BRISTOL-MYERS SQUIBB COMPANY

Synthesis
of
Carbon-Phosphorus
Bonds

Author

Robert Engel, Ph.D.
Professor of Chemistry and Biochemistry
Queens College, C.U.N.Y.
Flushing, New York

CRC Press, Inc.
Boca Raton, Florida

Library of Congress Cataloging-in-Publication Data

Engel, Robert.
  Synthesis of carbon-phosphorus bonds.

  Bibliography: p.
  Includes index.
  1. Organophosphorous compounds.   2. Chemistry,
Organic—Synthesis.   I. Title.
QD412.P1E59 1988   547'.07   87-22392
ISBN 0-8493-4930-3

Direct all inquiries to CRC Press, Inc., 2000 Corporate Blvd., N.W., Boca Raton, Florida, 33431.

© 1988 by CRC Press, Inc.

International Standard Book Number 0-8493-4930-3

Library of Congress Card Number 87-22392
Printed in the United States

PREFACE

Interest in the synthesis of organophosphorus compounds was at one time relatively limited, being of concern chiefly to those involved in the preparation of materials of certain commercial interest (e.g., insecticides, flame retardants, and detergents). With this situation existing, it has been relatively rare to find university courses being taught on the topic of synthesis of these materials.

However, recent developments have made an understanding of organophosphorus compounds and their syntheses of significantly more general value. Not only have there been found applications for organophosphorus compounds in a wide range of commercial applications, but also these materials have become of great utility for the facilitation of other organic transformations. Yet, the university-trained organic chemist is provided with little guidance to the performance of organophosphorus chemical conversions or to the organophosphorus literature. It is in an attempt to provide direction to the organic chemist lacking specific training in organophosphorus chemistry that this book is presented.

A major portion of the effort of preparing this manuscript was performed while I was on sabbatical leave from Queens College at the Rohm and Haas Company in Spring House, Pa. I wish to thank Queens College for the time and the Rohm and Haas Company for their hospitality, both of which were important for this work. In addition, I would dedicate this work to my children, Cheryl and Erik.

Robert Engel, 1987

THE AUTHOR

Robert Engel, Ph.D., is Professor of Chemistry and Biochemistry at Queens College of the City University of New York. Dr. Engel received his B.S. from Carnegie Institute of Technology in 1963 and his Ph.D. from The Pennsylvania State University in 1966.

From 1966 to 1968 Dr. Engel served in the U.S. Army at the U.S. Army Edgewood Arsenal Chemical Research Laboratories. Upon completion of his tour of duty, Dr. Engel took up his position on the faculty at Queens College. For two years he served as Chairman of the Chemistry Department.

Dr. Engel has worked in several areas of chemistry. These have included the investigation of reaction mechanisms and the spectroscopic study of complexes of paramagnetic transition metal ions. The major portion of his work has been concerned with studies of organophosphorus compounds, their synthesis, reaction mechanisms associated with them, and their use as specific metabolic regulators. His current research interests continue in all of these latter areas.

Dr. Engel is a member of the American Chemical Society, the New York Academy of Sciences, the Royal Society of Chemistry, and Sigma Xi. He has published over 70 articles and has given presentations at 20 scientific meetings.

TABLE OF CONTENTS

Chapter 1

INTRODUCTION

In recent years there has been a broadly expanding interest in the synthesis of organophosphorus compounds, i.e., those bearing a carbon atom bound directly to a phosphorus atom. This interest has resulted from the recognition of the value of such materials for a variety of industrial, biological, and chemical synthetic uses. Several categories of such compounds should be noted as being of particular significance for modern organic chemistry. These include: (1) quinquevalent [P(V)] phosphorus oxyacids and their derivatives as end products for a variety of applications;

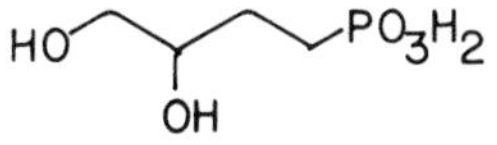

HO ⌒ ⌒ PO$_3$H$_2$
OH

$\left(Cl \frown \frown O \right)_2$ $\overset{\overset{O}{\|}}{P}$ $\frown$ Cl

3,4-Dihydroxybutyl-l-phosphonic Acid

$\left[\text{Metabolic Probe} \right]$

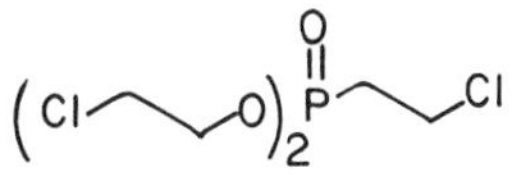

bis(2-Chloroethyl) Chloroethylphosphonate

$\left[\text{Flame Retardant} \right]$

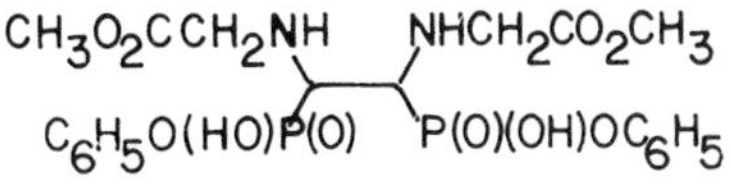

CH$_3$O$_2$CCH$_2$NH   NHCH$_2$CO$_2$CH$_3$
C$_6$H$_5$O(HO)P(O)   P(O)(OH)OC$_6$H$_5$

Diphenyl 1,2-Di(methoxycarbonylmethylamino)-

ethylene-l,2-diphosphonate

$\left[\text{Herbicide} \right]$

(2) trivalent neutral [P(III)] species, often used as intermediates in the synthesis of other organophosphorus compounds with direct applications;

CH$_3$P(OC$_2$H$_5$)$_2$

(C$_6$H$_5$)$_2$POCH$_3$

Diethyl Methylphosphonite

Methyl Diphenylphosphinite

and (3) tetravalent cationic [P(IV)] species of utility for general synthetic organic applications.

Br$^-$

C$_6$H$_5$CH$_2$$\overset{+}{P}$(C$_6H_5$)$_3$

Triphenylbenzylphosphonium Bromide

Examples of the utility of compounds in the first category are numerous. Esters of phosphonic acids (phosphonates) or phosphinic acids (phosphinates) find significant application in a wide range of areas. These materials have been demonstrated to have industrial use as flame-retardants, either by admixture with polymeric materials[1] or by copolymerization with other monomers,[2,3] inhibitors of oxidation in lubricants,[4] and as surfactants.[5]

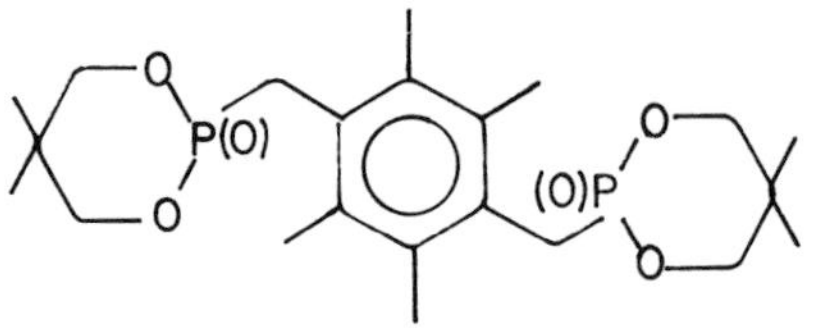

$(C_2H_5O)_2P(O)C_6H_5$

$C_9H_{19}O(CH_2CH_2O)_8-P(O)(OC_2H_5)C_2H_5$

These applications, among many others of commercial significance, have been reviewed recently.[6]

Applications of this category of P(V) organophosphorus materials for biological regulation have received considerable attention in recent years. Many compounds of this category have been synthesized for use in the regulation of plant growth. Included here are series of compounds related to glyphosphate[7] and diphenyl ethers,[8] as well as morphactins.[9]

$HO_2CCH_2NHCH_2PO_3H_2$

N –(Phosphonomethyl) glycine

[Glyphosate]

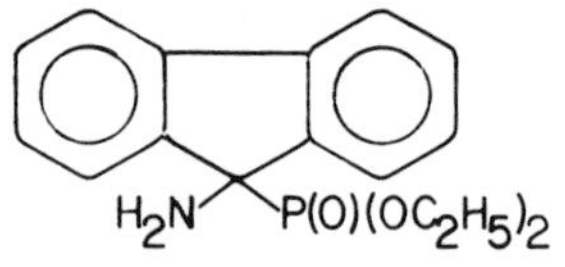

H_2N $P(O)(OC_2H_5)_2$

Diethyl
9–Aminofluorene –9– phosphonate

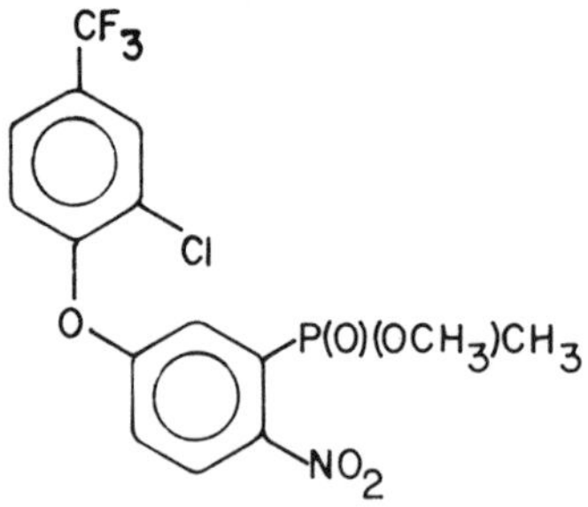

Methyl

2–Nitro–5–(2'–chloro–4'–

trifluoromethylphenoxy)phenylmethylphosphinate

Significant effort has also been devoted to the synthesis of P(V) organophosphorus compounds, particularly phosphonates, for medicinal applications. Major programs of laboratory synthesis of the antibiotic phosphonomycin (fosfomycin)[10] and structurally related molecules,[11] as well as potential antiviral materials in the phosphonoacetate and phosphonoformate series[12,13] have been undertaken.

$$HO_2CCH_2PO_3H_2 \qquad\qquad HO_2CPO_3H_2$$

Phosphonoacetic Acid Phosphonoformic Acid

Studies of the use of such materials in both of these areas have been reviewed recently[14,15] and will not be discussed further here.

A further biological application of organophosphorus compounds of the P(V) category has been for the mechanistic probing of metabolic processes. Compounds which are structural analogues of natural phosphate esters, contain a C-C-P linkage in place of the natural C-O-P linkage, might be anticipated to be stable to natural hydrolytic enzymes and could be used to provide information regarding biological mechanisms.

Beginning with the methylene analogue of GTP, of use in the determination of the individual steps of protein biosynthesis,[16] data regarding details of a variety of metabolic processes have been obtained.[17,18]

Guanosine 5'-(α,β-methylene)

Triphosphate

It should be noted that the syntheses necessary to allow these biological investigations to be performed are often quite complex. Usually the introduction of the carbon-phosphorus linkage needs to be made into species rich in reactive chemical functionalities. The available reactions for forming the desired carbon-phosphorus bond(s) must be of sufficient variety to allow its introduction in the presence of a wide range of such functionalities.

Compounds in the second category [P(III) species] generally have associated with them their own significant reactivity and are of greatest interest as precursors of the other two categories of compounds. Exceptions to this generalization are found with oxyacids of interest as analogues of carboxylic acid species in biological systems.

Finally, there should be noted the utility of organophosphorus compounds of the third category [P(IV) species] for the facile performance of chemical transformations, without which long and arduous synthetic routes would often be required. The Wittig reaction opened for organic chemists a one-step method for the substitution of a carbonyl oxygen by a multiply-bonded carbon function. The general utility of the Wittig approach was broadened through the Horner and the Wadsworth-Emmons modifications, which use stabilized phosphonate species, derived from compounds of the P(V) category.[19-22]

Quite recently, investigations of carbon-phosphorus bond formation reactions have been spurred by the potential application of these materials as *umpolung* reagents. Compounds bearing the phosphoryl linkage accompanied by a suitable functionality on an adjacent carbon atom conveniently provide a reactivity equivalent to a carbonyl function of polarity inverted from the normal situation.[23] This application of organophosphorus chemistry to general organic synthetic procedures has relied greatly on the use of silylated derivatives of P(V) phosphorus oxyacids, reagents themselves of relatively recent vintage.

The purpose of this volume is not to address the application of organophosphorus compounds to the attack of a variety of problems, but rather to survey the recent advances in the procedures for their preparation. While particular emphasis is given to that work reported since 1970, earlier efforts of fundamental significance will be noted. This volume is intended to serve as a manual for the general synthetic organic chemist in the design and performance of organophosphorus syntheses. Attention will be given to compounds in categories 1 and 2 as noted previously. Attention will *not* be given to the synthesis of phosphonium species of category 3 nor to their applications.

Carbon-phosphorus bond formation in the present discussion is viewed with regard to both the type of reaction used to produce the linkage and the nature of the bond to be generated. In all, carbon-phosphorus bond formation is viewed here in five parts. First, displacement reactions at aliphatic carbon by a phosphorus ester [P(III) species] is considered. In each instance a (formally) trivalent phosphorus atom is converted to a quinquevalent [P(V)] species in the process. A comparison of the feasibility of the several approaches available will be made with regard to the target organophosphorus product. Particular attention is given to the recently developed use of silylated phosphorus reagents.

Second, reactions are reviewed in which treatment of an organometallic reagent with a phosphorus halide is involved. Several oxidation and coordination states of phosphorus are considered, including mixed ester-halides. The alkylation of phosphorus halides mediated by Lewis acids are also reviewed, along with several other methods.

Polar addition reactions at unsaturated carbon are considered in two categories. The first of these involves addition at carbonyl-type carbon sites, and the second is concerned with conjugate addition processes at olefinic sites. Again, particular attention is given to silylated phosphorus reagents for these reactions.

Finally, attention is given to the variety of methods available for the direct generation of carbon-phosphorus bonds at vinylic and aromatic sites. Secondary approaches to such materials, starting with aliphatic carbon-phosphorus linkages, are not reviewed.

In each instance examples of experimental procedures are included as well as a tabulation of syntheses performed using the approaches. These tabulations are not encyclopedic, but rather give representative examples of the use of the various procedures.

It is in order here to provide a brief survey of organophosphorus nomenclature. This is done in an outline-tabular form organized with regard to the oxygen coordination about the phosphorus atom in the free-acid state. The naming system provided will be used throughout this review and is the standard notation for the classes of materials in the English-language literature. It should be noted that where an ''R'' group is indicated and described as an ''alkyl'' function, aromatic or vinylic functions may also exist and be named in an analogous manner.

$P(OH)_2R$	$OP(OH)_2R$	$P(OH)_3$	$OP(OH)_3$
Phosphonous Acid	Phosphonic Acid	Phosphorous Acid	Phosphoric Acid
$P(OH)(OR')R$	$OP(OH)(OR')R$	$P(OH)_2OR'$	$OP(OH)_2OR'$
Monoalkyl Phosphonite	Monoalkyl Phosphonate	Monoalkyl Phosphite	Monoalkyl Phosphate
$P(OR')_2R$	$OP(OR')_2R$	$P(OH)(OR')_2$	$OP(OH)(OR')_2$
Dialkyl Phosphonite	Dialkyl Phosphonate	Dialkyl Phosphite	Dialkyl Phosphate
		$P(OR')_3$	$OP(OR')_3$
		Trialkyl Phosphite	Trialkyl Phosphate
PR_3	OPR_3	$P(OH)R_2$	$OP(OH)R_2$
Phosphine	Phosphine Oxide	Phosphinous Acid	Phosphinic Acid
		$P(OR')R_2$	$OP(OR')R_2$
		Alkyl Phosphinite	Alkyl Phosphinate

REFERENCES

1. **Hoffman, J. A.,** Cyclic Diphosphonates, U.S. Patent 4,268,459, 1981.
2. **Kleiner, H.-J., Linke, F., and Dursch, W.,** Carbamoyl-oxyalkyl-phosphinic Acid Derivatives, U.S. Patent 4,173,601, 1979.
3. **Block, H.-D. and Frohlen, H.-G.,** Preparation of Alkane Phosphonic and Phosphinic Acid Aryl Esters, U.S. Patent 4,377,537, 1983.
4. **Martin, D. J.,** Catalyzed Process for Producing Pentavalent Phosphorus Derivatives, U.S. Patent 3,705,214, 1972.
5. **Walz, K., Nolte, W., and Muller, F.,** Surface-active Phosphonic Acid Esters, U.S. Patent 4,385,000, 1983.
6. **Drake, G. L., Jr. and Calamari, T. A., Jr.,** Industrial Uses of Phosphonates, in *The Role of Phosphonates in Living Systems,* Hilderbrand, R. L., Ed., CRC Press, Boca Raton, Fla., 1983, chap. 7.
7. **Rupp, W., Finke, M., Bieringer, H., Langeluddeke, P., and Kleiner, H.-J.,** Herbicidal Agents, U.S. Patent 4,168,963, 1979.
8. **Maier, L.,** Herbicidally Active 2-Nitro-5-(2'-chloro-4'-trifluoromethylphenoxy)phenylphosphinic Acid Derivatives, U.S. Patent 4,434,108, 1984.
9. **Gancarz, R. and Wieczorek, J. S.,** Synthesis of Phosphonic Analogues of Morphactines, *J. Prakt. Chem.,* 322, 213, 1980.

10. **Hendlin, D., Stapley, E. O., Jackson, M., Wallick, H., Miller, A. K., Wolf, F. J., Miller, T. W., Chaiet, L., Kahan, F. M., Foltz, E. L., Woodruff, H. B., Mata, J. M., Hernandez, S., and Mochales, S.,** Phosphonomycin, a New Antibiotic Produced by Strains of *Streptomyces, Science,* 166, 122, 1969.
11. **Glamkowski, E. J., Rosas, C. B., Sletzinger, M., and Wantuck, J. A.,** Process for the Preparation of *cis*-1-Propenylphosphonic Acid, U.S. Patent 3,733,356, 1973.
12. **Castaner, J. and Hopkins, S. J.,** Phosphonoacetic Acid, *Drugs Future,* 2, 677, 1978.
13. **Helgstrand, A. J. E., Johansson, K. N., Misiorny, A., Noren, J. O., and Stening, G. B.,** Phosphonoformic Acid Esters and Pharmaceutical Compositions Containing Same, U.S. Patent 4,368,081, 1983.
14. **Overby, L. R., Duff, R. G., and Mao, J. C.-H.,** Antiviral Properties of Phosphonoacetic Acid, *Ann. N.Y. Acad. Sci.,* 284, 310, 1977.
15. **Hilderbrand, R. L.,** The Effects of Synthetic Phosphonates on Living Systems, in *The Role of Phosphonates in Living Systems,* Hilderbrand, R. L., Ed., CRC Press, Boca Raton, Fla., 1983, chap. 6.
16. **Hershey, J. W. B. and Monro, R. E.,** Competitive Inhibitor of the Guanosine Triphosphate Reaction in Protein Synthesis, *J. Mol. Biol.,* 18, 68, 1966.
17. **Engel, R.,** Phosphonates as Analogues of Natural Phosphates, *Chem. Rev.,* 77, 349, 1977.
18. **Engel, R.,** Phosphonic Acids and Phosphonates as Antimetabolites, in *The Role of Phosphonates in Living Systems,* Hilderbrand, R. L., Ed., CRC Press, Boca Raton, Fla., 1983, chap. 5.
19. **Maercker, A.,** The Wittig Reaction, *Org. React.,* 14, 270, 1965.
20. **Horner, L., Hoffman, H., Wippel, H. G., and Kluhre, G.,** Phosphine Oxides as Olefination Reagents, *Chem. Ber.,* 92, 2499, 1959.
21. **Wadsworth, W. S., Jr. and Emmons, W. D.,** The Utility of Phosphonate Carbanions in Olefin Synthesis, *J. Am. Chem. Soc.,* 83, 1733, 1961.
22. **Wadsworth, W. S., Jr.,** Phosphonate-anions, *Org. React.,* 25, 73, 1977.
23. **Evans, D. A., Takacs, J. M., and Hurst, K. M.,** Phosphonamide Stabilized Allylic Carbanions. New Homoenolate Equivalents, *J. Am. Chem. Soc.,* 101, 371, 1979.

Chapter 2

NUCLEOPHILIC DISPLACEMENT AT CARBON BY PHOSPHORUS

I. INTRODUCTION

Displacement of a suitable leaving group from carbon by a nucleophilic trivalent phosphorus species is the most commonly used method for the formation of carbon-phosphorus bonds. In addition to those reactions which lead to phosphoryl-type compounds, such as the Michaelis-Becker and Michaelis-Arbuzov reactions, phosphonium ion and phosphine generation are usually performed in this manner. The present review is concerned only with a background summary and recent advances for the formation of phosphoryl-type compounds. Several recent reviews of phosphonium ion formation are available.[1-5]

The discussion here will be divided into two major parts. The first of these will be concerned with the nucleophilic reactions of the anionic forms of tricoordinated phosphorus species, such as phosphite diesters, phosphonite monoesters, and phosphinites.

$$(RO)_2\overset{\overset{O}{\|}}{P}{:}^- \qquad RO\overset{\overset{O}{\|}}{\underset{R'}{P}}{:}^- \qquad R'_2\overset{\overset{O}{\|}}{P}{:}^-$$

Such reactions, generally known as Michaelis-Becker reactions, will be reviewed with a particular emphasis on recent advances toward increasing their efficiency in the generation of phosphoryl-type products and their practical application to syntheses.

The second part of this discussion will center on the nucleophilic displacement reactions (Michaelis-Arbuzov reactions) of neutral trivalent phosphorus species. Although some variants will be noted, the compounds of interest will be fully esterified forms of phosphorous, substituted phosphonous, and substituted phosphinous acids.

$$(RO)_3P{:} \qquad (RO)_2\ddot{P}R' \qquad RO\ddot{P}R'_2$$

Particular attention will be given to recent advances in the use of silyl esters of phosphorus acids and the variety of leaving groups from carbon.

II. MICHAELIS-BECKER REACTIONS AND MECHANISM

The displacement of halide ion from saturated carbon by a dialkyl phosphite anion has been known for quite some time.[6]

$$(RO)_2\overset{\overset{O}{\|}}{P}{:}^- \quad + \quad R'X \quad \longrightarrow \quad (RO)_2P(O)R' \quad + \quad X^-$$

Much of the important early work has been summarized in a review by Crofts[7] and in further reviews of the Michaelis-Arbuzov reaction.[8]

Several mechanisms have been postulated to account for the overall conversion. In spite of the complex kinetics and product variation with structure that has been noted,[9] it would appear that initial attack is by phosphorus rather than oxygen on carbon. A possibility considered at one time involved displacement of halide by oxygen, followed by Michaelis-Arbuzov reaction of the fully esterified species.

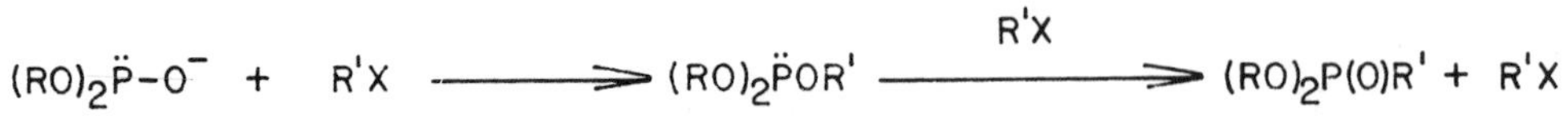

$$(RO)_2\ddot{P}-O^- \quad + \quad R'X \quad \longrightarrow \quad (RO)_2\ddot{P}OR' \quad \xrightarrow{\quad R'X \quad} \quad (RO)_2P(O)R' \quad + \quad R'X$$

This route is eliminated as a possibility for simple alkylations upon consideration of the Michaelis-Becker reaction of methyl phenylphosphonite with d_3-methyl iodide.[10] A nearly quantitative yield of methyl d_3-methylphenylphosphinate is isolated, indicating that direct attack by phosphorus on carbon occurs, presumably in a classically S_N2 manner with regard to the carbon center. Displacement by oxygen has been reported in one instance in reaction involving an imidoyl chloride.[11]

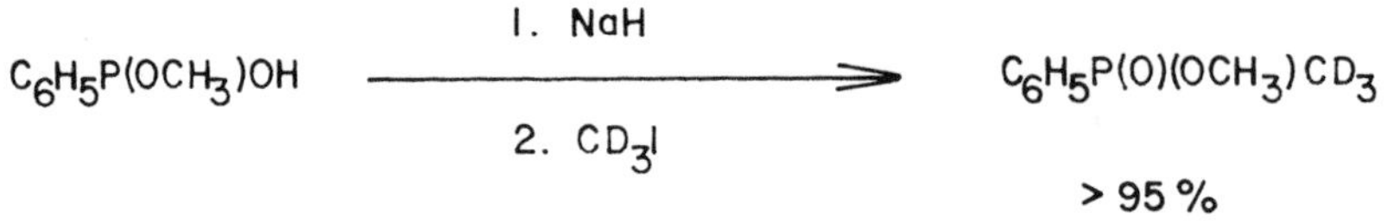

$$C_6H_5P(OCH_3)OH \quad \xrightarrow[\text{2. CD}_3\text{I}]{\text{1. NaH}} \quad C_6H_5P(O)(OCH_3)CD_3$$

$$> 95\%$$

However, there remains uncertainty regarding the coordination about phosphorus in this reaction. An understanding of the process with regard to the phosphorus center has been facilitated by the preparation in recent years of phosphonite monoesters with a chiral phosphorus center. The first preparation of such species[12] proceeded via the Raney-nickel desulfurization of a chiral O-monoester of a phosphonothioic acid, obtained in optically active form by a standard resolution technique.[13]

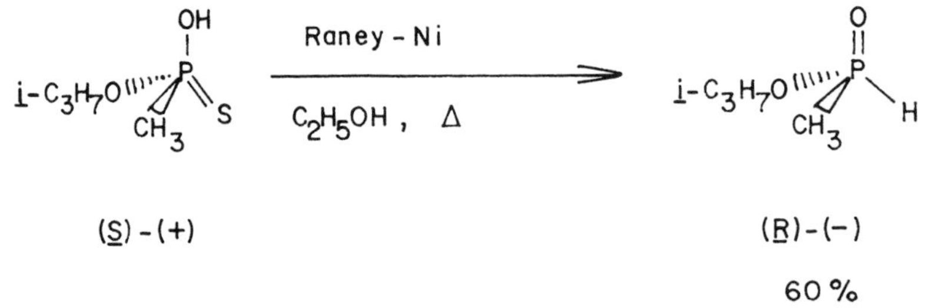

$$\text{(S)}-(+) \qquad\qquad\qquad \text{(R)}-(-)$$

$$60\%$$

Within 3 months of the initial report, two other approaches were described which allowed isolation of chiral phosphonite monoesters. One involved the fractional crystallization of diastereoisomeric forms of menthyl phenylphosphonite, prepared from an achiral phosphorus reagent (phenylphosphonous dichloride) and a chiral alcohol, (−)-menthol.[14] The other utilized the selective inclusion of the levorotatory enantiomer of isopropyl methylphosphonite in cyclohepta-amylose,[15] a reagent previously explored as a resolving agent.[15,16]

In solution, the "phosphoryl" form is dominant in the equilibrium with the "hydroxyl" form. The stereochemical integrity of the phosphorus center is maintained in this equilibration, and appears to be stable to acid-catalyzed racemization.[12]

Upon generation of the anion using sodium hydride, followed by addition of an alkyl halide, alkylation of the phosphorus occurs with net retention of configuration at the chiral center.[10,18]

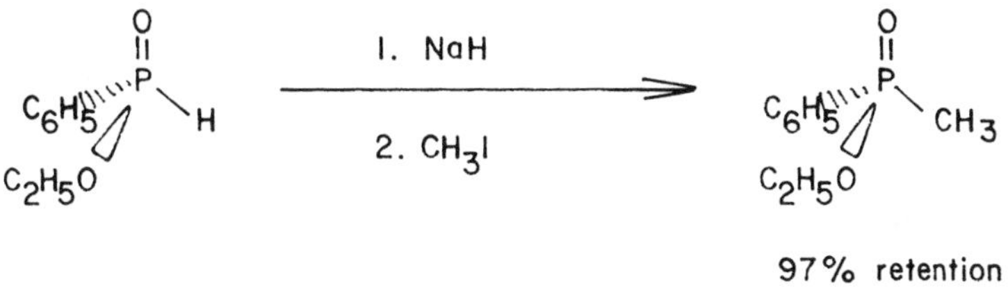

97% retention

From this result one may conclude that the anion maintains significant stereochemical integrity in solution, and that the halide displacement occurs with direct generation of a tetracoordinated phosphorus species[10]

Early efforts would indicate that standard procedure in the performance of the Michaelis-Becker reaction involves formation of the sodium salt of the monobasic trivalent phosphorus acid (phosphite diester, phosphonite monoester, or secondary phosphine oxide). In much of the current work this is certainly reinforced, although the use of sodium metal has been replaced by sodium hydride. However, other approaches to the anionic species have been developed which are quite useful.

Tertiary amines have been found to be quite suitable as bases for the Michaelis-Becker reaction. This is particularly the situation with highly reactive substrates. Triethylamine has been used successfully with dialkyl phosphites in their reaction with imidoyl chlorides,[11] carboxylic acid chlorides,[19] and chloroformates.[20,21]

$$[(CH_3)_3SiO]_2P(O)H \;+\; ClCO_2C_2H_5 \quad \xrightarrow{(C_2H_5)_3N} \quad [(CH_3)_3SiO]_2P(O)CO_2C_2H_5$$

89%

Tertiary amines have also been used to serve the dual purpose of base and leaving group. With diethyl phosphite, symmetrical cyclic 1,1-diaminomethane derivatives have been used for the preparation of 1-aminoalkylphosphonates.[22,23] Reaction presumably occurs by displacement of a protonated nitrogen from the central carbon.

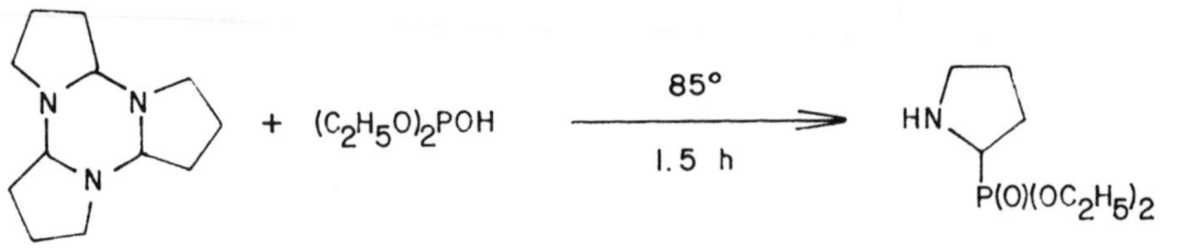

The α-alkoxyalkylureas represent an interesting related class of compounds for reaction with dialkyl phosphites.[24] Simply warming the substituted urea and the dialkyl phosphite results in reaction, the urea serving both as base and substrate.

A similar reaction is found between the more acidic O,O-diethyl thiophosphite and tertiary o-hydroxybenzylic amines.[25] Displacement of dialkylamine from the benzylic carbon leads to formation of substituted benzylthiophosphonates, albeit in only fair yield. Use of this particular reaction for synthetic purposes is plagued by side reactions involving the sulfur and the phenolic site.

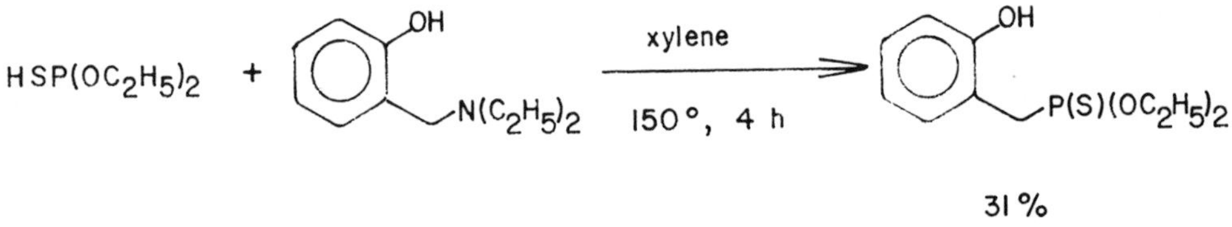

31 %

Of course, the use of amines to generate the anionic forms of dialkyl phosphites has been used for quite some time in the Todd reaction for the preparation of phosphorochloridates.[26-28] Sodium salts may also be used, but the tertiary amines provide a clean and experimentally convenient alternative. The use of amines rather than sodium or sodium hydride with dialkyl phosphites to generate the anions is one approach to help overcome an important difficulty often found in performing Michaelis-Becker reactions. That difficulty is the relatively low solubility of the salts in a variety of compatible solvents. Another recent approach to overcoming this difficulty has been the use of phase transfer catalysts. Biphase solvent systems with quaternary ammonium salts added have been used to good result in several instances.[29-31] Potassium carbonate with catalytic amounts of 18-crown-6 has also been used in a solid-liquid two phase system for the alkylation of dialkyl phosphites.[32]

Preformed heavy-metal salts of dialkyl phosphites have been used in several syntheses of phosphonate derivatives of carbohydrates. Facilitated by silver[33] and mercury[34-36] salts, dialkyl phosphites undergo addition to acylonium ions derived from 2-O-acetylglycosyl-bromides to generate dioxolanephosphonates.

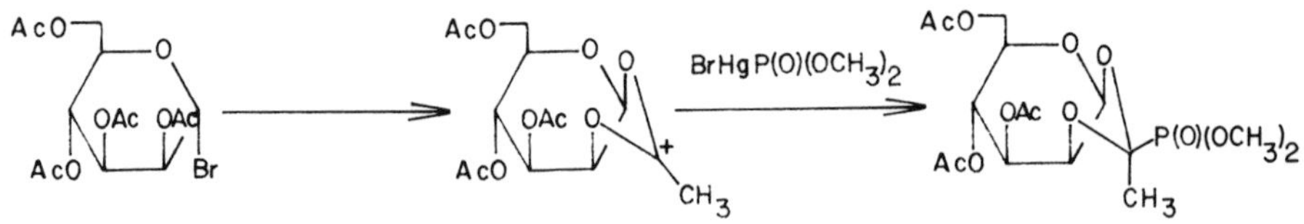

While the vast majority of Michaelis-Becker reactions which have been reported involve the action of dialkyl phosphites on simple alkyl halides,[37-45] a variety of other functionalities has been used for analogous reaction processes. For example, acetylenic halides give acetylenicphosphonates in fair to moderate yield. Interestingly, it is found that the yields are quite dependent upon the cation used with the dialkyl phosphite, sodium providing the significantly more favorable result.[46] Two reports have also been made recently of reaction involving imidoyl chlorides, although the detailed course of the reaction appears to differ between them.[11,47]

Reaction of dialkyl phosphites under Michaelis-Becker conditions with α-halocarbonyl compounds generally proceeds to give "Perkow-type" products rather than direct displacement of halide.[48-50] That is, products are isolated which are the result of phosphorus attack at the carbonyl position rather than the halogen bearing site.

$$(RO)_2\overset{\overset{O}{\|}}{P}{:}^- \quad + \quad ArCOCH_2X \quad \longrightarrow \quad (RO)_2P(O)OC(Ar){=}CH_2$$

An exception to this is found with ethyl 4-bromoacetylacetonate which gives the simple Michaelis-Becker product in moderate yield.[51] Similarly, α-haloesters proceed with simple displacement reaction.[52] The α-halophosphonates are also known to undergo displacement cleanly.[53]

A study of other haloketones under Michaelis-Becker conditions provides an insight into details of the competing possibilities of direct displacement and carbonyl attack.[48,49] For halogen located at either the γ- or δ-position relative to the carbonyl group, reaction proceeds to generate cyclic products, that is, substituted tetrahydrofuranyl- or tetrahydropyranyl-phosphonates, respectively. In some instances small amounts of direct substitution products are obtained as well. This would indicate that attack by the phosphorus reagent at the carbonyl site is a kinetically (but probably not a thermodynamically) favored process which can lead, if the geometry is suitable, to an intramolecular alkoxide displacement of halide.

$$RCO(CH_2)_nCl + (R'O)_2\overset{\overset{O}{\|}}{P}{:}^- \longrightarrow (R'O)_2P(O)\underset{R}{\overset{\overset{O^-}{|}}{C}}(CH_2)_nCl \longrightarrow (R'O)_2P(O)\underset{R}{\overset{\overset{O\frown}{|}}{C}}(CH_2)_n$$

$$CH_3CO(CH_2)_3Cl \quad + \quad NaOP(OC_2H_5)_2 \quad \longrightarrow$$

66%

Leaving groups other than halogen have also been used in Michaelis-Becker type reactions. In addition to the reports previously noted in which tertiary amino sites served as bases facilitating reaction at adjacent sites,[22,23,25] the sodium salt of diethyl phosphite has been reported to attack at a site adjacent to a quaternary ammonium center to displace a tertiary amine.[54] While the yield in this particular reaction was quite low, modification of structural features of the reaction system would likely lead to more favorable results.

$$CH_3COCH_2CH_2\overset{+}{N}(C_2H_5)_2CH_3 + NaOP(OC_2H_5)_2 \longrightarrow CH_3COCH_2CH_2P(O)(OC_2H_5)_2$$

$$I^-$$

18%

There has been reported recently a reaction under Michaelis-Becker conditions in which an acetate function is displaced.[55] The lithium salt of diphenylthiophosphinite, in the presence of catalytic amounts of tetrakis(triphenylphosphine)palladium, displaces acetate from allylic positions to generate tertiaryphosphine sulfides.

$$\text{OAc} + LiSP(C_6H_5)_2 \xrightarrow{Pd\left[P(C_6H_5)_3\right]_4} P(S)(C_6H_5)_2$$

In another instance, attack on the carboxyl carbon of an acetate has been reported, although alternative mechanistic pathways to the observed products cannot be definitively excluded.[56]

Labile three-membered rings are also susceptible to phosphonation under Michaelis-Becker conditions. Epoxides react readily with the sodium salts of dialkyl phosphites to generate the β-hydroxyphosphonates in fair yield.[57] The *N*-acylaziridines are also found to undergo reaction with dialkyl phosphites in the absence of added base.[58] Product is generated in moderate to good yield with heating over several days.

III. COMPARISON OF METHODS

For the performance of alkylations of phosphonyl anionic species, the fundamental experimental variables include the basic reagent and the solvent system used, two factors which are closely related.

Choosing an optimal solvent system has always posed a difficult experimental problem for the performance of Michaelis-Becker reactions. Metal salts of the phosphorus acids bear relatively low solubilities for aprotic solvents compatible with the performance of the reaction. Even using an excess of the phosphorus acid itself as the solvent works only in certain instances due to limited salt solubility.

Several relatively favorable situations may be noted, however. Aromatic solvents, such as toluene or xylene, appear to provide greater solubility than simple oxygenated aprotic media, such as diethyl ether or tetrahydrofuran. Dioxane has been reported to have superior solvent characteristics compared to the mono-oxygenated ethers. In each instance, the sodium salt of di-*n*-butyl phosphite exhibits greater solubility than the corresponding salts of other dialkyl phosphites. When using an excess of the acid as solvent, again di-*n*-butyl phosphite is superior. If the identity of the alkyl ester linkages of the target phosphonate is not critical to the overall synthesis, the *n*-butyl esters are experimentally the most convenient to use. It is anticipated that further investigations into the use of biphasic reaction systems with phase transfer catalysts or cationic binding agents will increase the variability of reaction conditions available for synthetic purposes.

In the choice of base for the performance of the reaction, the great majority of the prior efforts have used an alkali metal, usually sodium due to its availability and ease of handling. When comparisons have been made with other alkali metal salts, sodium has also been found to provide superior product yields. The use of sodium metal in the solvent systems described can at times pose difficulties due to slow formation of the anion. The use of sodium hydride, either dry or as a mineral oil dispersion, works very nicely to give the salt rapidly and cleanly.

Relatively little work has been performed on Michaelis-Becker alkylations using amines as the base. That work which has been reported as exhibited only moderate success. It is anticipated, however, that with the proper choice of amine (low nucleophilicity) and completely anhydrous solvent systems, improved results might be found.

IV. EXPERIMENTAL PROCEDURES

Diethyl 3,3-Diethoxypropynyl-1-phosphonate — Reaction of sodium salt of a dialkyl phosphite with an acetylenic halide.[46]

A solution of sodium diethylphosphite was prepared by the addition of sodium hydride (3.72 g, 0.155 mol) to diethyl phosphite (20.7 g, 0.15 mol) in tetrahydrofuran (200 mℓ). The resultant solution was cooled to $-70°C$ and stirred rapidly while 3,3-diethoxypropynyl-1-bromide (30.05 g, 0.15 mol) in tetrahydrofuran (20 mℓ) was added dropwise. The reaction mixture was stirred for 16 hr at $-70°C$, after which it was allowed to warm to room temperature. The reaction mixture was centrifuged and the solution decanted from the precipitated sodium bromide. The solid material was dissolved in water (40 mℓ) and extracted with diethyl ether (2 × 20 mℓ). The organic phases were combined, dried over magnesium sulfate, filtered, and evaporated under reduced pressure to give an oil. The oil was vacuum distilled (119 to 130°C/0.13 torr) to give the diethyl 3,3-diethoxypropynyl-1-phosphonate (17 g, 64%). Trace amounts of an impurity could be detected in this material. This could be removed by preparative scale thin layer chromatography using silica PF-254 eluting with hexane-acetone (9/1).

bis-(2,2-Dimethyltrimethylene)yl [(2,5-Dimethyl-1,4-phenylene)dimethylene] Diphosphonate — Reaction of the sodium salt of a cyclic phosphite diester with a bis-benzylic halide.[39]

A solution of the sodium salt of the cyclic 2,2-dimethyltrimethylene phosphite was prepared by the addition of a 57% mineral oil dispersion of sodium hydride (8.42 g, 0.2 mol) to a solution of the cyclic 2,2-dimethyltrimethylene phosphite (30 g, 0.2 mol) in dry dimethylformamide (150 mℓ) while the temperature was maintained below 30°C. To the resultant solution was added a solution of 1,4-bis(chloromethyl)-2,5-dimethylbenzene (20.3 g, 0.1 mol) in dry dimethylformamide (150 mℓ). After the addition was complete and the exotherm subsided, the reaction mixture was heated at 60 to 65°C for 15 hr. After the reaction mixture was cooled to room temperature, the solid which formed was filtered. The solid was washed with dimethylformamide and dried in a vacuum oven. The dried solid was washed with warm water, filtered, rinsed with water again and dried in a vacuum oven to give the bis-(2,2-dimethyltrimethylene)yl [(2,5-dimethyl-1,4-phenylene)dimethylene]diphosphonate (32 g, 74.4%) of mp 231 to 233°C.

di-*n*-Butyl *N,N,*-Diethylcarbamoylmethylphosphonate — Phase transfer catalyzed reaction of a dialkyl phosphite with an alkyl chloride.[29]

Into a 500-mℓ, three-necked, round-bottomed flask equipped with a thermowell, a 125-mℓ pressure-equalizing addition funnel, a mechanical stirrer, inert gas fittings, and a septum was placed a solution of *N,N*-diethylchloroacetamide (14.95 g, 0.1 mol) and methyltricaprylammonium chloride (0.5 g) in methylene chloride (75 mℓ), along with 50% sodium hydroxide (100 mℓ). The solution was stirred at 200 rpm under a gentle purge of nitrogen and maintained at 5 to 10°C while a solution of di-*n*-butyl phosphite (21.34 g, 0.11 mol) and methyltricaprylammonium chloride (0.5 g) in methylene chloride (75 mℓ) was added dropwise. The addition was complete in 1 hr After 2 hr additional, di-*n*-butyl phosphite (1.94 g, 0.01 mol) was added, the stirring was continued for 2 hr, and then the phases were separated. The aqueous layer was extracted with pentane (50 mℓ) and the combined organic solutions were washed with 50% aqueous methanol (3 × 50 mℓ) and saturated sodium chloride solution (50 mℓ). After the mixture was dried over anhydrous potassium carbonate

Table 1
FORMATION OF CARBON TO PHOSPHORUS BONDS *VIA* THE MICHAELIS-BECKER REACTION

Product	Phosphorus reagent	Method	Yield (%)	Refs.
$(C_2H_5O)_2P(O)CH_2CO_2H$	$(C_2H_5O)_2POH$	A	100	50
$(C_2H_5O)_2P(O)CH_2CH(OH)CH_3$	$(C_2H_5O)_2POH$	B	55	55
$(C_2H_5O)_2P(O)CH_2CH(OH)C_2H_5$	$(C_2H_5O)_2POH$	B	55	55
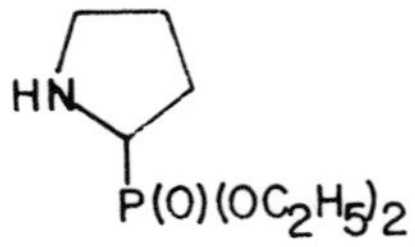	$(C_2H_5O)_2POH$	C	52	21
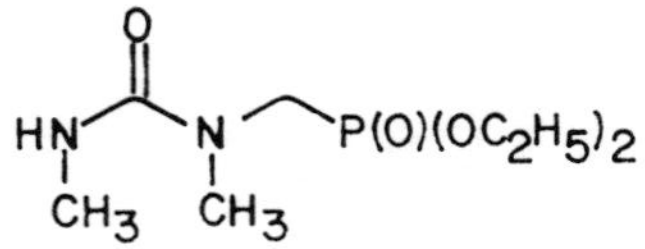	$(C_2H_5O)_2POH$	C	40	23
$(C_2H_5O)_2P(O)CH_2CH_2COC_2H_5$	$(C_2H_5O)_2POH$	D	40	46
$(C_2H_5O)_2P(O)CF_2P(O)(OC_2H_5)_2$	$(C_2H_5O)_2POH$	D	13	51
$(C_2H_5O)_2P(O)CH_2CH_2NHCO_2C_2H_5$	$(C_2H_5O)_2POH$	E	42	56
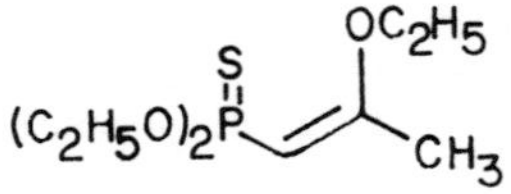	$(C_2H_5O)_2PSH$	D	40	47
$(TMSO)_2P(O)CO_2C_2H_5$	$(TMSO)_2POH$	F	89	19

Structure	Reagent	Method	Yield	Yield
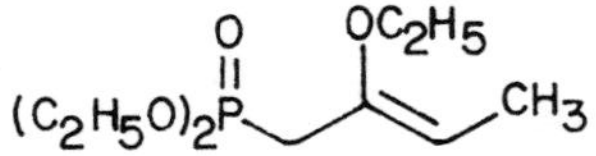 ![structure] $(C_2H_5O)_2P$ group with OC_2H_5, CH_3	$(C_2H_5O)_2POH$	D	60	47
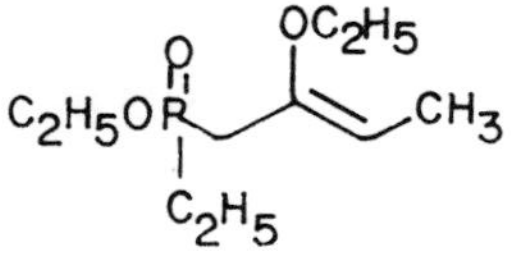 C_2H_5OR group with OC_2H_5, CH_3, C_2H_5	$C_2H_5P(OH)OC_2H_5$	D	62	47
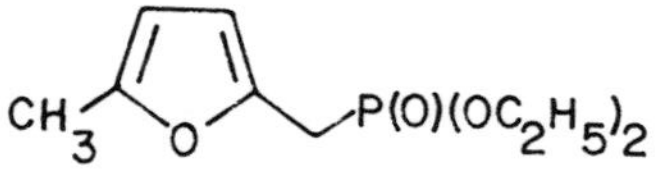 furan with CH_3, CH_2, $P(O)(OC_2H_5)_2$	$(C_2H_5O)_2POH$	D	72	43
$(C_2H_5O)_2P(O)CH_2COCH_2CO_2C_2H_5$ $(n-C_4H_9O)_2P(O)CH_2CH(OH)CH_3$	$(C_2H_5O)_2POH$ $(n-C_4H_9O)_2POH$	G B	50 35	49 55
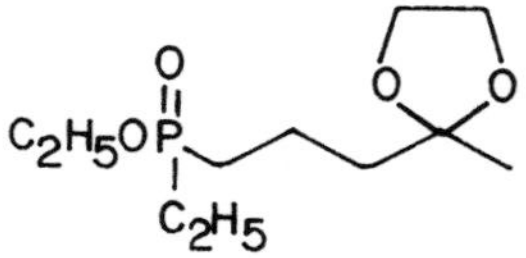 dioxolane structure with C_2H_5, C_2H_5OP, O	$C_2H_5P(OH)OC_2H_5$	D	74	47
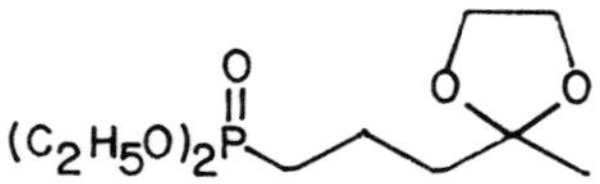 dioxolane structure with $(C_2H_5O)_2P$, O	$(C_2H_5O)_2POH$	D	80	47
$(C_2H_5O)_2P(O)-C\equiv C-CH(OC_2H_5)_2$	$(C_2H_5O)_2POH$	G	64	44

Table 1 (continued)
FORMATION OF CARBON TO PHOSPHORUS BONDS *VIA* THE MICHAELIS-BECKER REACTION

Product	Phosphorus reagent	Method	Yield (%)	Refs.
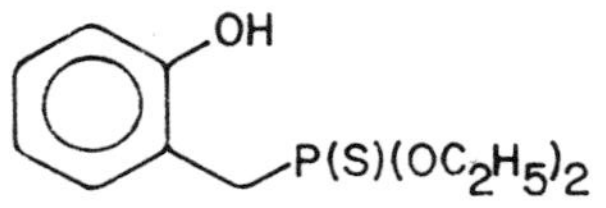 structure with OH and P(S)(OC$_2$H$_5$)$_2$	(C$_2$H$_5$O)$_2$PSH	C	31	24
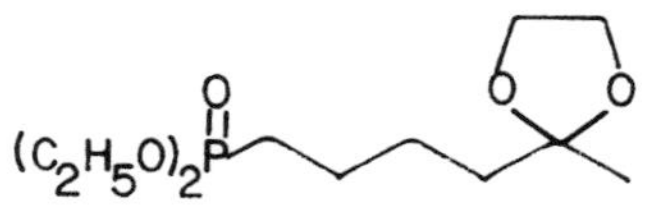 dioxolane structure, (C$_2$H$_5$O)$_2$P	(C$_2$H$_5$O)$_2$POH	G	72	47
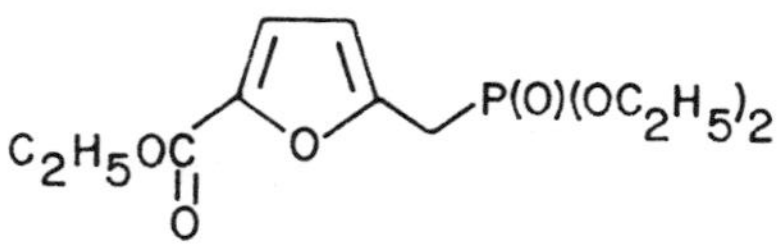 furan structure, C$_2$H$_5$OC, P(O)(OC$_2$H$_5$)$_2$	(C$_2$H$_5$O)$_2$POH	D	23	43
[(CH$_3$)$_2$CH]$_2$P(O)CH$_2$CH$_2$NHCO$_2$CH(CH$_3$)$_2$	[(CH$_3$)$_2$CH]$_2$POH	E	52	56
(C$_2$H$_5$O)$_2$P(O)C(=NCH$_3$)C$_6$H$_5$	(C$_2$H$_5$O)$_2$POH	F	38	11
(C$_2$H$_5$O)$_2$P(O)CH$_2$NHCH$_2$C$_6$H$_5$	(C$_2$H$_5$O)$_2$POH	C	68	22

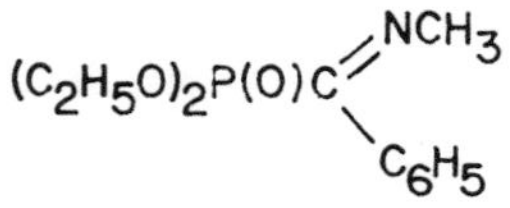

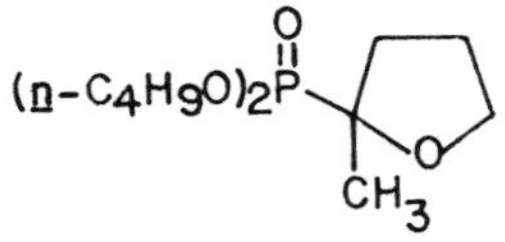

Structure	Reagent	Method	Yield	Yield
(n-C₄H₉O)₂P tetrahydrofuran, CH₃	(n-C₄H₉O)₂POH	D	66	46
dioxolane–C₂H₅P(O)–OC₂H₅ chain	C₂H₅P(OH)OC₂H₅	D	72	47
(C₂H₅O)₂P(O)CH₂CH₂NHCOC₆H₅	(C₂H₅O)₂POH	E	77	56
(n-C₄H₉O)₂P pyran structure	(n-C₄H₉O)₂POH	D	30	46
quinoline-CH₂P(O)(OC₂H₅)₂	(C₂H₅O)₂POH	D	94	39
quinoline-P(O)(OC₂H₅)₂	(C₂H₅O)₂POH	D	78	39

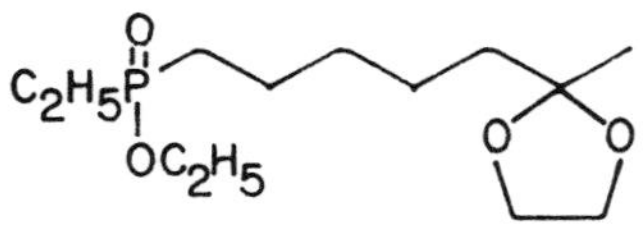

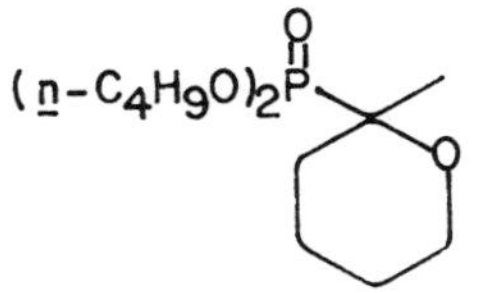

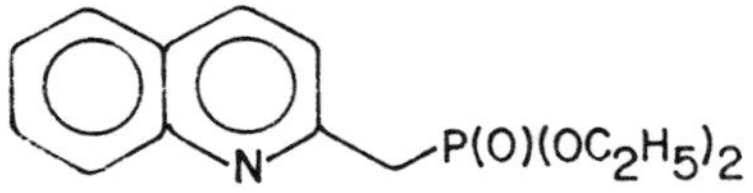

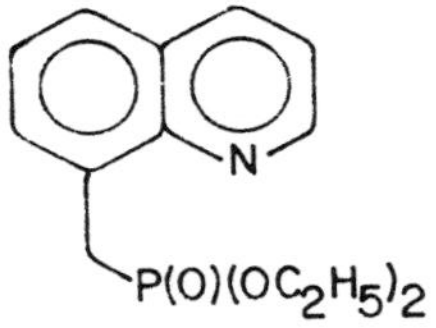

Table 1 (continued)
FORMATION OF CARBON TO PHOSPHORUS BONDS *VIA* THE MICHAELIS-BECKER REACTION

Product	Phosphorus reagent	Method	Yield (%)	Refs.
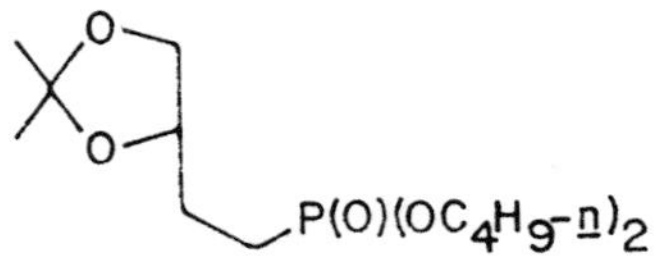 $P(O)(OC_4H_9\text{-}n)_2$	$(n\text{-}C_4H_9O)_2POH$	D	45	41
$CH_3CH=CHCH_2P(S)(C_6H_5)_2$	$(C_6H_5)_2PSH$	H	80	53
$(n\text{-}C_4H_9O)_2P(O)CF_2P(O)(OC_4H_9\text{-}n)_2$	$(n\text{-}C_4H_9O)_2POH$	D	47	51
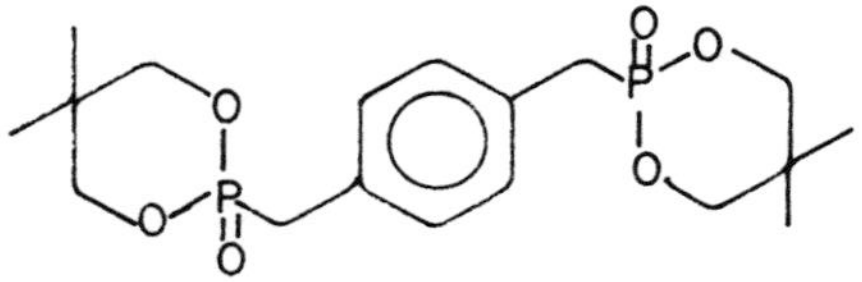	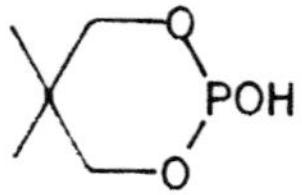 POH	G	90	38
$(n\text{-}C_4H_9)_2P(O)CH_2CON(C_2H_5)_2$	$(n\text{-}C_4H_9)_2POH$	I	87	28
	$(C_6H_5)_2PSH$	H	85	53
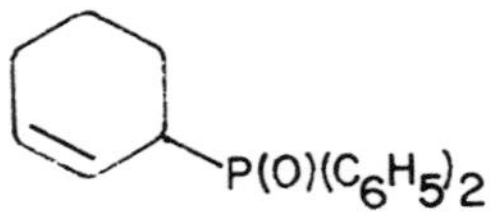 $P(O)(C_6H_5)_2$	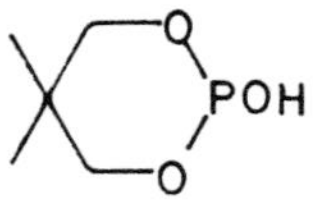 POH	G	74	38

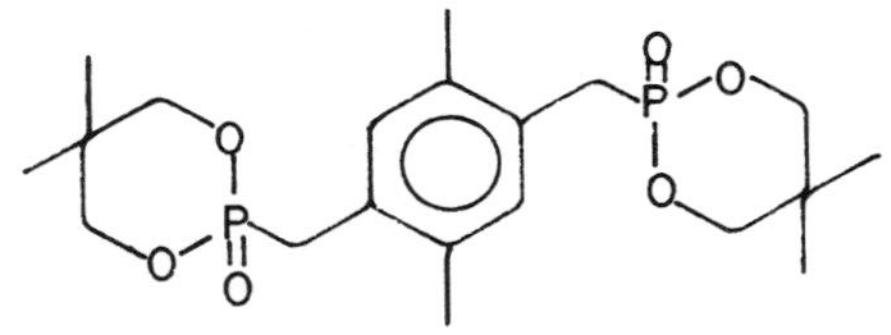

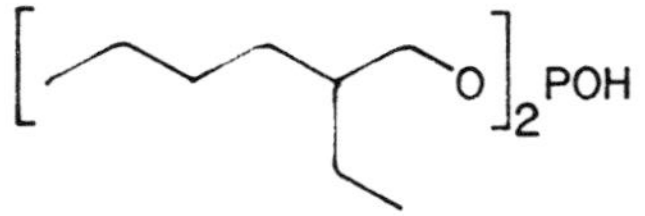

Structure			
$[$ ⌬ ring structure with two P(O) dioxaphosphorinane groups $]$	$\left[\begin{array}{c}\\ O\end{array}\right]_2 POH$	I	75	28
$[CH_3(CH_2)_3CH(C_2H_5)CH_2O]_2P(O)CH_2CON(CH_3)_2$	$(C_2H_5O)_2POH$	G	63	40
SO_2N—O—$CH_2C_6H_5$ ... $P(O)(OC_2H_5)_2$				
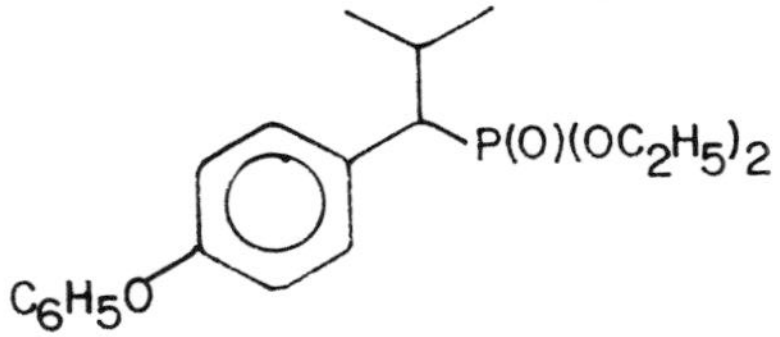 C_6H_5O ... $P(O)(OC_2H_5)_2$	$(C_2H_5O)_2POH$	G	35	42
$(n\text{-}C_8H_{17})_2P(O)CH_2CON(C_2H_5)_2$	$(n\text{-}C_8H_{17})_2POH$	I	87	28
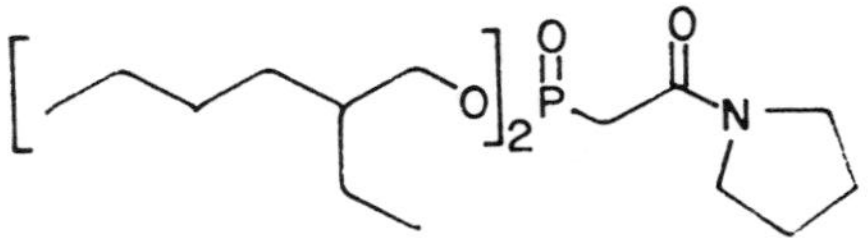 pyrrolidine amide phosphonate structure	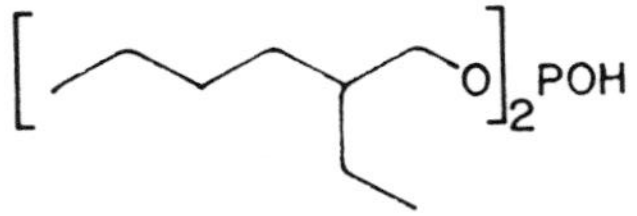 $\left[\begin{array}{c}\\ O\end{array}\right]_2 POH$	I	85	28
$[CH_3(CH_2)_3CH(C_2H_5)CH_2O]_2P(O)CH_2CON(C_2H_5)_2$	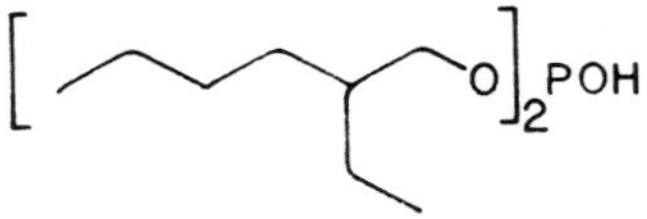 $\left[\begin{array}{c}\\ O\end{array}\right]_2 POH$	I	85	28

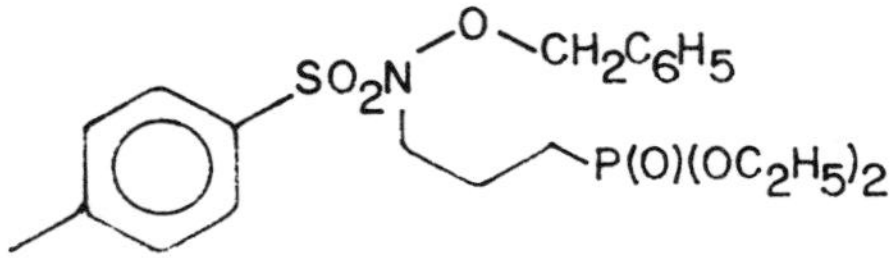

Table 1 (continued)
FORMATION OF CARBON TO PHOSPHORUS BONDS *VIA* THE MICHAELIS-BECKER REACTION

Note: A, sodium alkoxide used as base; organic halide used as substrate; B, sodium metal used as base; epoxide used as substrate; C, internal amine site acting as base; D, sodium metal used as base; organic halide used as substrate; E, aziridine used as substrate; F, amine used as base; organic halide used as substrate; G, sodium hydride used as base; organic halide used as substrate; H, lithium metal used as base; organic acetate used as substrate; I, phase transfer catalysis; organic halide used as substrate.

and filtered, the filtrate was evaporated under reduced pressure to give the di-*n*-butyl *N,N*-diethylcarbamoylphosphonate (27.9 g, 91%).

V. MICHAELIS-ARBUZOV REACTIONS

A. Reactions and Mechanism

The direct displacement of a suitable leaving group from carbon by a fully esterified trivalent phosphorus acid remains as the most common method for the formation of carbon to phosphorus bonds in quinquevalent organophosphorus compounds. This is the situation due not only to the relative availability of the required starting materials, but also to the relative ease of the performance of the reaction. In most instances, little more than heating the reagents together in the absence of solvent and purification of the product by distillation is required. With this simplicity, even the fact that yields are often only in the fair to moderate range does not cause one to look for other methods.

Since the earliest work using methyl iodide and triphenyl phosphite,[59] the reaction system has been broadened to allow wide variation in the nature of the trivalent phosphorus reagent and the carbon-centered substrate. As noted previously, this review is concerned primarily with developments in the use of the Michaelis-Arbuzov reaction since 1970. Much of the work reported in this period, of course, involves rather straightforward extensions of earlier developments. This work will be noted with relatively little comment. However, several areas of more major advances will receive extensive consideration.

For earlier synthetic accomplishments and compound tabulations, the reader is referred to several prior reviews. Of particular note are two early reviews by Kosolapoff,[60,61] the 1964 review by Harvey and de Sombre,[62] and two relatively recent treatments in the American[63] and Russian literature.[64]

Viewing the overall Michaelis-Arbuzov reaction, it is evident that more than one step is involved.

$$RX + (R'O)_3P \longrightarrow RP(O)(OR')_2 + R'X$$

Thereby, consideration of its mechanism must be undertaken in several stages. The first of these stages has as its concern the nature of the addition of phosphorus to the carbon center. An understanding of this process in detail involves three separate factors: (1) the electronic nature of the approach of phosphorus to carbon; (2) the stereochemical fate of the central carbon atom; and (3) the stereochemical fate of the phosphorus center.

Following this, a second stage in understanding the overall mechanism is concerned with the nature of the intermediate species generated upon initial formation of the carbon-phosphorus bond. And finally, there is a concern for both the electronic and stereochemical aspects of the formation of the by-product alkyl halide in the reaction.

It is clear that, with certain peculiar exceptions, the initial association of phosphorus with carbon is a polar process rather than radical in nature. Within the spectrum of possible polar processes, however, a definitive statement is not simple to make as the process appears to vary with the nature of the alkyl halide substrate.

The fundamental kinetic observation with simple alkyl halides would tend to indicate that the attack by phosphorus on carbon is of a classic S_N2 nature.[62] That is, maintaining other factors constant, iodides react faster than bromides, which in turn react faster than chlorides. Moreover, methyl and other primary halides react significantly faster than secondary halides, the latter usually resulting principally in elimination reaction. Ordinary tertiary halides fail to give substitution, usually reacting via an elimination route. The critical stereochemical experiment, involving use of a chiral secondary alkyl halide, has been attempted.[65] However, elimination rather than substitution occurs with the secondary halide, precluding a definitive answer to the question.

Variation of the substituents of the trivalent phosphorus reagent would also seem to support the concept of an initial S_N2 process. Replacement of ester linkages on phosphorus by alkyl or aryl groups increases the reactivity for Michaelis-Arbuzov reaction.[66-68] That is, phosphinites react more rapidly than do phosphonites, which in turn react faster than phosphites.

$$(C_2H_5)_2POC_2H_5 \quad > \quad C_2H_5P(OC_2H_5)_2 \quad > \quad P(OC_2H_5)_3$$

It is further observed that Michaelis-Arbuzov reaction occurs when substrates are used which serve as sources of relatively stable carbocations.[69-71] For those systems where carbocations may be generated with ease, reaction can proceed via an S_N1 route.

Thus, there is seen to be a significant variability possible in the initial step of the reaction. While kinetic studies indicate a favored role for a good nucleophile in this initial step pointing toward an S_N2 approach,[72,73] if a particularly favorable substrate is presented, reaction can proceed via an S_N1 approach. It would be of interest to see the carbon stereochemical result upon the use of a chiral primary substrate in the Michaelis-Arbuzov reaction. Such a substrate has been available for quite some time with the chiral 1-butylbrosylate-1-*d*.[74]

An understanding of the stereochemistry about the phosphorus center in the initial attack has also been the target of numerous investigations. Due to the potential of phosphorus to form bonds to more than four other atoms, the possibility of phosphorus binding to both halogen and carbon more or less simultaneously to generate a phosphorane intermediate must be considered. The two possibilities of phosphorane intermediate and simple nucleophilic displacement of halide would be anticipated to yield different stereochemical results at phosphorus. If a phosphorane were generated, pseudorotation[75] would lead to a loss of stereochemical integrity at the phosphorus center.

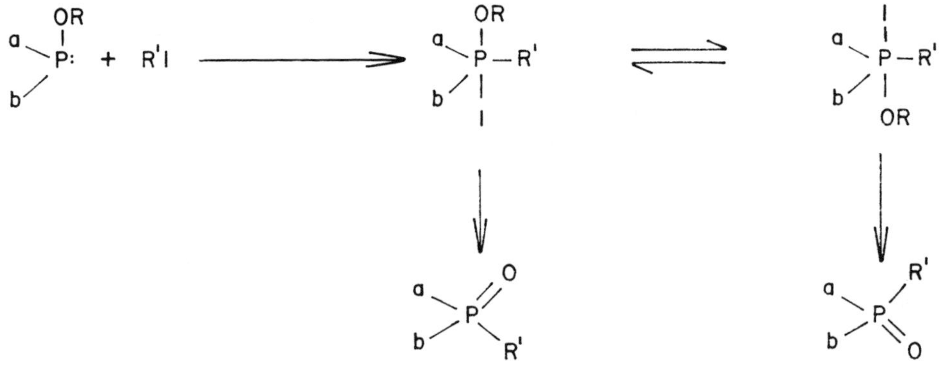

If the process were one of simple nucleophilic displacement leading to a tetracoordinated phosphorus center, retention of configuration at phosphorus would be anticipated.

This question has been investigated through the use of cyclic trialkyl phosphites, 2-ethoxy- and 2-isopropoxy-4-methyl-1,3,2-dioxaphosphorinane, in reaction with both alkyl halides and a source of trityl cation.[76,77]

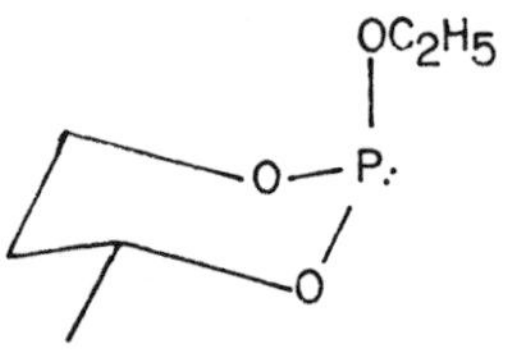

Assuming that this cyclic phosphite maintains its stereochemical integrity prior to undergoing Michaelis-Arbuzov reaction, observation of a single product, as contrasted with a pair of products, would identify the stereochemical course at phosphorus as involving retention of configuration.

In reactions using ordinary alkyl iodides, nearly complete loss of stereochemical integrity at phosphorus was reported.[76] However, the opposite result was observed when trityl tetrafluoroborate was used,[77] leading to the conclusion that while a phosphorane species was involved in the former reaction, only a quasi-phosphonium ion was generated in the latter "S_N1-type" reaction. The reaction of the same cyclic phosphite with methyl iodide was later investigated by another group and found also to proceed with retention of configuration at phosphorus.[78] These workers concluded that the lack of stereospecificity reported in the initial investigation was the result of a preliminary inversion of the phosphite before it reacted with the alkyl halide.

When the structurally related 2-methoxy-5-t-butyl-1,3,2-dioxaphosphorinane was used in reaction with methyl iodide, clearly a single phosphonate product isomer was obtained with retention of configuration at phosphorus.[79,80]

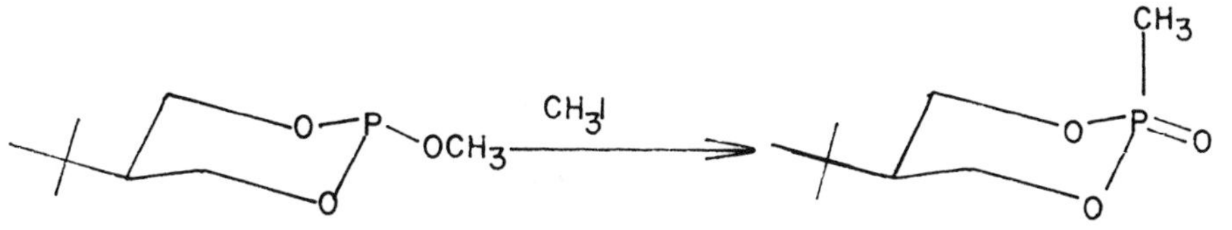

Both of these systems depended on the formation of a particular geometric isomer or isomer pair to understand the stereochemistry of the reaction. Two reports have also been made wherein chiral phosphorus centers were involved in the Michaelis-Arbuzov reaction. A chiral silylated phosphonite, produced from the chiral partial ester of phenylphosphonous acid, reacted cleanly with ethyl or methyl iodide to generate the chiral phosphinate with virtually complete retention of configuration at phosphorus.[18]

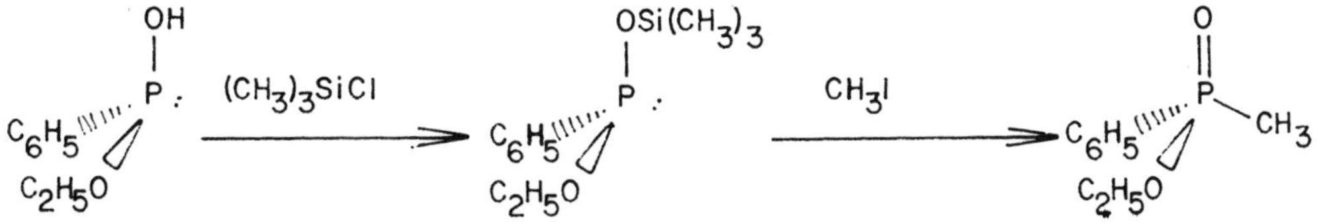

Similarly, the Michaelis-Arbuzov reaction of a series of chiral phosphinites has been reported to proceed with complete retention of configuration.[81]

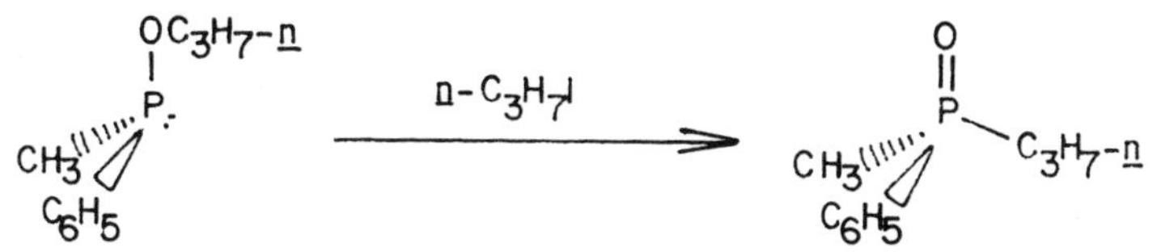

Although intermediate species in the Michaelis-Arbuzov reaction had been observed and isolated even in the earliest studies,[59] until recently only indirect and equivocal evidence had been gained as to their nature. With dialkyl phosphonites, particulartly reactive trivalent phosphorus species, intermediates are formed without heating and can be isolated readily.[82,83] Similarly, adducts of triphenyl phosphite or phenyl phosphonites with alkyl halides form readily and decompose slowly, thereby being excellent candidates for NMR investigation. Both ^{31}P and ^{1}H NMR spectra of such adducts clearly indicate a phosphonium ion structure rather than a phosphorane structure to be involved.[84-88]

Nuclear magnetic resonance studies have also been performed using intermediates derived from two trialkyl phosphites. Trineopentyl phosphite, a reagent which undergoes the second step of the Michaelis-Arbuzov reaction rather slowly, gives an intermediate exhibiting a ^{31}P signal indicating clearly phosphonium ion character to be dominant.[88,89] Likewise, trimethyl phosphite taken in reaction with methyl triflate, a reagent generating the Michaelis-Arbuzov intermediate without heating, yields the same conclusion upon NMR analysis.[90]

Finally, an X-ray analysis of the intermediate from reaction of dineopentyl phenylphosphonite with methyl iodide indicates the phosphorus center to be bound with a tetrahedral geometry.[91] Thus, under a variety of experimental conditions, the intermediate is indicated to have a phosphonium ion type structure. A review of the ''quasi-phosphonium ions'' produced by the Michaelis-Arbuzov reaction and other methods has recently been presented.[5]

The evidence regarding the dealkylation (final) step of the Michaelis-Arbuzov reaction remains in a conflicting state. The clearest piece of evidence comes from the stereochemical work reported over 30 years ago by Gerrard and Green.[92] Chiral tri-2-octyl phosphite, which was prepared from (*S*)-2-octanol, underwent reaction with methyl iodide to generate (*R*)-2-iodooctane, evidence of a clean inversion of configuration at carbon in the dealkylation step, and an S_N2 nature to the reaction.

Early qualitative kinetic studies do not appear to be in accord with this concept, however. Reactions of simple alkyl halides with trialkyl phosphites derived from secondary alcohols appear to proceed significantly faster than those using trialkyl phosphites derived from primary alcohols.[62,93] In recent years, careful kinetic measurements have been performed on the decomposition of an isolated intermediate quasi-phosphonium ion, diisobutyloxymethylethylphosphonium iodide.[94] The reaction was found to exhibit variable kinetics, first order at high concentration and second order at low. The reaction was also found to be significantly faster when performed in methylene chloride as compared to acetonitrile.

The classical S_N2 route for dealkylation would be impossible for intermediates derived from reaction of triadamantyl phosphite. Yet, both the tri(1-adamantyl) and tri(2-adamantyl) phosphites react with a variety of halide reagents and proceed to completion in moderate to good yield.[95] Presumably a variability in mechanism is possible depending on the reagent presented.

Historically, the Michaelis-Arbuzov reaction has involved the heating of a trivalent phosphorus ester, either alkyl or aryl, with an alkyl halide for the generation of the quinquevalent phosphorus product. While there have been numerous efforts in recent years to expand the scope and utility of the reaction by the introduction of fundamental structural changes in the reagents and modification of the reaction conditions, many have found the simple reaction system to continue to be of use for syntheses of quinquevalent organophosphorus materials. The variety of these recent synthetic efforts will be reviewed here briefly.

The reaction of trialkyl phosphites with simple aliphatic halides has been undertaken to generate phosphonate diesters for a variety of applications in recent years, but particularly for biologically related purposes. One application has involved the preparation of alkylphosphonate diesters for further use as reagents for Horner-type reactions. Included here are the syntheses of *cis*-jasmone,[96] and other natural products.[41,97]

The fundamental Michaelis-Arbuzov reaction has been used in several recent efforts for the preparation of phosphorus-containing analogues of natural products. These include precursors to phosphonic and phosphinic acid analogues of natural phosphates in nucleotide,[98-102] carbohydrate,[103-106] amino acid,[107,108] glycerol,[109-111] isoprenoid,[112] and lipid[113-116] series.

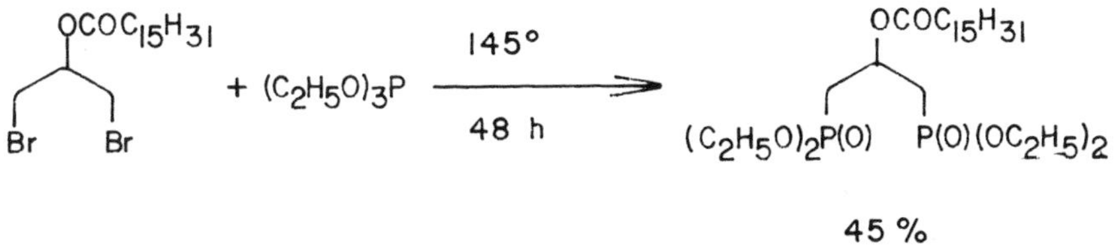

Several antiviral agents are also prepared directly by Michaelis-Arbuzov reaction of primary alkyl halides and trialkyl phosphites.[117-119]

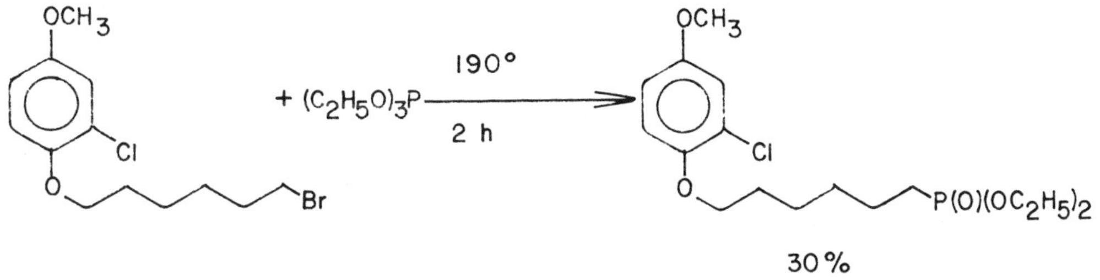

Other investigations have been performed which use the reaction of trialkyl phosphites with simple alkyl halides to generate phosphonates containing singular structural features. One particularly interesting category includes those trivalent phosphorus esters which undergo ''internal'' reaction, that is, contain an alkyl halide functionality within the ester linkage.[120-123] The products are of utility as precursors for flame retardant materials. Also interesting are reactions involving the formation of cyclopropylphosphonates[124] and cyclic phosphonoesters.[125]

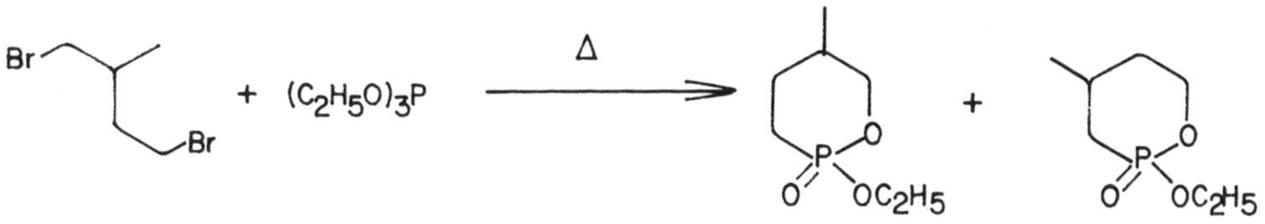

In addition to the trialkyl phosphites, numerous applications of the Michaelis-Arbuzov reaction of dialkyl alkylphosphonites with alkyl halides have been reported recently. Included here are syntheses of analogues of natural amino acids,[108] and of the natural phosphinate antibiotic, phosphinothricin[126,127] and its analogues.[127]

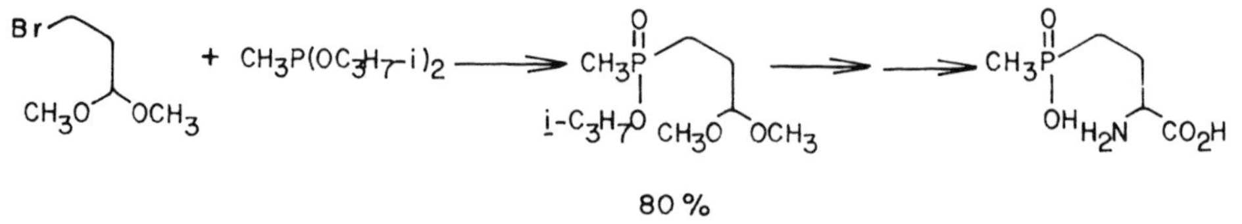

80%

Similarly, analogues of phospholipids,[128,129] portions thereof,[130] and other surfactants[131] are prepared using phosphonite diesters in these reactions.

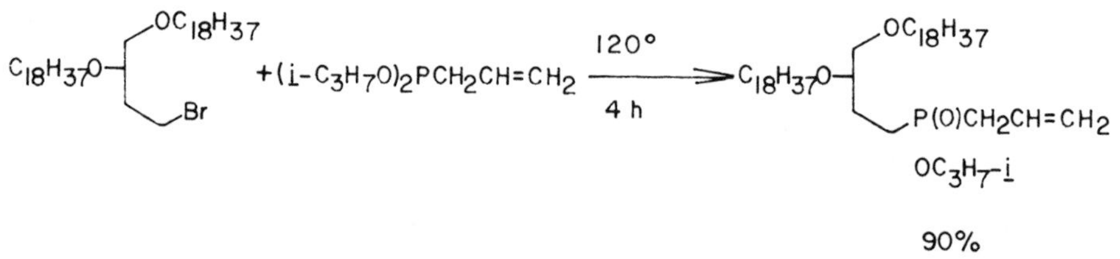

90%

An ingenious reaction system for the generation of more highly alkylated phosphorus centers involves the Michaelis-Arbuzov reaction of trivalent phosphorus species generated *in situ.* Quinquevalent phosphorus esters undergo facile reduction to the trivalent esters by reaction with bis(2-methoxyethoxy) aluminum hydride (vitride).[132] In a one-pot procedure, a phosphonate diester can be reduced to the corresponding phosphonite diester, to which the alkyl halide is added directly for the Michaelis-Arbuzov reaction.[133] This approach has been used for the synthesis of a series of macrocyclic diphosphines of interest as potential bidentate ligands for transition metals.[134]

Allylic halides exhibit particularly high reactivity in the Michaelis-Arbuzov reaction, and generally give high yields of phosphonate diesters with trialkyl phosphites. Several reports of such reactions have been made in recent years.[135-138] One report of particular interest for biological application is concerned with the synthesis of an isosteric phosphonic acid analogue of phosphoenolpyruvate.[139]

71%

Similarly, benzyl- and substituted benzylphosphonate diesters have been prepared by this route.[140-143] The early report of Kosolapoff[144] in which 9-chloroacridine was used as a substrate for the Michaelis-Arbuzov reaction is a special example in this category. Two reports have been made which compare directly the Michaelis-Becker and Michaelis-Arbuzov routes to these substituted benzylphosphonates.[145,146] In both instances the Michaelis-Arbuzov route was found to give superior yields.

Two other reaction systems are worthy of mention here. Multiple reaction is observed with poly(halomethyl)aromatics. A tetraphosphonate is formed on reaction of triethyl phosphite with 2,2′,6,6′-tetra(bromomethyl)biphenyl,[147] although the material decomposed when distillation was attempted. Also, Michaelis-Arbuzov reaction is found to proceed in high yield at a benzylic chloride site in the presence of a sulfonyl fluoride linkage.[148]

80 %

As expected, reaction proceeds quite readily between benzylic halides and phosphinous esters.[149] A series of analogues of phosphinothricin has been prepared from benzylic halides and diethyl 2-chloroethylphosphonite.[150] The benzylic halide added is significantly more reactive than the internal halide, allowing high yields of target material to be isolated when nickel salts were added as catalyst.

Reaction of trialkyl phosphites with acetylenic halides proceeds readily and in a controlled manner. By judicious choice of reaction conditions, mono- or disubstitution can be accomplished with dihaloacetylene.[151,152] Both symmetrical and mixed acetylenediphosphonates can be prepared readily.[152]

$$(\underline{i}\text{-}C_3H_7O)_3P \quad + \quad ClC\equiv CCl \xrightarrow[\text{room temp}]{1\ h} (\underline{i}\text{-}C_3H_7O)_2P(O)C\equiv CCl$$

77 %

Alkylphosphonate diesters bearing heteroatom substituents at the position adjacent to phosphorus have gained particular significance in recent years as organic reagents. One approach to the preparation of such materials is through the Michaelis-Arbuzov reaction of trialkyl phosphites with 1-haloethers. Several reports have appeared on the preparation of such materials by this route for their use as *umpolung*[153-156] and Wadsworth-Emmons[157] reagents. Moderate to excellent yields are obtained in these reactions.

$$C_6H_5CH(OCH_3)Cl \quad + \quad (C_2H_5O)_3P \longrightarrow C_6H_5CH(OCH_3)P(O)(OC_2H_5)_2$$

87 %

Similarly high yields are obtained using ethyl diphenylphosphinite to generate reagents for Wittig-Horner type reactions.[158]

Geminal dichloro compounds have been used in several instances to give phosphonates highly functionalized at the carbon adjacent to phosphorus.[159,160] Halomethyl sulfides have also been reported to participate readily in Michaelis-Arbuzov reactions with trialkyl phosphites. The resultant thiomethylphosphonates have been used in Horner-Wittig reactions[161] and as specific enzyme inhibitors.[162]

A further interesting reaction system with chloromethyl ethers involves the use of methyl phosphorodichloridite as the trivalent phosphorus reagent.[163,164] Catalyzed by boron trifluoride, a Michaelis-Arbuzov reaction is observed to occur generating the phosphonic dichloride. While only fair to good yields are obtained, the presence of the phosphonic dichloride function provides a useful versatility for further chemical transformations.

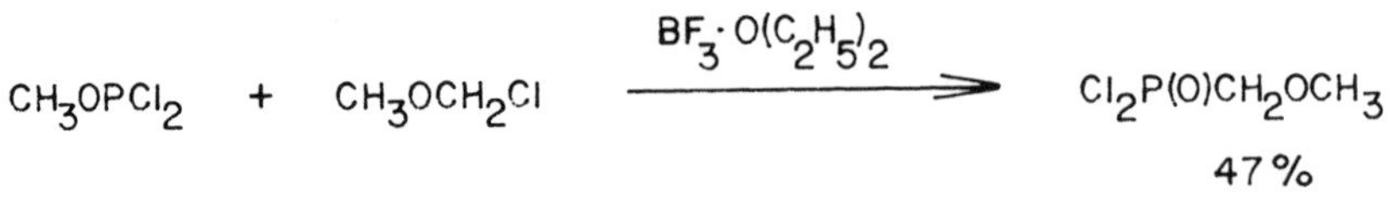

$$CH_3OPCl_2 \quad + \quad CH_3OCH_2Cl \quad \xrightarrow{\quad BF_3 \cdot O(C_2H_5)_2 \quad} \quad Cl_2P(O)CH_2OCH_3$$

$$47\%$$

Catalysis of reaction of methyl phosphorodichloridite using ferric choride has been used with α-haloisocyanates to give the phosphonic dichloride in good yield.[165]

Haloepoxides, a special category of 1-haloethers, exhibit a singular reaction with trialkyl phosphites. Moderate to excellent yields of 2-ketophosphonates are obtained from them under relatively mild conditions.[166]

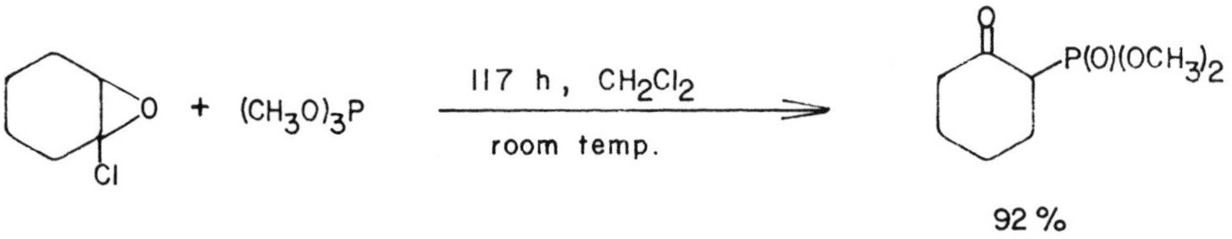

$$92\%$$

A series of *N*-halomethylamides have also been demonstrated to undergo Michaelis-Arbuzov reaction easily and in good yield.[24] Two patents have been issued using this route for the preparation of the herbicide *N*-phosphonomethylglycine and its analogues.[167,168] In addition, a cyclic amide recently has been reported to participate well in the same type of reaction.[169]

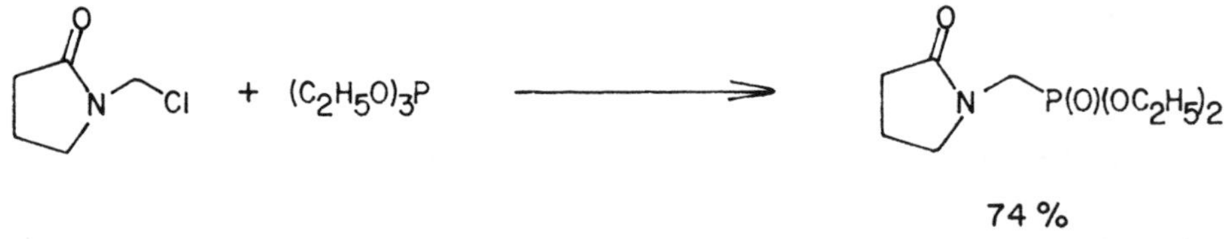

$$74\%$$

The use of the haloepoxides for the synthesis of 2-ketophosphonates represents an important synthetic achievement.[166] A route that might be anticipated to generate such compounds, Michaelis-Arbuzov reaction of α-haloketones, generally leads instead to phosphate esters via the Perkow reaction (*vide infra*). Only in a few instances are the direct reactions of 1-haloketones with trialkyl phosphites feasible for the synthesis of the 2-ketophosphonates.[170-172] Usually a preliminary masking of the carbonyl function is required prior to reaction with the trivalent phosphorus reagent.[49,173-176]

The closely related α-halocarboxylate derivatives are not troubled by the alternative Perkow reaction pathway. These materials undergo Michaelis-Arbuzov reaction with trialkyl phosphites readily to generate 1-phosphonocarboxylates in good yield. The 1-phosphono-carboxylates generated in this manner have found use as Wittig-Horner reagents,[177-179] herbicides,[166] and precursors to analogues of natural amino acids.[180-182] Similarly, 2-chloroimidoyl chlorides,[183] α-haloamides,[152] and α-halonitriles[184-187] readily undergo this reaction. Yields vary widely, but generally are in the fair to good range.

$$(C_2H_5O)_3P \ + \ \text{Cl}\!-\!\!\text{CH}_2\!\!-\!\!\overset{H}{N}\!\!-\!\!\text{P(O)(OC}_6\text{H}_5)_2 \ \xrightarrow{\Delta} \ (C_2H_5O)_2\overset{O}{\underset{}{P}}\!\!-\!\!\text{CH}_2\!\!-\!\!\overset{H}{N}\!\!-\!\!\text{P(O)(OC}_6\text{H}_5)_2$$

63 %

With the more reactive phosphonite diesters[188] and phosphinite esters,[189] generally higher yields are observed with the same categories of α-halocarboxylate derivatives. This increased yield is particularly convenient if the materials are to be used in Wittig-Horner type reactions wherein the phosphorus component of the material is ultimately eliminated and its exact structural nature is relatively unimportant.

The formation of 1,1-bisphosphonates and related materials via the Michaelis-Arbuzov reaction should also be noted. The preparation by this route and the utility of the methylenebisphosphonic acids has been known for quite some time.[190] In recent years, several reports indicating improvements in yields have been made.[191,192] Moreover, successful applications to syntheses of analogues of nucleotides[193] and phospholipids[194] normally bearing pyrophosphate linkages have been reported.

$$+(C_2H_5O)_3P \ \xrightarrow[\text{48 h}]{150°}$$

54 %

Acyl halides constitute a category of extremely reactive substrates for the Michaelis-Arbuzov reaction which have received particularly strong attention in recent years. Since the report of the preparation of dimethyl acetylphosphonate,[195] numerous acid chlorides have been used as facile substrates for studies of the scope of the reaction. After the utility of the products as organic reagents and biologically interesting materials was established, an even greater interest in the products of the reaction developed.

In some reports, the reaction conditions as stated include heating the mixed reagents at relatively high temperatures for several hours. This is generally unnecessary. It is our experience that excellent results can be obtained by dropwise addition of the acyl halide to the trivalent phosphorus reagent with vigorous stirring in the absence of solvent. The addition rate is controlled to allow a mild exotherm to occur which sustains the reaction as the trivalent phosphorus reagent becomes consumed. If possible, the by-product alkyl halide should be vented from the reaction system. Some of the early syntheses have been reinvestigated with modified conditions to provide products in higher yield.[196-202]

The use of acyl chlorides in Michaelis-Arbuzov reactions has been spurred by activity in several areas of biological chemistry. In the expansion of studies of biological regulation by structural analogues of natural substrates, interest developed for studies involving the replacement of the carboxylic acid function of the α-amino acids by acidic phosphorus sites.

As a ketonic function could easily be converted to an amino function, the synthesis of α-ketophosphonic acids for this application is of great significance. Reports have been made on the synthesis of analogues of tryptophan,[203] phenylalanine,[204,205] leucine,[204] isoleucine,[204] valine,[204] cysteine,[206] and proline[207] by this route.

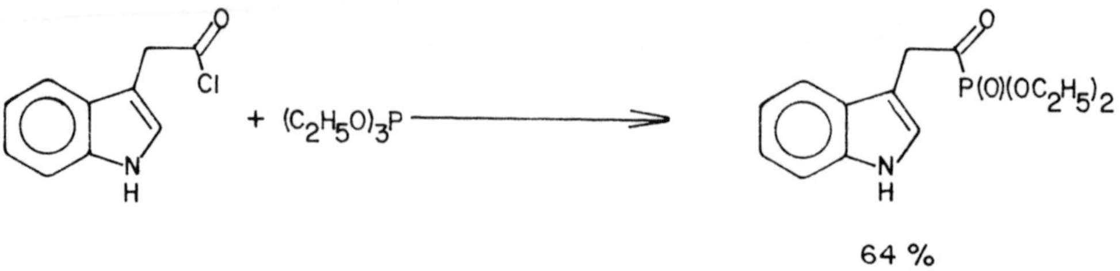
Cl
+ (C2H5O)3P
P(O)(OC2H5)2
N
H
N
H
64 %

The α-ketophosphonates have also been prepared in this manner for use as herbicides,[208,209] inhibitors of alanine racemase,[47] and as analogues of coumarin derivatives.[210]

Chloroformates constitute a particularly interesting class of acyl halide. Being both esters and acid chlorides, they bring a wealth of functionality to the products of the Michaelis-Arbuzov reaction. The first report of the synthesis of a member of this class of compounds, triethyl phosphonoformate, was made at a very early date[211] and improvements were later made on the procedure.[212] The recognition of phosphonoformic acid as an antiviral agent (while being investigated in a routine screening procedure) resulted in a series of synthetic efforts using these reactive acyl chlorides.[20,213-216]

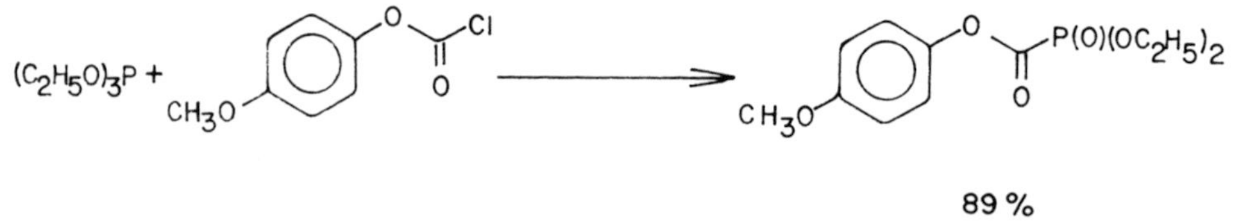
(C2H5O)3P +
CH3O
O
Cl
O
CH3O
O
P(O)(OC2H5)2
O
89 %

The thermal decomposition of these phosphonoformate esters has recently been investigated.[217]

The facile formation of phosphonates which are functionalized at the α-position also makes this an important route for the preparation of useful organic reagents. After reduction or amination, the derivatives of α-ketophosphonates can serve in a variety of reactions as acyl-anion equivalents, *umpolung* reagents. To this end syntheses of numerous α-ketophosphonates have been performed.[218-223] The α-iminophosphonates can also be generated directly by the Michaelis-Arbuzov reaction using imidoyl chlorides.[11]

Cl3CCH2OCOCl + (C2H5O)3P
Cl3CCH2OCOP(O)(OC2H5)2
71 %

Numerous other reports have been made regarding the preparation of α-ketophosphonates by this route for a wide range of purposes.[224-234]

Two interesting reports have also appeared in which products of carbon-phosphorus bond formation are not isolated. The reaction of acyl halides with triphenyl phosphite is reported to lead to the formation of phenyl carboxylates and diphenyl phosphorochloridite.[235] Presumably, choride ion attacks phosphorus rather than carbon of the intermediate quasi-phosphonium ion.

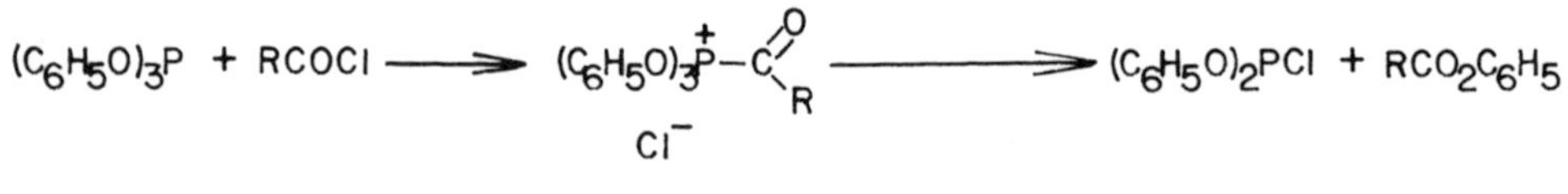
(C6H5O)3P + RCOCl
(C6H5O)3P–C
O
R
Cl−
(C6H5O)2PCl + RCO2C6H5

Benzoyl chloride was used as an example of a particularly reactive species participating in the Michaelis-Arbuzov reaction to demonstrate the lack of reactivity of bridgehead-phosphorus esters.[236] Even with the highly reactive substrate, no reaction was observed due to the stereoelectronic effect.

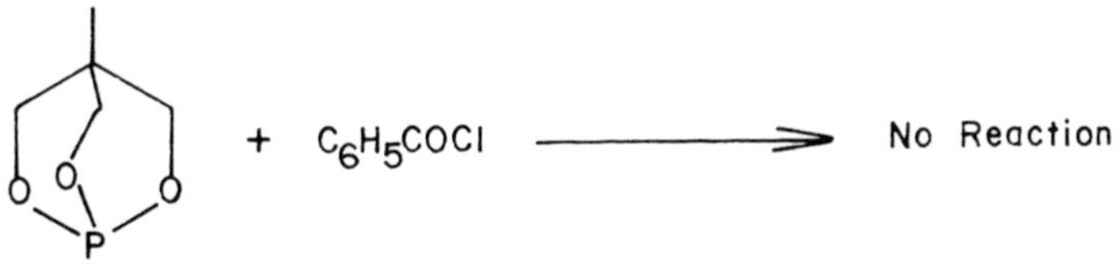

The Michaelis-Arbuzov reaction has been observed to occur with leaving groups from carbon other than halide. For example, alkyl quaternary ammonium compounds can undergo reaction in much the same way as the alkyl halides. A tertiary amine is displaced which performs the subsequent dealkylation. In several instances where direct comparison is possible, yields by this approach were found to be superior to those obtained using alkyl halides.[54,237-239]

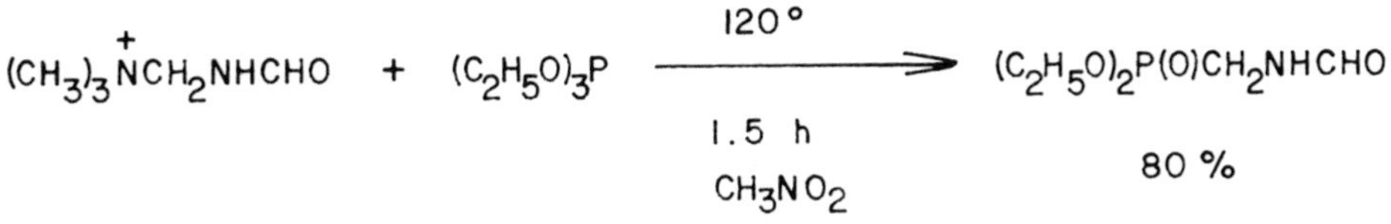

The reaction proceeds in good yield with benzylic ammonium species as well.[240] *N*-Acylisoquinoline salts undergo phosphorus addition (rather than displacement) at the 1-position.[241,242]

In a few instances, acetate has been used successfully as a leaving group from alkyl carbon. With triphenyl phosphite in the presence of acetic acid, *N*-(acetoxymethyl)-benzylcarbamate reacts to give a substituted phosphorus analogue of glycine.[243.] This route bears some practical advantages over other approaches to the same material.

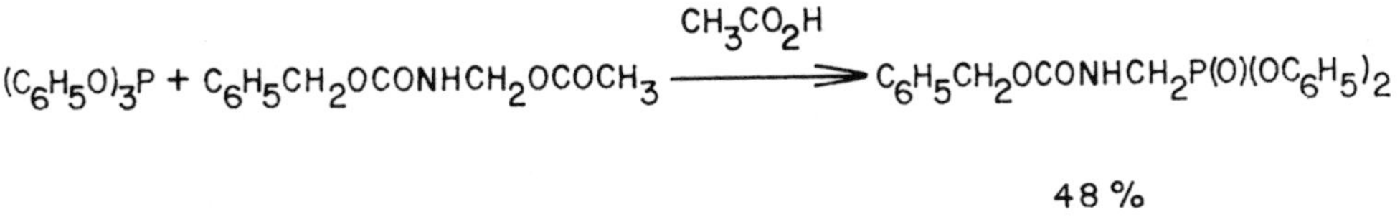

Michaelis-Arbuzov reaction is found to proceed reasonably well with 4-acetoxyazetidin-2-ones, providing yet another approach to phosphorus analogues of natural amino acids.[244,245]

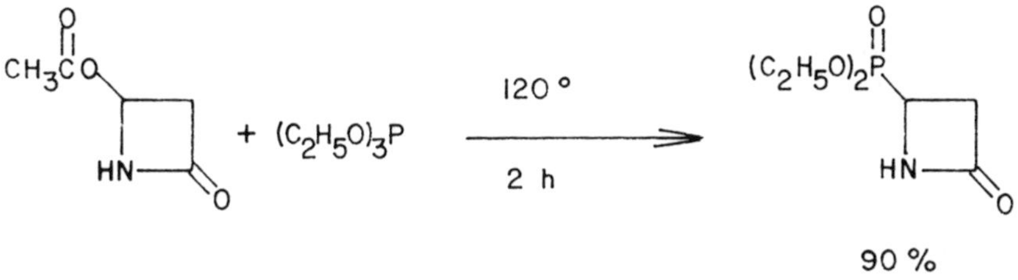

Reaction is also observed with *N*-acylaziridines proceeding with nucleophilic opening of the strained ring.[58,246] Dealkylation is accomplished by nitrogen of the intermediate zwitterion, although the inter- or intramolecular nature of this dealkylation is not defined.

$$(C_2H_5O)_3P \ + \ \underset{NCOC_6H_5}{\triangle} \quad \xrightarrow[\text{15 h}]{110°} \quad (C_2H_5O)_2\overset{O}{\overset{\parallel}{P}} \diagdown N \diagup \overset{O}{\underset{\parallel}{C}} C_6H_5$$

66 %

Several reports have appeared recently in which acids have been used to facilitate reaction. With a derivatized glycosyl halide, antimony pentachloride was used to generate a carbocation which subsequently underwent Michaelis-Arbuzov reaction with triethyl phosphite.[35] Acetals of substituted benzaldehydes undergo Michaelis-Arbuzov reaction after treatment with boron trifluoride etherate.[247] A carbocation is generated to which the trialkyl phosphite adds readily.

$$Cl\diagdown \bigcirc \overset{OCH_3}{\underset{OCH_3}{\diagdown}} + (C_2H_5O)_3P \quad \xrightarrow{F_3B \cdot O(C_2H_5)_2} \quad Cl\diagdown \bigcirc \overset{P(O)(OC_2H_5)_2}{\underset{OCH_3}{\diagup}}$$

80 %

This represents yet another approach to phosponates with heteroatom substitution adjacent to phosphorus, compounds of use for *umpolung* processes. Organic acids have also been used to facilitate the rearrangement of dialkyl phosphonites to phosphinate esters[248] and the reaction of alcohols with triphenyl phosphite.[249] The latter reaction generates Michaelis-Arbuzov type products under relatively mild conditions.

A major advance in the use of the Michaelis-Arbuzov reaction has come with the preparation of silyl esters of the trivalent phosphorus acids. Their application to syntheses of biologically significant materials has recently been reviewed.[250] Several methods for their preparation are now available.[251-255] Of these, the last[255] is the most convenient for the preparation of tris(trimethylsilyl) phosphite. The use of silyl esters provides a particularly convenient approach for the preparation of the free quinquevalent phosphorus acids in the Michaelis-Arbuzov reaction. They participate readily in the reaction, and the ester linkages may be hydrolyzed under extremely mild conditions to generate the free acids. Moreover, with mixed ester trivalent phosphorus reagents, the silyl ester linkage is that which is always transferred in the dealkylation process rather than alkyl or aryl ester functions.

Once the tris(trialkylsilyl) phosphites had been prepared, their use in a variety of Michaelis-Arbuzov reactions followed quickly. Relatively high yields for the valence expansion reaction were reported using simple alkyl halides.[251,256]

Tris(trimethylsilyl) phosphite is found to be a particularly convenient reagent for the preparation of phosphonolipids, phosphonic acid analogues of natural phospholipids. In the syntheses of phosphonolipids, liberation of the free phosphonic acid subsequent to carbon-phosphorus bond formation, without disturbing other functionalities present, can be a difficult experimental task. With the phosphonate being formed bearing silyl esters, hydrolysis can be achieved under extremely mild conditions, stirring with aqueous tetrahydrofuran. Using this approach numerous phosphonolipids have been prepared.[257-261]

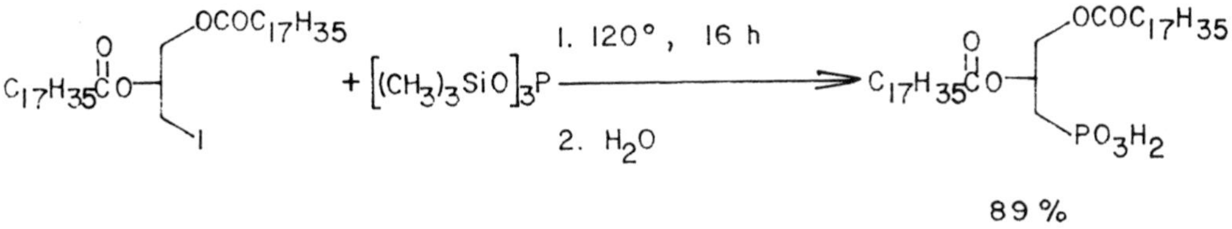

89 %

Of course, the silyl phosphites undergo reaction readily with acid chlorides, providing a facile route to the free α-ketophosphonic acids.[218,224,225,262,263]

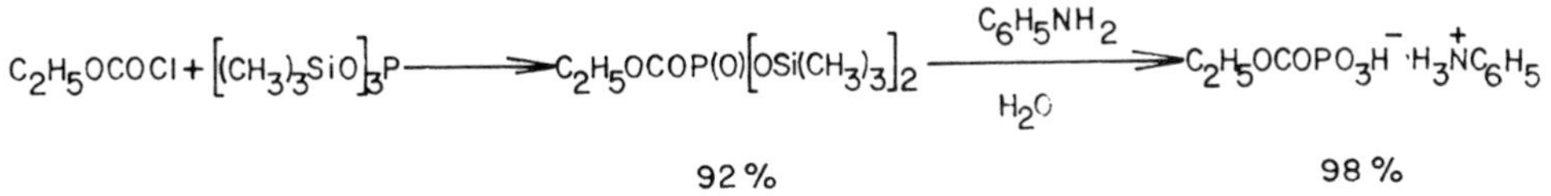

92 %   98 %

Silyl esters of other trivalent phosphorus acids have been used in the Michaelis-Arbuzov reaction. These include phosphinous esters,[264,265] phosphonous esters,[266,267] and mixed phosphites.[221] When mixed ester reagents are used, it is only the silyl ester function which is lost from phosphorus in the dealkylation step. A notably convenient reagent for organophosphorus syntheses is the bis(trimethylsilyl) trimethylsiloxymethylphosphonite,[266] produced easily from readily available reagents.[268] The material undergoes facile Michaelis-Arbuzov reaction to generate an α-functionalized phosphinate, useful for further derivitization.

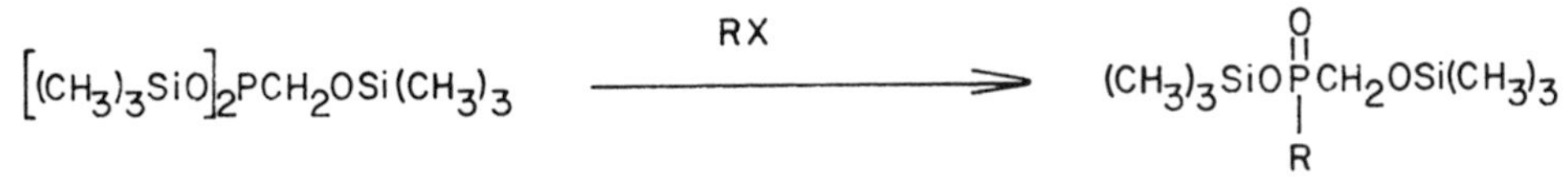

In the past, simple alkyl esters of the trivalent phosphorus acid were generally used for the performance of the Michaelis-Arbuzov reaction. Although vinylic and aromatic esters readily undergo the first step of the reaction, dealkylation does not occur easily with the aromatics and the reaction is not at all useful with the vinylics.[269,270] In recent years, efforts have been made to facilitate reactions of aromatic esters, particularly triphenyl phosphite.[249,271-274] Aside from the use of higher than "normal" temperatures, above 250°C, success has been obtained through the addition of nickel[272,273] and strong organic acids.[249]

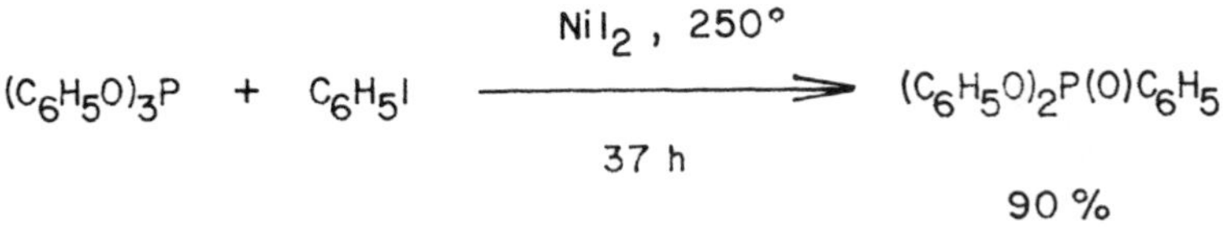

37 h

90 %

An anomalous reaction has been noted in the reaction of triphenyl phosphite with carboxylic acid chlorides.[235] Presumably a carbon-phosphorus bond is formed initially, but is ultimately broken. Isolated are the phenyl ester of the carboxylic acid and diphenyl phosphorochloridite.

Some effort has been devoted to the general catalysis of the Michaelis-Arbuzov reaction. In addition to those reaction systems to be discussed later in Chapter 6 and concerned with Michaelis-Arbuzov type reactions of aromatic and vinylic halides, several reports have been made using common Lewis acids as catalytic agents. These include nickel salts,[272,275] ferric salts,[165] and zinc salts.[276] Raney nickel has also been used successfully,[277] as has been palladium acetate,[278] and protic acids when alcohols are the substrate.[248,249]

Brief mention has been made already of "abnormal" reactions which are found to occur under Michaelis-Arbuzov conditions. The most prominent "abnormal" process is the Perkow reaction, the formation of vinylic phosphate esters rather than 2-ketophosphonates upon reaction of α-haloketones with trialkyl phosphites. For example, in an attempt to prepare a phosphonic acid analogue of ribose 5-phosphate, a vinylic phosphate ester was produced upon treatment of the related chloromethylketone with trimethyl phosphite.[279]

The Perkow route is generally found to be a process competitive with the Michaelis-Arbuzov route whenever there is an α-haloketone or α-haloaldehyde involved. Several studies of this competition as a function of the electronic nature of interacting substituents have been reported.[280-282] It would appear that strongly electron withdrawing functionalities result in a favoring of the Perkow route, whereas electron donating groups give predominantly Michaelis-Arbuzov type product. This conclusion is in keeping those of other mechanistic investigations of the reaction[283-285] which suggest that the reaction involves reversible addition of phosphorus to the carbonyl function. (See Chapter 3, the Abramov and Pudovik reactions.) Immediate loss of halide leaves phosphorus bound to the carbonyl oxygen and results in the formation of the olefinic linkage. With halogen further removed from the carbonyl linkage, other "abnormal" reactions involving initial phosphorus attack at the carbonyl position are found.[48,112]

Several reports have been made of abnormal reactions involving trivalent phosphorus amides.[286,287] These can best be understood as proceeding via nitrogen rather than phosphorus attack at carbon. In another report,[124] significant amounts of reduction are noted to occur when the Michaelis-Arbuzov reaction is performed with a 1,1-dibromocyclopropane in the presence of a base and a hydroxylic solvent.

In reactions wherein a secondary alkyl halide is used, elimination often competes effectively with substitution by phosphorus. In one instance using an α-haloether, elimination is reported to be the dominant process.[288]

As noted previously, a bridgehead phosphite triester failed to give Michaelis-Arbuzov reaction.[236] However, this material is observed to yield anticipated product in reaction with halogen, a route proceeding through a similar quasi-phosphonium ion intermediate.[289,290] The reaction of trivalent phosphorus esters with carbon tetrachloride is also to be categorized as "abnormal", or at least unusual. Isolation of the simple monosubstitution product can be accomplished, although the product mixture is complicated by the presence of multiple substitution products and phosphorus acid chlorides, and yields are often low.[191,291,292] From several studies it is concluded that initial attack is by phosphorus on chlorine.[293,294] There may subsequently be involved a true phosphorane intermediate.

B. Comparison of Methods

The variables available to the experimenter in the performance of Michaelis-Arbuzov reactions relate principally to the substitutents of the reagents, that is, the ester linkages of the trivalent phosphorus reagent and the leaving group at carbon which is to become bound to phosphorus.

With regard to the ester linkages at phosphorus, several alternatives are available. For the isolation of esterified products, where the identity of the ester linkage is not critical and may be varied, particular advantages are to be found with both methyl and isopropyl groups.

The principal advantage of having methyl ester linkages present in the product species is that they may be removed with peculiar ease and selectively compared to other alkyl linkages, through the use of chlorotrimethylsilane.[296] Thus, with particularly reactive carbon substrates, use of trimethyl phosphite or other methyl esters of trivalent phosphorus acids is recommended. Several cautionary notes, however, are in order. In the use of trimethyl phosphite, or other methyl ester, methyl halide by-product will be generated. This by-product is more reactive than other alkyl halide reagents in the Michaelis-Arbuzov reaction and will compete with the substrate for remaining phosphorus reagent. In instances where acyl halides or benzylic halides are involved, this difficulty will be minimized, but deleterious competition may be anticipated with other simple alkyl halides.

This difficulty is also obviated if mixed esters are used wherein methyl and silyl ester linkages are present. Dealkylation will occur at the silyl ester site leaving product bearing methyl esters. On the other hand, where mixed alkyl esters are used, the methyl ester linkage will compete most effectively in the dealkylation step and be eliminated as methyl halide.

The use of isopropyl esters in the phosphorus reagent, such as with triisopropyl phosphite, has other advantages. The major advantage is the lack of reactivity of the isopropyl halide by-product. Most substrates will be able to compete quite effectively with this by-product. Isopropyl halide by-product can be removed by evaporation at the completion of the reaction. A difficulty with the isopropyl esters of quinquevalent phosphorus compounds, however, has to do with their cleavage. Both hydrolysis and chlorotrimethylsilane mediated cleavage are relatively slow. Isopropyl esters may be removed, however, thermolytically.[297] This becomes an option when the remaining portion of the compound is free of thermolabile functions.

If the target quinquevalent phosphorus material is to be obtained as the free phosphorus acid, the ester linkages of choice are silyl esters. These may be introduced easily into the appropriate starting phosphorus reagent by chlorotrimethylsilane treatment of the corresponding trivalent phosphorus acid. They are removed under extremely mild conditions which will leave most other functionalities present intact and exhibit preferential removal in the dealkylation of the Michaelis-Arbuzov reaction.

With regard to the carbon substrate for the Michaelis-Arbuzov reaction, general favorability is found for iodide as compared to bromide for the leaving group. Normally, bromide is found to be better than chloride for the purpose as well. An inversion of these leaving group preferences is reported in certain instances with silyl ester reagents.[258]

Relatively little work has been reported where there are good comparisons between the leaving facilities of halide ions as compared to amines. The few instances where comparisons are possible indicate the amines (quaternary ammonium substrates) to be quite suitable for the reaction. In a given synthetic procedure, the use of the quaternary ammonium species is well worth investigating if it is available conveniently.

VI. EXPERIMENTAL PROCEDURES

Trisodium Phosphonoformate — Reaction of a chloroformate with a trialkyl phosphite and cleavage of the ester linkages.[212]

Triethyl phosphonoformate was prepared by the dropwise addition of ethyl chloroformate (10.85 g, 0.1 mol) to triethyl phosphite (16.6 g, 0.1 mol) with vigorous stirring at ambient temperature. After the exotherm had subsided, the volatile materials were removed under reduced pressure and the residue vacuum distilled to give pure triethyl phosphonoformate (17.2 g, 82%) of bp 86 to 88°C/0.25 torr which exhibited NMR, IR, and mass spectra in accord with the proposed structure.

To the resultant triethyl phosphonoformate (105 g, 0.5 mol) was added sodium hydroxide solution (250 mℓ, 10 *M*) over a period of 15 min at ambient temperature. The solution became hot and ethanol by-product boiled off from the reaction mixture. Upon cooling, a precipitate formed which was recrystallized from water to give colorless crystals of the pure trisodium phosphonoformate hexahydrate (27.5 g, 19%) which exhibited IR and NMR spectra and X-ray crystal analysis in accord with the proposed structure.

Tris(trimethylsilyl) Phosphite — Preparation of a silyl ester of a trivalent phosphorus acid for Michaelis-Arbuzov reaction.[255]

An aqueous solution (100 mℓ) of trisodium phosphonoformate (2.83 g, 9.43 mmol) was eluted through a column (3 × 20 cm) of Diaion SK 1B in the pyridinium form using a pyridine:water mixture (1:4, 500 mℓ). The eluents were evaporated under reduced pressure, coevaporated with dry pyridine several times, and dissolved in dry THF (30 mℓ). Triethyl amine (3.2 g, 31.1 mmol) and chlorotrimethylsilane (3.4 g, 31.1 mmol) were added and the mixture was stirred vigorously at room temperature. After 2 hr, dry ether (15 mℓ) was added and the resulting precipitate was removed by filtration and washed with dry ether (10 mℓ). The filtrate and washings were combined, evaporated under reduced pressure, and the residue was vacuum distilled (76 to 77°C/10 torr) to give pure tris(trimethylsilyl) phosphite (2.27 g, 81%).

2,3-Dioleoyloxypropylphosphonic Acid — Reaction of an alkyl halide with a silyl phosphite ester.[258]

To 2,3-dioleoyloxy-1-iodopropane (3.65 g, 5 mmol) was added tris(trimethylsilyl) phosphite (15.05 g, 50 mmol) along with a trace of butyl hydrogen phthalate. The reaction mixture was stirred under a static nitrogen atmosphere with heating at 125°C for 16 hr. After this time excess tris(trimethylsilyl) phosphite and iodotrimethylsilane were removed by high vacuum distillation (bath 100°C) to leave a colorless oil. This residue was dissolved in THF:water (9:1, 50 mℓ) and allowed to stand in the dark at room temperature for 12 hr. The solvent was removed under reduced pressure and the residue dried by repeated azeotropic distillation with 2-propanol at reduced pressure. The residual oil was dissolved in chloroform and chromatographed on a column of silicic acid, eluted first with chloroform (500 mℓ), and then chloroform:methanol (9:1, 500 mℓ). The eluents were evaporated under reduced pressure and the viscous residue was dissolved in a minimum of chloroform and passed through a Gelman Metricel (0.45 μm) filter to removed suspended silicic acid. Removal of the solvent under reduced pressure gave the pure 2,3-dioleoyloxypropylphosphonic acid (2.8 g, 81%) as a viscous oil.

Dimethyl (4-Methoxybenzyl)phosphonate — Reaction of a benzyl halide with a trialkyl phosphite.[140]

Trimethyl phosphite (12.4 g, 0.1 mol) and *p*-methoxybenzyl chloride (15.6 g, 0.1 mol) were heated at reflux under a nitrogen atmosphere for 20 hr. The residue was vacuum distilled (141°C/0.45 torr) to give the pure dimethyl (4-methoxybenzyl)phosphonate (12.6 g, 59%) as an oil which exhibited spectra in accord with the proposed structure.

5-(3-Benzoylpropionyl)-3-deoxy-3-diisopropoxyphosphinylmethyl-1,2-di-*O*-acetyl-D-ribofuranose — Reaction of an alkyl bromide in a carbohydrate series with triisopropyl phosphite.[106] A solution of 5-*O*-(3-benzoylpropionyl)-3-deoxy-3-bromomethyl-1,2-di-*O*-acetyl-D-ribofuranose (2.35 g, 4.89 mmol) in triisopropyl phosphite (17 g, 81 mmol) was heated at 160 to 180°C with exclusion of moisture for 72 hr. Volatile materials were removed under reduced pressure and the residue was purified by chromatography on a silica gel column (48 × 2.7 cm) eluting with chloroform:ethyl acetate (1:1). From the 150 to 500 mℓ eluent there was isolated the pure 5-(3-benzoylpropionyl)-3-deoxy-3-diisopropoxyphosphinylmethyl-1,2-di-*O*-acetyl-D-ribofuranose (1.73 g, 68%) as an oil which exhibited IR and NMR spectra and analyses in accord with the proposed structure.

Triethyl 2-Phosphonobutanoate — Reaction of a trialkyl phosphite with a 2-halocarboxylate ester.[180]

A reaction flask was equipped with stirrer, thermometer, dropping funnel, and a steam jacketed condenser which would allow ethyl bromide to pass through while condensing higher boiling materials. To the flask was added ethyl 2-bromobutanoate (195 g, 1 mol) and it was heated to 160°C. The triethyl phosphite (199 g, 1.2 mol) was added dropwise over a period of 2 hr. After completion of the addition the temperature was increased to 190°C and maintained there until the evolution of ethyl bromide ceased. The mixture was distilled rapidly below 3 torr, and then redistilled using a 24-in. spinning band column. In this manner was isolated the pure triethyl 2-phosphonobutanoate (219 g, 87%) of bp 117 to 118°C/0.6 torr.

Dimethyl 1-Isopropoxy-2-methylpropylphosphonate — Reaction of a trialkyl phosphite with a 1-chloroether.[157]

Trimethyl phosphite (24.8 g, 0.2 mol) was added dropwise at room temperature to 1-isopropoxy-1-chloro-2-methylpropane (30.1 g, 0.2 mol). As heating was begun, methyl chloride began to be evolved. After heating at 100°C for 2 hr the reaction mixture was vacuum distilled. There was thus isolated pure dimethyl 1-isopropoxy-2-methylpropylphosphonate (27.8 g, 62%) of bp 94 to 95°C/0.8 torr.

Diphenyl Benzyloxycarbonylaminomethanephosphonate — Reaction of a triaryl phosphite with an *N*-acetoxymethylcarbamate generated *in situ*.[243]

A mixture of benzylcarbamate (30.6 g, 0.2 mol), paraformaldehyde (6 g, 0.2 mol), acetic anhydride (25.5 g, 0.25 mol) and acetic acid (20 mℓ) was stirred for 3 hr at 60 to 70°C. Triphenyl phosphite (62.1 g, 0.2 mol) was then added and the mixture was stirred for 2 hr at 110 to 120°C. After evaporation of acetic acid and remaining acetic anhydride under reduced pressure, the residue was dissolved in methanol (150 mℓ). The mixture was allowed to stand at −10°C for 4 hr, after which the precipitate was isolated by suction filtration, washed with methanol, and air-dried. The precipitate was recrystallized from chloroform:methanol to give pure diphenyl benzyloxycarbonylaminomethanephosphonate (38 g, 48%) of mp 114 to 116°C which exhibited IR and NMR spectra and analyses in accord with the proposed structure.

Diethyl Isobutyrylphosphonate — Reaction of a trialkyl phosphite with a carboxylic acid chloride.[204]

Isobutyryl chloride (25.9 g, 0.24 mol) was added dropwise to triethyl phosphite (43.9 g, 0.26 mol) with stirring under a nitrogen atmosphere and the temperature was maintained at 30 to 40°C. After the addition was complete, the reaction mixture was allowed to stand overnight at room temperature. The reaction mixture was vacuum distilled to give the pure diethyl isobutyrylphosphonate (43.9 g, 88%) of bp 75 to 83°C/3-4 torr which exhibited IR spectral characteristics in accord with the assigned structure.

(Diethyl Phosphonomethyl) Acetyl Sulfide — Reaction of a trialkyl phosphite with a 1-halosulfide.[162]

Bromomethyl acetyl sulfide (26.8 g, 0.16 mol) and triethyl phosphite (28.4 g, 0.17 mol) were combined in a flask fitted with a Dean-Stark trap. The mixture was stirred at 130°C for 2.5 hr during which time ethyl bromide collected in the Dean-Stark trap. The reaction mixture was vacuum distilled to give the pure bromomethyl acetyl sulfide (23.5 g, 65%) as a clear oil of bp 105 to 106°C/0.03 torr which exhibited NMR spectra in accord with the proposed structure.

Diethyl 1-Oxo-2-(3-indolyl)ethanephosphonate — Reaction of a trialkyl phosphite with a carboxylic acid chloride in solution.[203]

Triethyl phosphite (16.6 g, 0.1 mol) was added dropwise to a solution of 3-indolylacetyl choride (19.35 g, 0.1 mol) in dry ether (200 ml) at 0 to 5°C. The reaction was stirred for 1 hr at this temperature and a further 4 hr at room temperature. The resulting precipitate was filtered, washed with dry ether, and vacuum dried to give pure diethyl 1-oxo-2-(3-indolyl)ethanephosphonate (18.9 g, 64%) of mp 107 to 110°C.

Diethyl 1-ethoxybenzylphosphonate — Reaction of a trialkyl phosphite with an acetal in the presence of boron trifluoride.[247]

Benzaldehyde diethyl acetal (5.94 g, 32.9 mmol) and triethyl phosphite (5.45 g, 32.9 mmol) were dissolved in dichloromethane (60 ml) under an inert atmosphere and cooled to −20°C. Boron trifluoride etherate (3.57 g, 35.2 mmol) was then added dropwise. The resulting mixture was allowed to return to ambient temperature over an 18 hr period. The reaction was quenched by the addition of water (10 ml) with stirring for 5 min. The organic layer was separated, dried over magnesium sulfate, and the volatile materials were evaporated under reduced pressure. The residue was dissolved in a small amount of chloroform and applied to a silica gel column in chloroform. After removing unreacted starting material by chloroform elution, the column was eluted with 20% ethyl acetate in chloroform. In this manner was isolated pure diethyl 1-ethyoxybenzylphosphonate (6.52 g, 73%) of bp 108 to 110°C/0.2 torr which exhibited NMR and mass spectra in accord with the proposed structure.

1,10-Diphenyl-1,10-diphosphacyclooctadecane 1,10-Dioxide — Reaction of an alkyl halide with a phosphinite ester generated *in situ* by reduction of a phosphinate.[134]

To a well-stirred solution of diisopropyl octamethylenebis(phenylphosphinate) (2.48 g, 5 mmol) in benzene (60 ml) in a 2 ℓ round-bottomed flask at 25°C was added dropwise a solution of sodium bis(2-methoxyethoxy)aluminum hydride (2.82 g, 14 mmol) in benzene (50 ml). Vigorous evolution of hydrogen took place throughout the addition. When all the material had been added, the reaction mixture was stirred for 5 min until the evolution of hydrogen ceased. The reaction mixture was then diluted with benzene to a total volume of 1ℓ, and then there was added 1,8-dibromooctane (1.36 g, 5 mmol). Precautions were taken to prevent oxygen entry to the reaction system during dilution and addition. The reaction mixture was heated at 80°C for 12 hr. After cooling and concentrating the reaction mixture to 500 ml, water (3 ml) was added at 25°C and the mixture was filtered. The precipitate was washed with benzene (100 ml) and the combined filtrate and washings were evaporated under reduced pressure. The residue was subjected to thin-layer chromatography from which was isolated the two isomeric forms of 1,10-diphenyl-1,10-diphosphocyclooctadecane 1,10-dioxide (combined 310 mg, 12.8%).

Table 2
FORMATION OF CARBON TO PHOSPHORUS BONDS VIA THE MICHAELIS-ARBUZOV REACTION

Product	Phosphorus reagent	Method	Yield (%)	Refs.
$CH_3OP(O)(CH_3)_2$	$CH_3P(OCH_3)_2$	CH_3COOH	63	248
$C_2H_5OCH_2P(O)Cl_2$	CH_3OPCl_2	RX, BF_3	56	163
$Cl_2P(O)CH(NCO)CH_2Br$	CH_3OPCl_2	RX, $FeCl_3$, heat	73	165
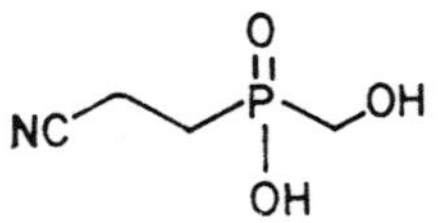 (structure: phosphonic acid bearing NC group)	$(CH_3)_3SiOCH_2P[OSi(CH_3)_3]_2$	RX, 135°C, 1 hr	91	266
$n\text{-}C_3H_7OCH_2P(O)Cl_2$	CH_3OPCl_2	RX, BF_3	52	163
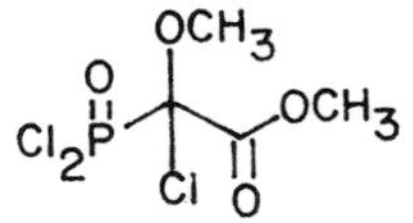 (structure: methyl ester with PCl_2 group)	CH_3OPCl_2	RX	80	164
$(CH_3O)_2P(O)CH_2CH_2Br$	$(CH_3O)_3P$	RX, 140°C, 3 hr	67	116
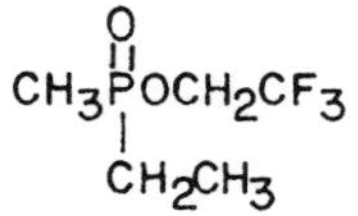 (structure: $CH_3P(O)OCH_2CF_3$ / CH_2CH_3)	$C_2H_5P(OCH_2CF_3)_2$	RX, 12 hr, 60°C	74	83
$s\text{-}C_4H_9OCH_2P(O)Cl_2$	CH_3OPCl_2	RX, BF_3	25	163
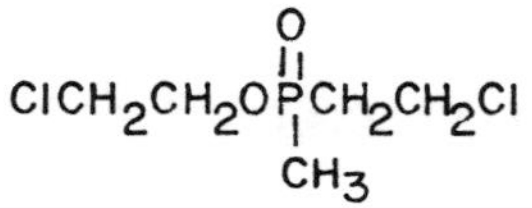 (structure: $ClCH_2CH_2OP(O)CH_2CH_2Cl$ / CH_3)	$CH_3P(OCH_2CH_2Cl)_2$	160°C, 1 hr	89	123

Table 2 (continued)

FORMATION OF CARBON TO PHOSPHORUS BONDS VIA THE MICHAELIS-ARBUZOV REACTION

Product	Phosphorus reagent	Method	Yield (%)	Refs.
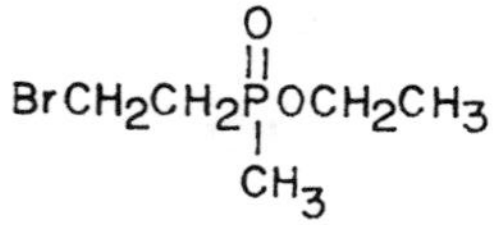 $BrCH_2CH_2P(O)(CH_3)OCH_2CH_3$	$CH_3P(OC_2H_5)_2$	RX, 2 hr, 80°C	13	130
$CH_3COP(O)(OC_2H_5)_2$	$(C_2H_5O)_3P$	AcX, $1/2$ hr, r.t.	87	200
$CH_3CH = CHCOP(O)(OCH_3)_2$	$(CH_3O)_3P$	AcX, r.t.	53	230
$CH_3OCH_2P(O)(OC_2H_5)_2$	$(C_2H_5O)_3P$	RX, 155°C, 3 hr	74	154
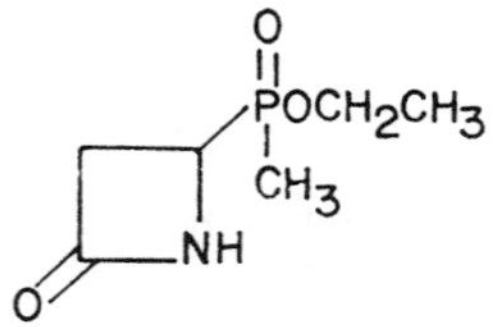	$CH_3P(OC_2H_5)_2$	ROAc, 1 hr, 60°C	89	245
$(C_2H_5O)_2P(O)CH_2NHCHO$	$(C_2H_5O)_3P$	RN +	90	238
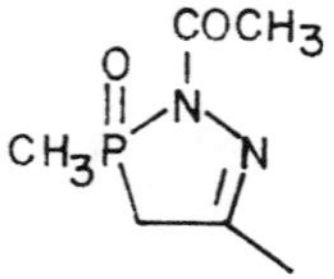	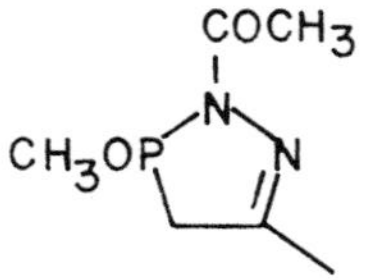		71	298
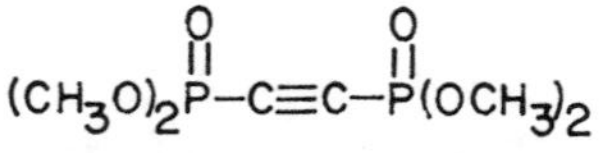 $(CH_3O)_2P—C{\equiv}C—P(OCH_3)_2$ (with $\overset{O}{\parallel}$ on each P)	$(CH_3O)_3P$	RX	87	152

Product	Reagent	Conditions	Yield (%)	Ref.
$CH_3SCH_2P(O)(OC_2H_5)_2$	$(C_2H_5O)_3P$	RX, 130°C, 16 hr	49	161
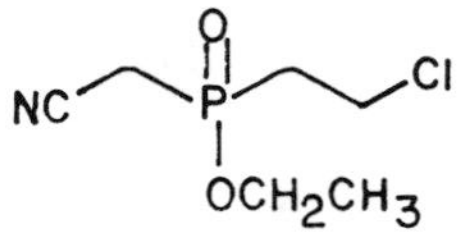 structure: NC–CH₂CH₂–P(O)(OCH₂CH₃), Cl	$(C_2H_5O)_2PCH_2CH_2Cl$	RX, NiCl₂	67	150
$(C_2H_5O)_2P(O)CCl_2CN$	$(C_2H_5O)_3P$	RX, 20°C, 1/2 hr RX	80 42	185 187
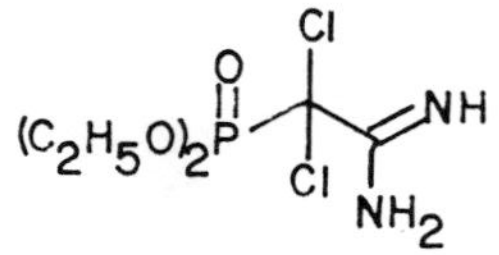 structure: $(C_2H_5O)_2P$, Cl, Cl, =NH, NH₂	$(C_2H_5O)_3P$	RX	84	184
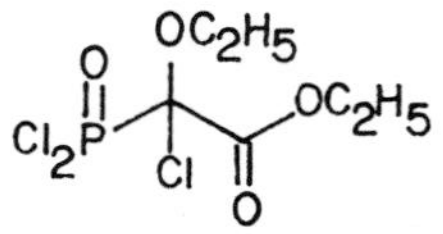 structure: $Cl_2P(O)$–, OC₂H₅, OC₂H₅, Cl	CH_3OPCl_2	RX	50	164
$CH_2=CHCH_2P(O)(OC_2H_5)$	$(C_2H_5O)_3P$	RX, hydroquinone, 120°C, Ni(O)	96	277
$(CH_3)_2CHCOCH_2P(O)(OCH_3)_2$	$(CH_3O)_3P$	Haloepoxide, r.t., 100 hr	89	166
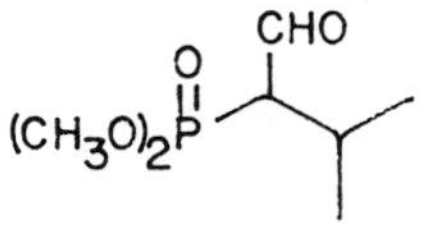 structure: $(CH_3O)_2P(O)$–, CHO	$(CH_3O)_3P$	Haloepoxide, 14 hr	36	166
$C_2H_5OCOP(O)(OC_2H_5)_2$ $(C_2H_5O)_2P(O)CH_2NHCOCH_3$	$(C_2H_5O)_3P$ $(C_2H_5O)_3P$	AcX RN+, 120°C, 1.5 hr RX, 80°C, 5 hr	82 80 8	212 237 237

41

Table 2 (continued)

FORMATION OF CARBON TO PHOSPHORUS BONDS VIA THE MICHAELIS-ARBUZOV REACTION

Product	Phosphorus reagent	Method	Yield (%)	Refs.
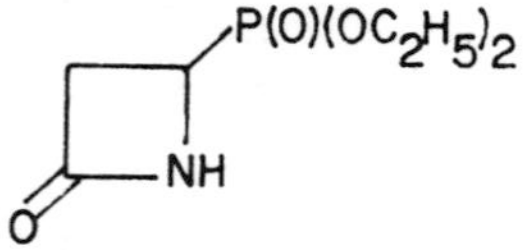 (β-lactam, P(O)(OC$_2$H$_5$)$_2$ substituted azetidinone with NH)	(C$_2$H$_5$O)$_3$P	ROAc, 120°C, 2 hr	90	245
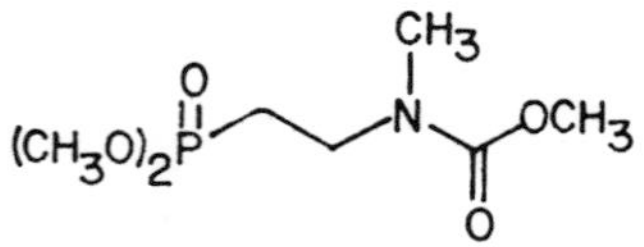 CH$_3$–N(C(O)OCH$_3$)–CH$_2$CH$_2$–P(O)(OCH$_3$)$_2$	(CH$_3$O)$_3$P	Acylaziridine, 50°C, 7 days	21	58
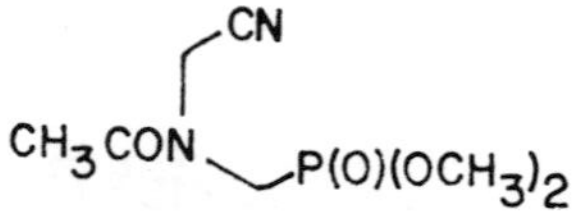 NC–CH$_2$CH$_2$CON–CH$_2$–P(O)(OCH$_3$)$_2$	(CH$_3$O)$_3$P	RX, 12 hr, r.t.	79	168
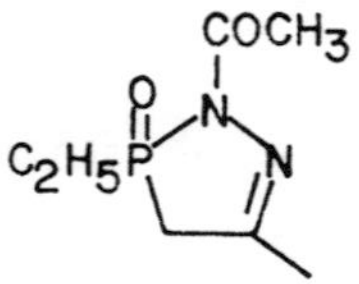 (imidazoline ring, COCH$_3$ on N, C$_2$H$_5$OP(O), CH$_3$)	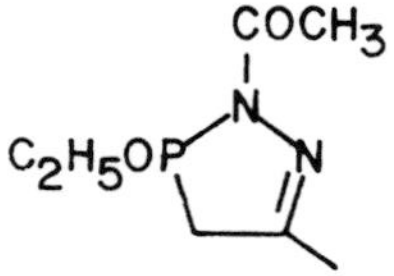 (imidazoline ring with COCH$_3$ on N, C$_2$H$_5$OP(O), CH$_3$)	RX, r.t.	50	298
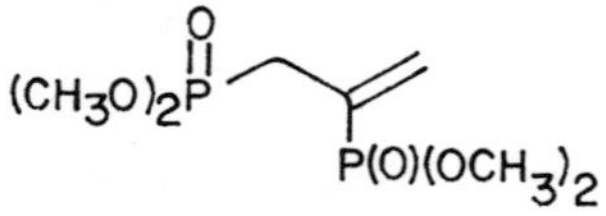 (CH$_3$O)$_2$P(O)CH$_2$–C(=CH$_2$)–CH$_2$–P(O)(OCH$_3$)$_2$	(CH$_3$O)$_3$P	RX, NiCl$_2$	41	275

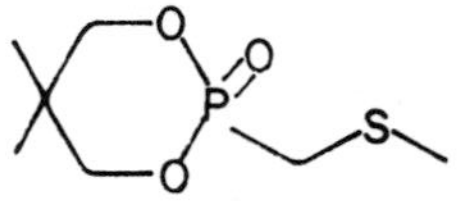

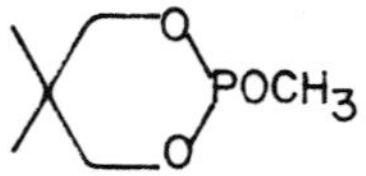

Reactant / Product	P-reagent	Conditions	Yield (%)	Ref.
[cyclic 2-oxo-1,3,2-dioxaphosphorinane, 5,5-dimethyl, with S–CH2CH3]	[5,5-dimethyl-2-methoxy-1,3,2-dioxaphosphorinane, POCH3]	RX, 125°C, 4 hr	61	161
(CH3COSCH2P(O)(OC2H5)2	(C2H5O)3P	RX, 130°C, 2.5 hr	65	162
(C2H5O)2P(O)CH2CH = CHCl	(C2H5O)3P	RX, NiCl2, 10 hr, 142°C	49	113
Cl3CCH2OCOP(O)(OC2H5)2	(C2H5O)3P	AcX	71	220
CH2 = CHCH2CH2P(O)(OC2H5)2	(C2H5O)3P	RX, heat	68	109
[cyclic 2-oxo-1,3,2-dioxaphosphorinane, 5,5-dimethyl, 2-CH3]	[5,5-dimethyl-2-methoxy-1,3,2-dioxaphosphorinane, POCH3]	RX, −15°C, 12 hr	58	80
C2H5O–P(O)(CH3)–CH2CH2–CO2C2H5	CH3P(OC2H5)2			
[2-oxocyclohexyl–P(O)(OCH3)2]	(CH3O)3P	RX	82	108
[CHO, tert-butyl substituted –P(O)(OCH3)2]	(CH3O)3P	Haloepoxide, r.t., 117 hr	92	166
(CH3)2CHCOP(O)(OC2H5)2	(CH3O)3P	Haloepoxide, 45 hr, 105°C	45	166
CH3COCH2P(O)(OC2H5)2	(C2H5O)3P	AcX, 40°C	94	204
(C2H5O)2P(O)CH2OCH2CH2CH2OCH3	(C2H5O)3P	RN+, heat	72	140
	(C2H5O)3P	RX, 155°C	77	153

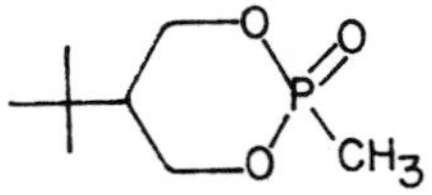

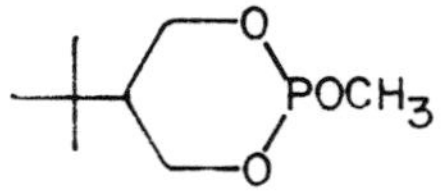

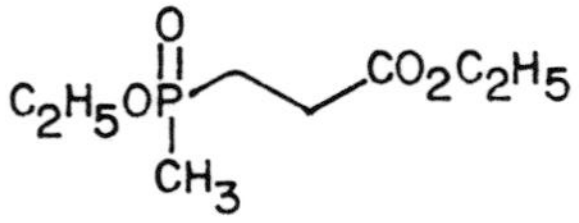

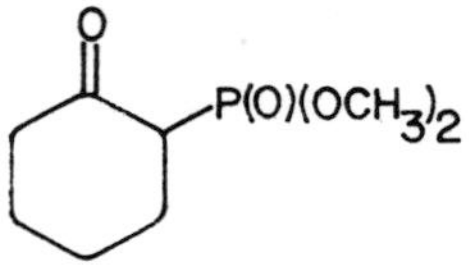

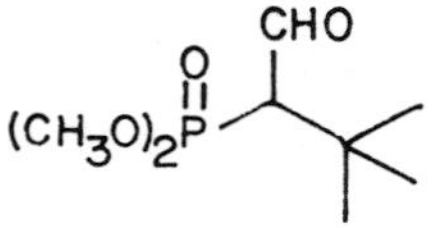

Table 2 (continued)
FORMATION OF CARBON TO PHOSPHORUS BONDS VIA THE MICHAELIS-ARBUZOV REACTION

Product	Phosphorus reagent	Method	Yield (%)	Refs.
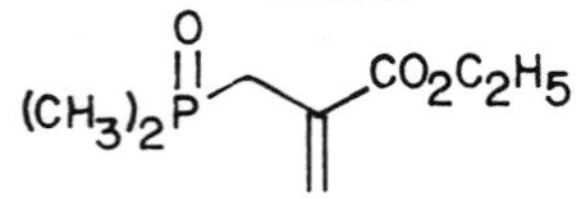	$(CH_3O)_3P$	RX, heat	71	139
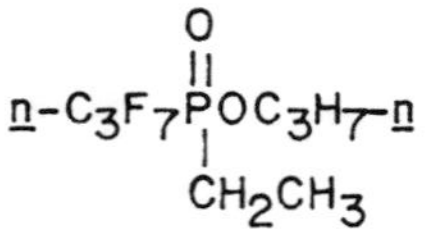	$C_2H_5P(OC_4H_9\text{-}n)_2$	RX, 150°C, 4 hr	42	299
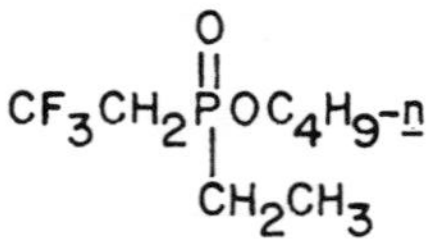	$C_2H_5P(OC_3H_7\text{-}n)_2$	RX, 150°C, 4 hr	32	299
$[(CH_3)_3SiO]_2P(O)COCF_3$	$[(CH_3)_3SiO]_3P$	AcX	46	226
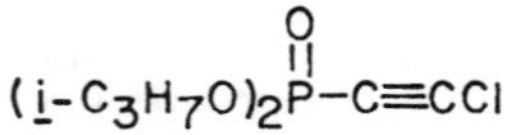				
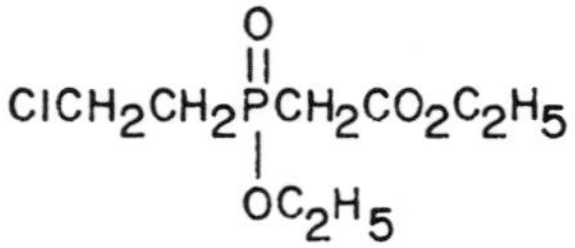	$(i\text{-}C_3H_7O)_3P$	RX	77	151

Product	Reagent	Conditions	Yield (%)	bp
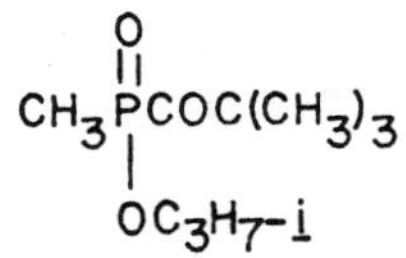 $CH_3PCOC(CH_3)_3$, $OC_3H_7\text{-}i$ (O=P)	$(C_2H_5O)_2PCH_2CH_2Cl$	RX, $NiCl_2$	93	150
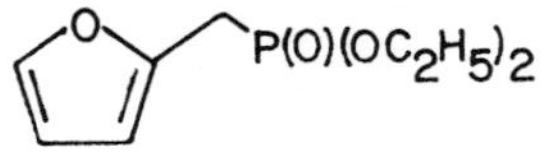 furyl-$CH_2P(O)(OC_2H_5)_2$	$CH_3P(OC_3H_7\text{-}i)_2$	AcX, 50°C, 1 hr	80	232
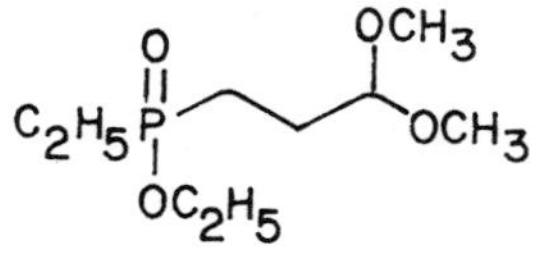 C_2H_5 (O=P) OC_2H_5, OCH_3 OCH_3	$(C_2H_5O)_3P$	RN+, 150°C, 4 hr	70	240
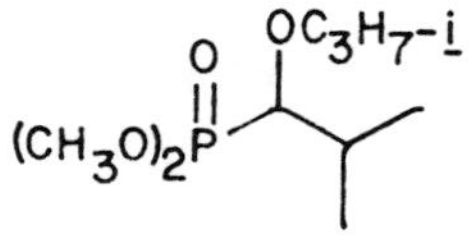 $(CH_3O)_2P$ (O=), $OC_3H_7\text{-}i$	$C_2H_5P(OC_2H_5)_2$	RX, heat	60	127
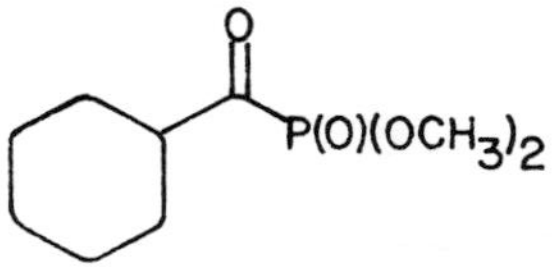 cyclohexyl-$C(O)P(O)(OCH_3)_2$	$(CH_3O)_3P$	RX	62	157
	$(CH_3O)_3P$	AcX	80	198
$CH_3COCH_2CH_2CH_2P(O)(OC_2H_5)_2$	$(C_2H_5O)_3P$	1. RX, heat; 2. aq. acid	38	112
$(C_2H_5O)_2P(O)CH_2CH_2CH(OCH_3)_2$	$(C_2H_5O)_3P$	RX, heat	65	127

Table 2 (continued)

FORMATION OF CARBON TO PHOSPHORUS BONDS VIA THE MICHAELIS-ARBUZOV REACTION

Product	Phosphorus reagent	Method	Yield (%)	Refs.
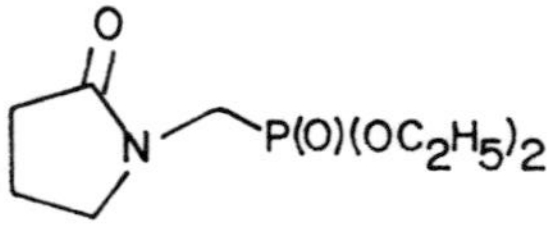 (structure: pyrrolidinone with N-CH$_2$-P(O)(OC$_2$H$_5$)$_2$)	(C$_2$H$_5$O)$_3$P	RX	74	169
(structure: n-C$_3$F$_7$-P(O)(O-CH$_2$CH(CH$_3$)CH$_2$CH$_3$))	C$_2$H$_5$P[OCH$_2$CH(CH$_3$)$_2$]	RX, 150°C, 4 hr	37	299
(C$_2$H$_5$O)$_2$P(O)CH$_2$CONHC(CH$_3$)$_2$PO$_3$H$_2$	(C$_2$H$_5$O)$_3$P	RX	29	300
(C$_2$H$_5$O)$_2$P(O)...CH(NH...)CO$_2$Na, PO$_3$H$_2$	(C$_2$H$_5$O)$_3$P	RX	42	300
(i-C$_3$H$_7$O)$_2$P(O)CH$_2$CH=CHCl	(i-C$_3$H$_7$O)$_3$P	RX, NiCl$_2$, 171°C, 15 hr	60	113
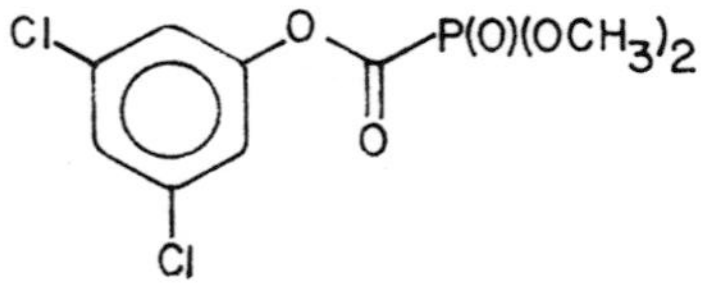 (structure: 2,4-dichlorophenyl ester of P(O)(OCH$_3$)$_2$-acetate)	(CH$_3$O)$_3$P	AcX, 100°C, 2 hr	64	216
[(C$_2$H$_5$O)$_2$P(O)]$_2$CCl$_2$ C$_6$H$_5$P(O)(OC$_2$H$_5$)$_2$	(C$_2$H$_5$O)$_3$P (C$_2$H$_5$O)$_3$P	RX ArX, Pd(II), 155°C, 3 hr	82 80	191 278

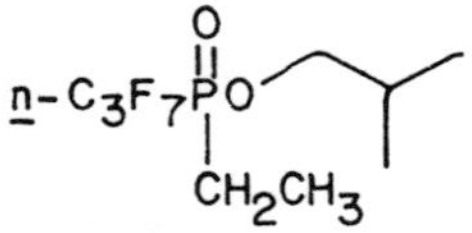

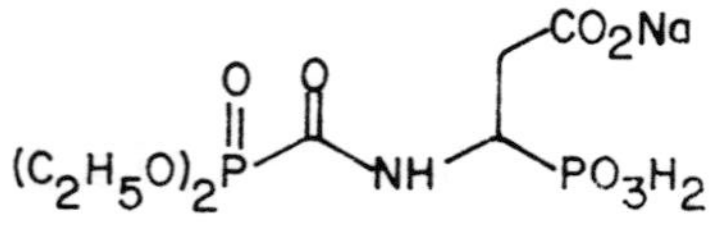

Product	Reagent	Conditions	Yield (%)	Ref.
CH₃CH₂CH₂CH(CH₃)CH₂P(O)(OC₂H₅)₂	(C₂H₅O)₃P	RX, heat	51	97
(4-CH₃O-C₆H₄)CH₂P(O)(OCH₃)₂	(CH₃O)₃P	RX	59	140
CH₃CO₂CH₂CH=CHCH₂P(O)(OC₂H₅)₂	(C₂H₅O)₃P	RX, heat	70	138
(C₂H₅O)₂P(O)C(CH₃)₂CO₂C₂H₅	(C₂H₅O)₃P	RX, 190°C	53	180
(C₂H₅O)₂P(O)CH(CH₂CH₂CO₂C₂H₅	(C₂H₅O)₃P	RX, 190°C	82	180
(C₂H₅O)₂P(O)CH(C₂H₅)CO₂C₂H₅	(C₂H₅O)₃P	RX, 190°C	87	180
(C₂H₅O)₂P(O)CH₂CH(OC₂H₅)₂	(C₂H₅O)₃P	RX, 145°C, 3 hr	91	174

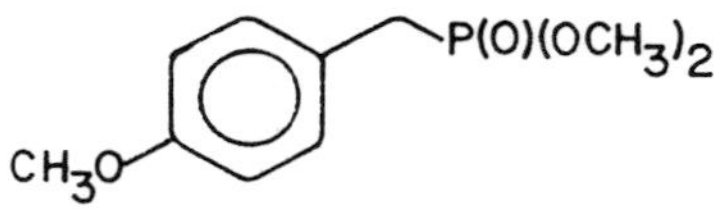

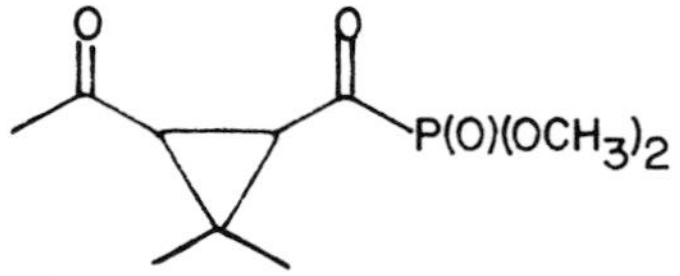

$CH_3CH_2CH_2CH(CH_3)CH_2P(O)(OC_2H_5)_2$ — (C₂H₅O)₃P — RX, heat — 51 — 97

$CH_3CO_2CH_2CH=CHCH_2P(O)(OC_2H_5)_2$

(C₂H₅O)₂P(O)C(CH₃)₂CO₂C₂H₅

(C₂H₅O)₂P(O)CH(CH₂CH₂CO₂C₂H₅

(C₂H₅O)₂P(O)CH(C₂H₅)CO₂C₂H₅

(C₂H₅O)₂P(O)CH₂CH(OC₂H₅)₂

Product	Reagent	Conditions	Yield (%)	Ref.
[cyclopropane, acetyl, P(O)(OCH₃)₂]	(CH₃O)₃P	AcX	72	228
(i-C₃H₇O)₂P(O)CH=CHCO₂CH₃	(i-C₃H₇O)₃P	VinX, 60—150°C	44	301
[(CH₃)₃Cl]₂P(O)CH₂CONH₂	[(CH₃)₃Cl]₂P(O)C₂H₅	120°C, RX, 2 hr	93	189
(C₂H₅O)₂P(O)CH₂CONHCH₂CO₂C₂H₅	(C₂H₅O)₃P	RX, 180°C, 3 hr	84	182
[aspartate structure, CO₂CH₃ / CO₂CH₃, NH, (CH₃O)₂P(O)]	(CH₃O)₃P	RX, 130°C, 3 hr	97	182
(CH₃)₃SiCH₂CH₂OCH₂P(O)(OC₂H₅)₂	(C₂H₅O)₃P	RX, 120°C, 2 hr	82	155
[(CH₃)₃SiO]₂P(O)C₄H₉-n	[(CH₃)₃SiO]₃P	RX, heat	90	251
		RX	89	256
[(C₂H₅O)₂P(O)]₂CHOCH₃	(C₂H₅O)₃P	RX, ZnCl₂	30	276
(n-C₄H₉O)₂P(O)CHClOCH₃	(n-C₄H₉O)₃P	RX, ZnCl₂, 150°C, 1 hr	62	276

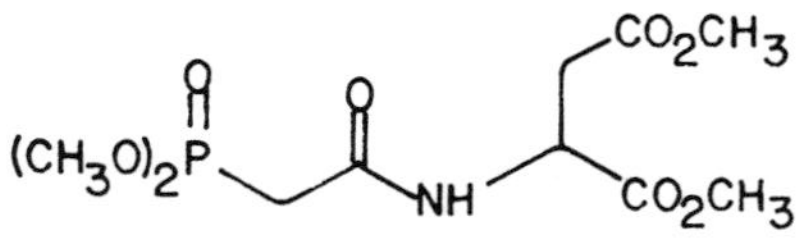

Table 2 (continued)

FORMATION OF CARBON TO PHOSPHORUS BONDS VIA THE MICHAELIS-ARBUZOV REACTION

Product	Phosphorus reagent	Method	Yield (%)	Refs.
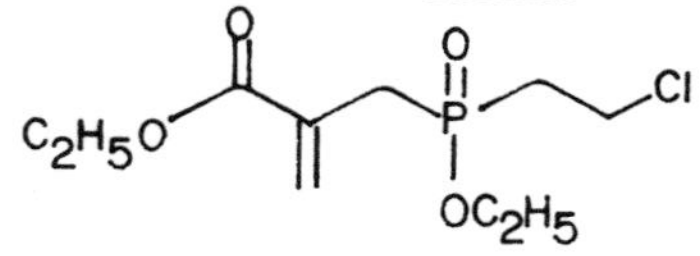	$ClCH_2CH_2P(OC_2H_5)_2$	RX, NiCl$_2$	79	150
$[(C_2H_5O)_2P(O)]_2CClCN$	$(C_2H_5O)_3P$	RX, 20°C, 0.5 hr	73	185
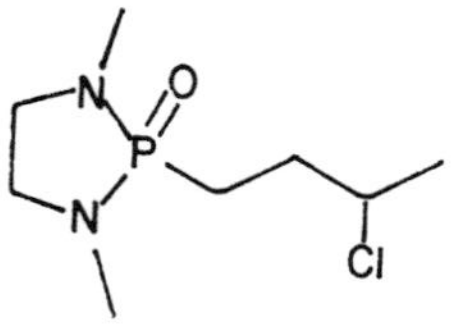	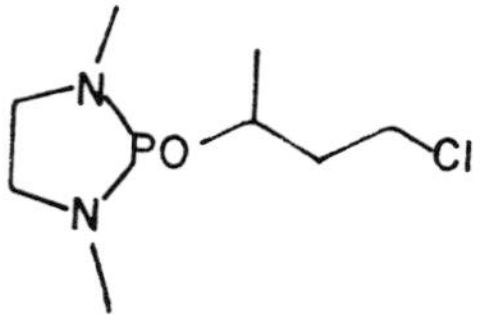	Heat	71	121
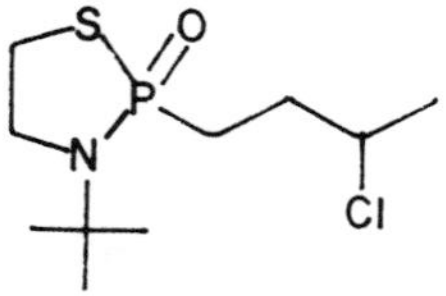	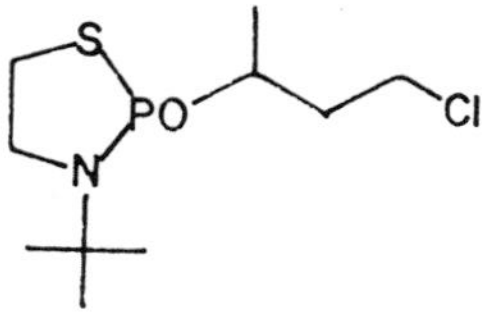	Heat	46	121
$[(CH_3)_3Cl_2P(O)CH_2COCH_3$ $C_6H_5CH_2P(O)(OC_2H_5)_2$	$[(CH_3)_3Cl]POC_2H_5$ $(C_2H_5O)_3P$	RX, 120°C, 2 hr $RN+$, 160°C, 4 hr	83 78	189 240
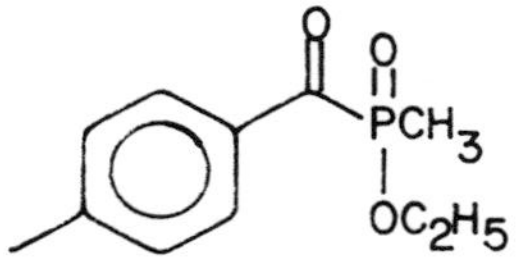	$CH_3P(OC_2H_5)_2$	AcX, 50°C, 1 hr	60	232

Product	Reagent	Conditions	%	Ref.
$C_6H_5CH=CHCOP(O)(OCH_3)_2$	$(CH_3O)_3P$	AcX, 0.5 hr, r.t.	38	222
$(CH_3O)_2P(O)CH(OC_2H_5)C_6H_5$	$(CH_3O)_3P$	RX	87	157
$(C_2H_5O)_2P(O)CH_2CH_2CH(OC_2H_5)_2$	$(C_2H_5O)_3P$	RX, heat	80	127
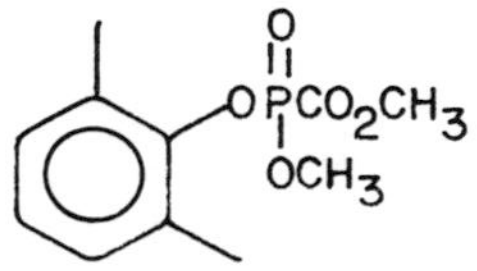 2,6-dimethylphenyl –O–P(O)(CO₂CH₃)(OCH₃) / –OP(OCH₃)₂	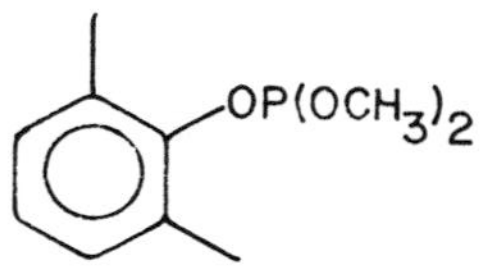 $(CH_3O)_3P$	AcX, 100°C, 4 hr	99	216
$(CH_3O)_2P(O)CH_2CO_2CH_2C_6H_5$	$(CH_3O)_3P$	RX, heat	83	179
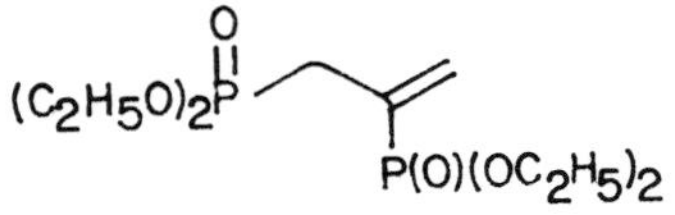 $(C_2H_5O)_2P(O)\,CH_2{=}C{-}P(O)(OC_2H_5)_2$	$(C_2H_5O)_3P$	VinX, NiCl₂	68	275
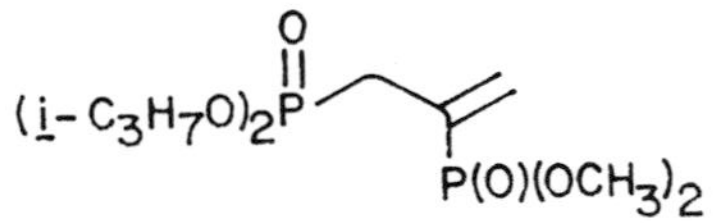 $(i{-}C_3H_7O)_2P(O)\,CH_2{=}C{-}P(O)(OCH_3)_2$	$(CH_3O)_3P$	VinX, NiCl₂	37	275
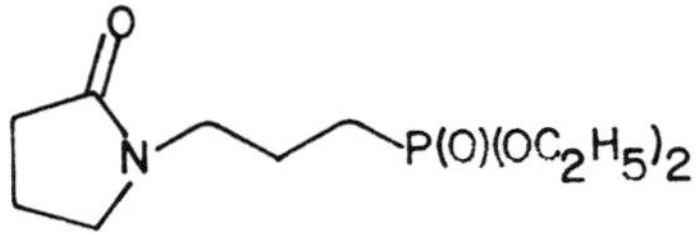 pyrrolidinone–$P(O)(OC_2H_5)_2$	$(C_2H_5O)_3P$	RX	29	169
$(C_2H_5O)_2P(O)NHCH_2CH_2CH_2P(O)(OC_2H_5)_2$	$(C_2H_5O)_3P$	RX, 156°C, 8 hr	80	107
$(C_2H_5O)_2P(O)NHCH(CH_3)CH_2P(O)(OC_2H_5)_2$	$(C_2H_5O)_3P$	RX, 156°C, 8 hr	68	107

Table 2 (continued)
FORMATION OF CARBON TO PHOSPHORUS BONDS VIA THE MICHAELIS-ARBUZOV REACTION

Product	Phosphorus reagent	Method	Yield (%)	Refs.
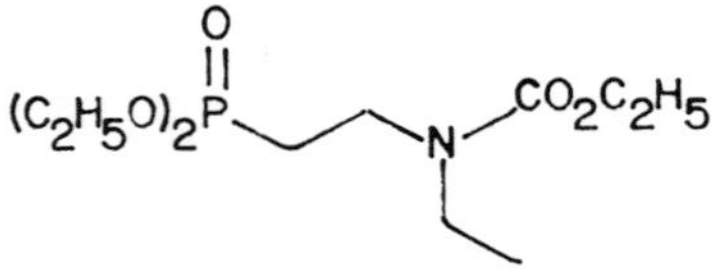 (structure: $(C_2H_5O)_2P(O)$–CH$_2$CH$_2$–N(C$_2$H$_5$)CO$_2$C$_2$H$_5$)	$(C_2H_5O)_3P$	Acylaziridine, 5 hr, 150°C	42	58
$(C_2H_5O)_2P(O)CH_2SC_6H_5$	$(C_2H_5O)_3P$	RX, 140°C, 16 hr	70	161
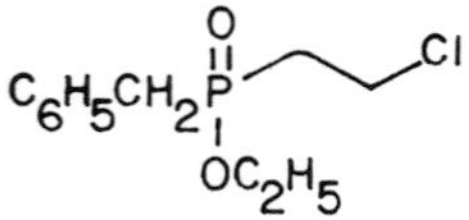 (structure: $C_6H_5CH_2P(O)(OC_2H_5)$–CH$_2$CH$_2$Cl)	$ClCH_2CH_2P(OC_2H_5)_2$	RX, NiCl$_2$	100	150
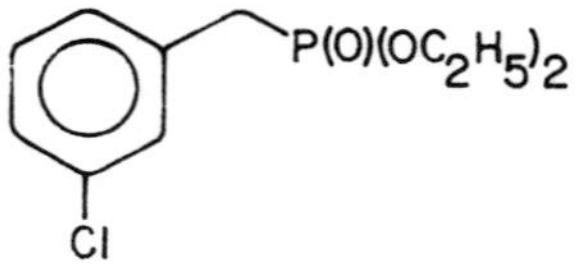 (structure: 3-chlorobenzyl $P(O)(OC_2H_5)_2$)	$(C_2H_5O)_3P$	RN+, 160°C, 4 hr	77	240
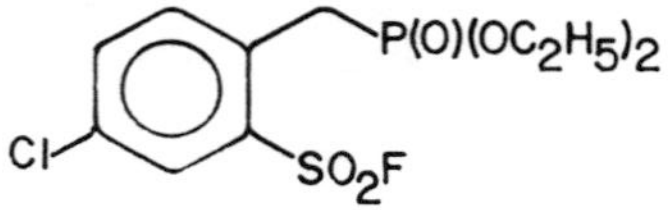 (structure: benzyl $P(O)(OC_2H_5)_2$ with SO$_2$F and Cl substituents)	$(C_2H_5O)_3P$	RX	80	148
$(i\text{-}C_3H_7O)_2P(O)CCl_2P(O)(OC_2H_5)_2$	$(i\text{-}C_3H_7O)_3P$	RX, 130°C, 2.5 hr	82	191

Structure	Reagent	Conditions		
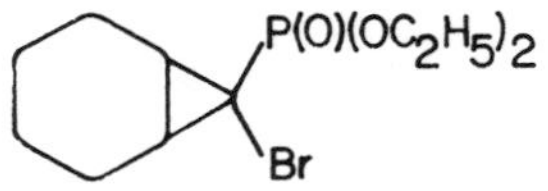	$(C_2H_5O)_3P$	RX	54	124
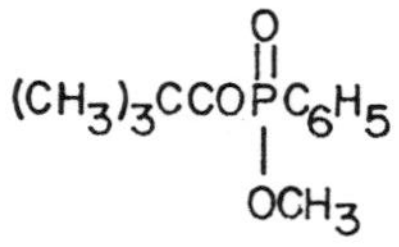	$C_6H_5P(OCH_3)_2$	AcX, 50°C, 1 hr	65	232
	$[(CH_3)_3Cl_2POC_2H_5$	RX, 120°C, 2 hr	94	189
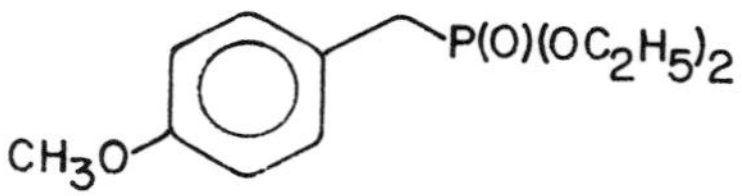	$(C_2H_5O)_3P$	RN+ , 170°C, 4 hr	88	240
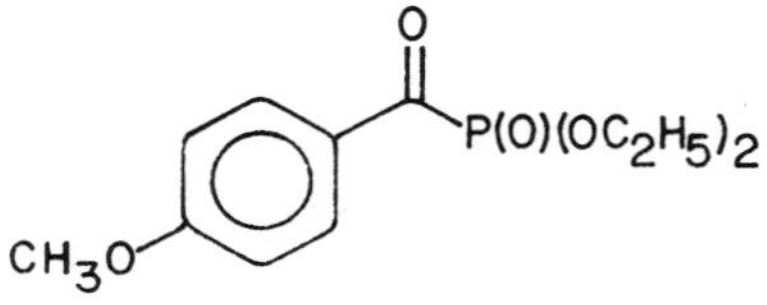	$(C_2H_5O)_3P$	AcX	79	224
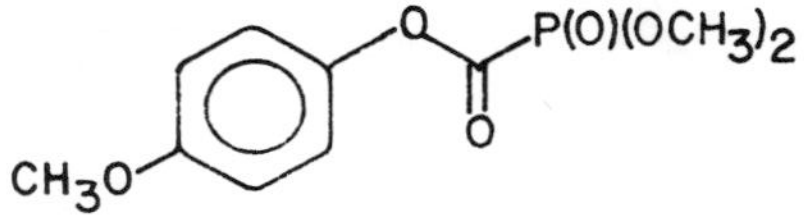	$(C_2H_5O)_3P$	AcX, 120°C, 1.5 hr	89	216
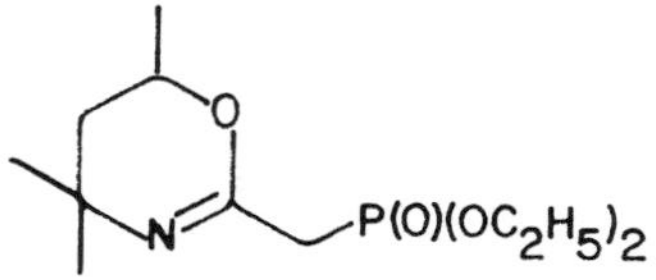	$(C_2H_5O)_3P$	RX, heat, 24 hr	80	177

Table 2 (continued)
FORMATION OF CARBON TO PHOSPHORUS BONDS VIA THE MICHAELIS-ARBUZOV REACTION

Product	Phosphorus reagent	Method	Yield (%)	Refs.
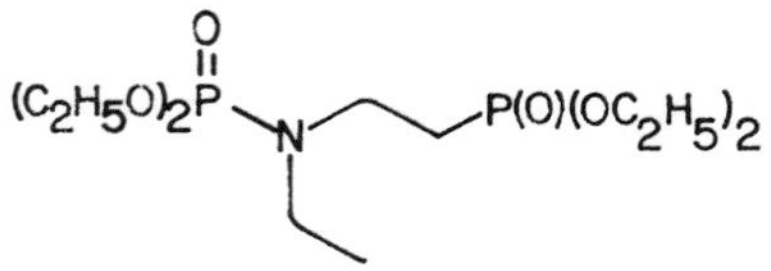	$(C_2H_5O)_3P$	RX, 156°C, 8 hr	57	107
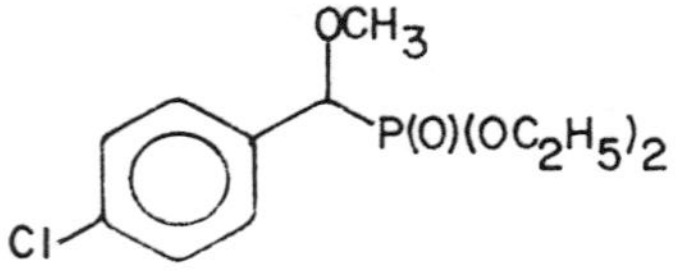	$(C_2H_5O)_3P$	BF$_3$, acetal	69	247
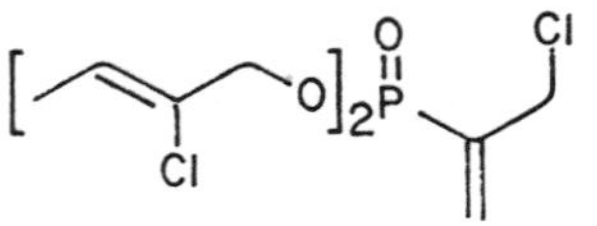	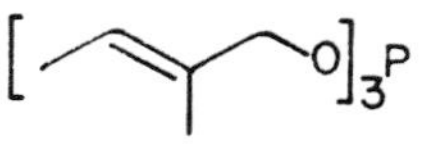	160°C, 5 hr	98	135
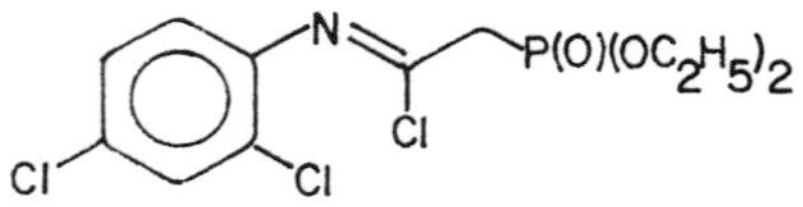	$(C_2H_5O)_3P$	RX	81	183
$(C_6H_5)_2P(O)CH_3$ $(C_6H_5O)_2P(O)CH_3$	$(C_6H_5)_2POSi(CH_3)_3$ $(C_6H_5O)_3P$	RX CH$_3$OH, acid RX, CH$_3$OH, 250°C, 2 hr ArX, (CH$_3$O)$_3$P, heat	93 78 92 95	264 249 271 274

Product	Reagent	Conditions	Yield (%)	Ref.
$(i\text{-}C_3H_7O)_2P(O)COC_6H_5$	$(i\text{-}C_3H_7O)_3P$	RX, 60°C, 0.75 hr AcX BF$_3$, acetal	72 55 73	196 224 247
$(C_2H_5O)_2P(O)CH(OC_2H_5)C_6H_5$	$(C_2H_5O)_3P$	RX, heat	15	110
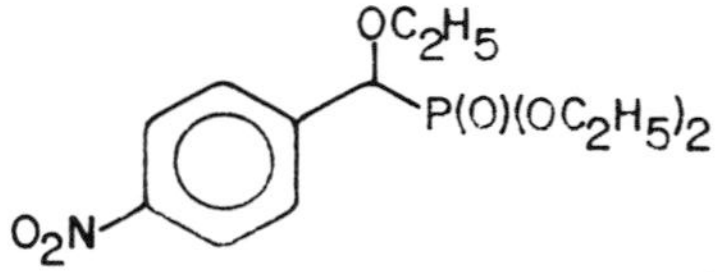 (structure: tetrahydropyranyloxy, CN, $P(O)(OC_2H_5)_2$)	$(C_2H_5O)_3P$	BF$_3$, acetal	78	247
(structure: p-nitrophenyl, OC_2H_5, $P(O)(OC_2H_5)_2$)	$[(CH_3)_3SiO]_3P$ $[(CH_3)_3SiO]_3P$ $[(C_2H_5)_3SiO]_3P$ $(C_2H_5O)_3P$	RX AcX RX, heat RX, -10°C	80 72 100 60	256 218 251 206
$[(CH_3)_3SiO]_2P(O)CH_2C_6H_5$ $[(CH_3)_3SiO]_2P(O)COC_6H_5$ $[(C_2H_5)_3SiO]_2P(O)CH_3$ $(C_2H_5O)_2P(O)COCH_2SCH_2C_6H_5$				
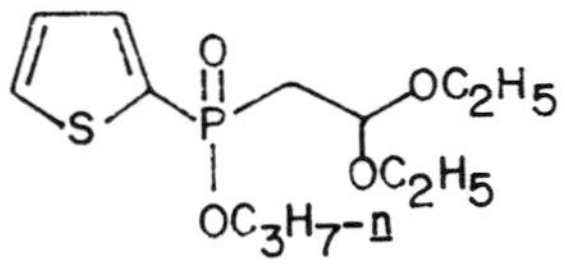 (structure: thiophene, $P(=O)$ with OC_2H_5, OC_2H_5, $OC_3H_7\text{-}n$)	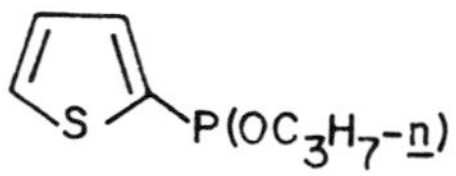 (structure: thiophene-$P(OC_3H_7\text{-}n)$)	RX, 120°C, 7 hr	82	176
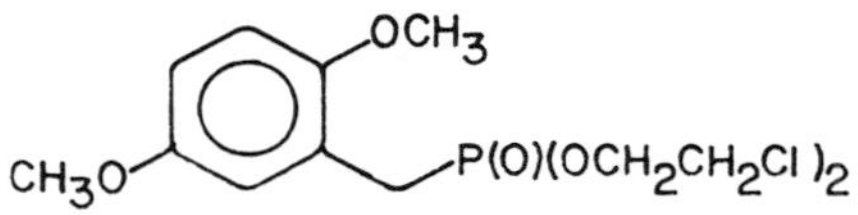 (structure: 2,5-dimethoxybenzyl $P(O)(OCH_2CH_2Cl)_2$)	$(ClCH_2CH_2O)_3P$	RX	84	145
$(C_6H_5)_2P(O)CH_2OCH_3$	$(C_6H_5)_2POC_2H_5$	RX	89	194

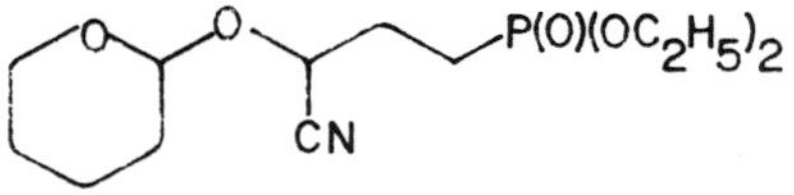

Table 2 (continued)
FORMATION OF CARBON TO PHOSPHORUS BONDS VIA THE MICHAELIS-ARBUZOV REACTION

Product	Phosphorus reagent	Method	Yield (%)	Refs.
[structure: Ph–(CH$_2$)$_n$–P(=O)(OC$_2$H$_5$)CO$_2$H]	[structure: $\overset{OC_2H_5}{|}$ P–OSi(CH$_3$)$_3$, Ph chain]	RX, r.t., 5 hr	97	267
[structure: C$_6$H$_5$P(=O)(OC$_3$H$_7$-i)CH$_2$CO$_2$C$_3$H$_7$-i]	C$_6$H$_5$P(OC$_3$H$_7$-i)	RX, 130°C, 0.5 hr	95	188
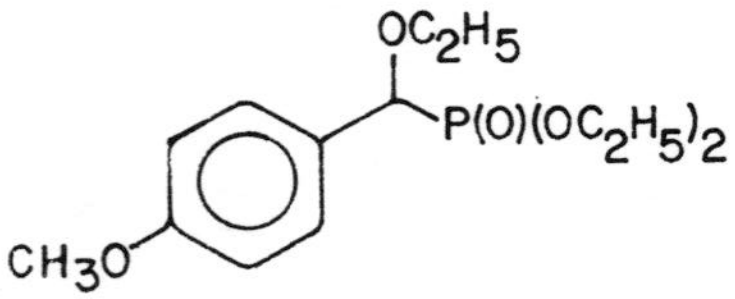 [structure: CH$_3$O–C$_6$H$_4$–CH(OC$_2$H$_5$)P(O)(OC$_2$H$_5$)$_2$]	(C$_2$H$_5$O)$_3$P	BF$_3$, acetal	84	241
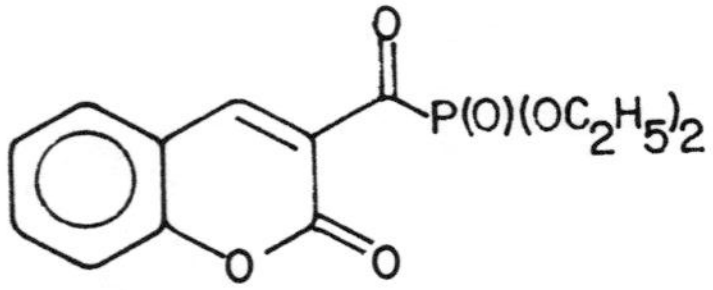 [structure: coumarin–C(=O)–P(O)(OC$_2$H$_5$)$_2$]	(C$_2$H$_5$O)$_3$P	AcX	80	210
[structure: NC–C$_6$H$_4$–CH$_2$P(O)(C$_3$H$_7$-n)$_2$]	(n-C$_3$H$_7$)$_2$POC$_2$H$_5$	RX	57	149

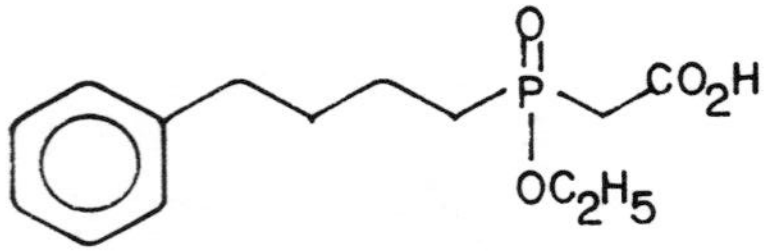

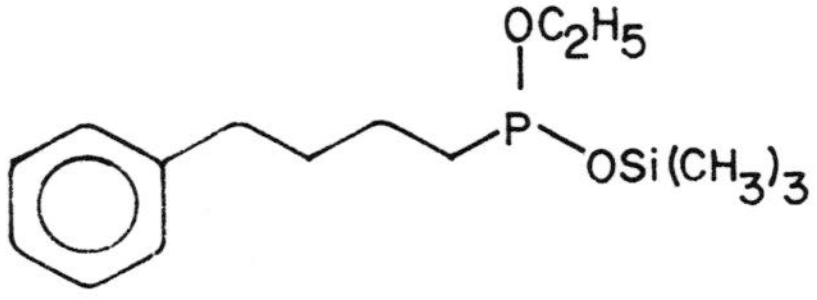

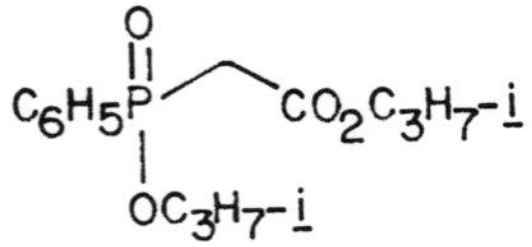

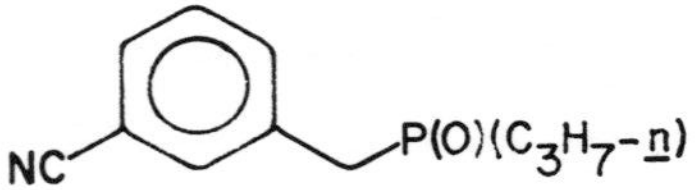

Structure	Reagent	Conditions	Yield (%)	No.
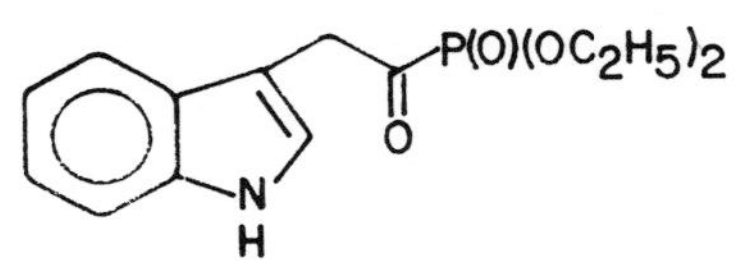 (indol-3-yl)CH₂C(O)P(O)(OC₂H₅)₂	$(C_2H_5O)_3P$	AcX, 5°C, 1 hr	64	203
$(C_2H_5O)_2P(O)CH_2C(O)NHCH(cyclopropyl)P(O)(OC_2H_5)_2$	$(C_2H_5O)_3P$	RX	26	296
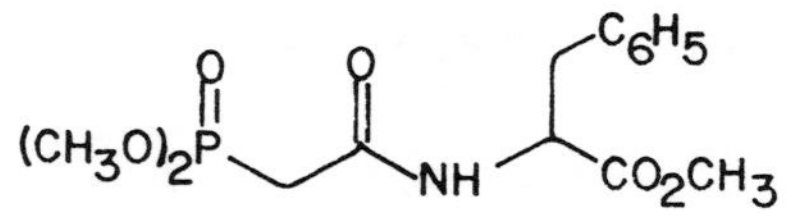 $(CH_3O)_2P(O)CH_2C(O)NHCH(CH_2C_6H_5)CO_2CH_3$	$(CH_3O)_3P$	RX, 130°C, 3 hr	97	182
$(C_2H_5O)_2P(O)CH_2C(O)NHCH(CH_2CH_2CO_2Na)P(O)(OC_2H_5)_2$	$(C_2H_5O)_3P$	RX	25	300
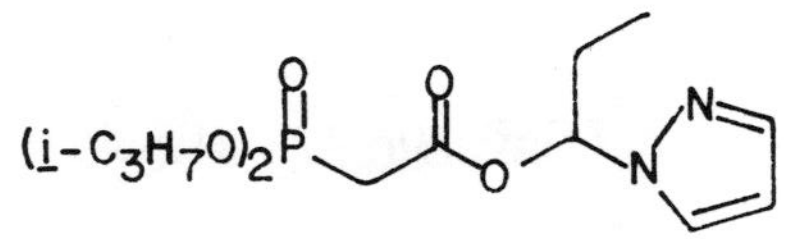 $(i-C_3H_7O)_2P(O)CH_2C(O)OCH(C_2H_5)(imidazol-1-yl)$	$(i-C_3H_7O)_3P$	RX, 4 hr, 140°C	85	302
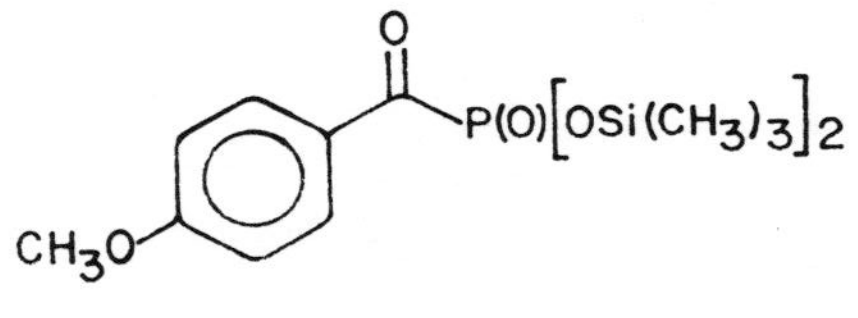 $(4-CH_3O-C_6H_4)C(O)P(O)[OSi(CH_3)_3]_2$	$[(CH_3)_3SiO]_3P$	AcX	72	218

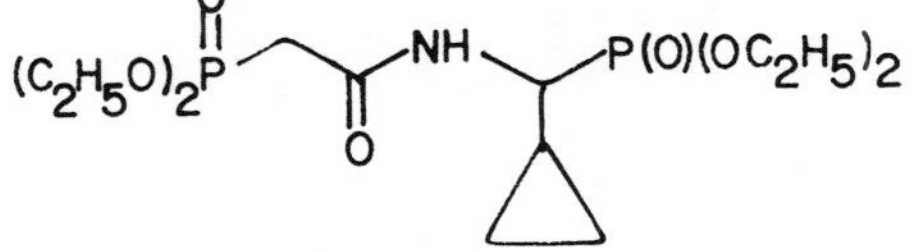

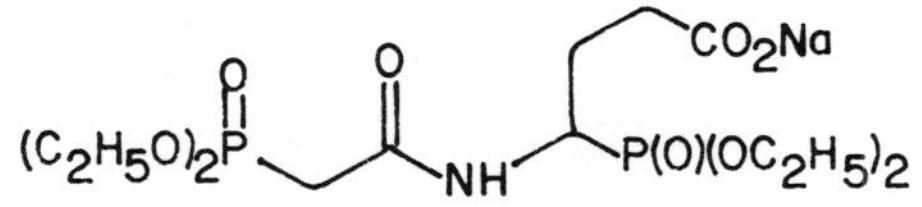

Table 2 (continued)

FORMATION OF CARBON TO PHOSPHORUS BONDS VIA THE MICHAELIS-ARBUZOV REACTION

Product	Phosphorus reagent	Method	Yield (%)	Refs.
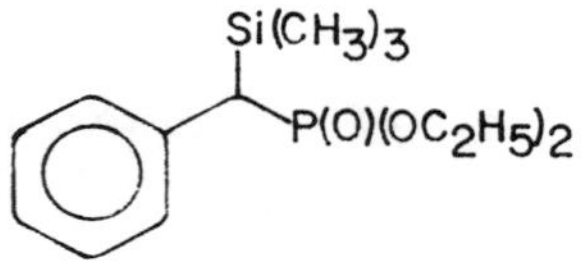 Si(CH$_3$)$_3$ / C$_6$H$_5$CH[P(O)(OC$_2$H$_5$)$_2$]	(C$_2$H$_5$O)$_3$P	RX	56	146
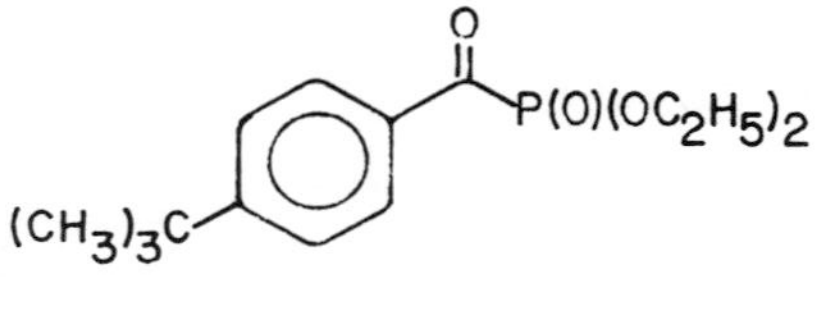 (CH$_3$)$_3$C–C$_6$H$_4$–C(O)–P(O)(OC$_2$H$_5$)$_2$	(C$_2$H$_5$O)$_3$P	AcX	29	224
C$_6$H$_5$COP(O)[OCH(CH$_3$)CH$_2$CH$_3$]$_2$	[CH$_3$CH$_2$CH(CH$_3$)O]$_3$P	AcX	59	199
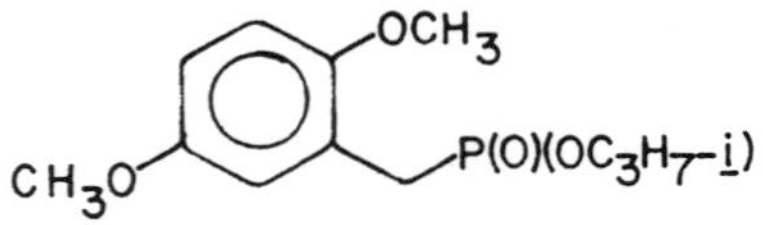 OCH$_3$ / CH$_3$O–C$_6$H$_3$–CH$_2$–P(O)(OC$_3$H$_7$-i)	(i-C$_3$H$_7$O)$_3$P	RX	54	145
(C$_2$H$_5$O)$_2$P(O)–CH$_2$CH$_2$–N(CO$_2$C$_6$H$_5$)(C$_2$H$_5$)	(C$_2$H$_5$O)$_3$P	Acylaziridine, 15 hr, 110°C	66	58

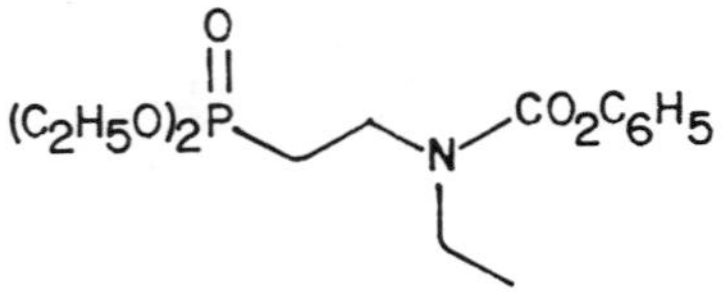

Product	Reagent	Conditions	Yield (%)	(°C)
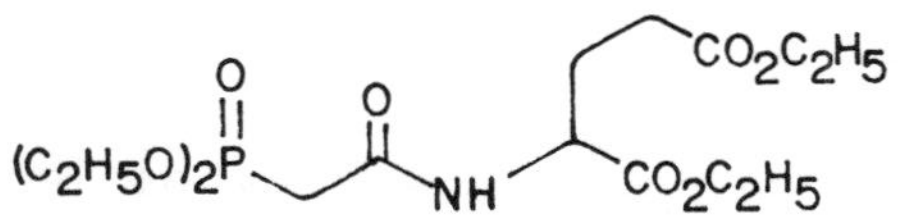 $(C_2H_5O)_2P$, $CO_2C_2H_5$, NH, $CO_2C_2H_5$ structure	$(C_2H_5O)_3P$	RX, 130°C, 3 hr	84	182
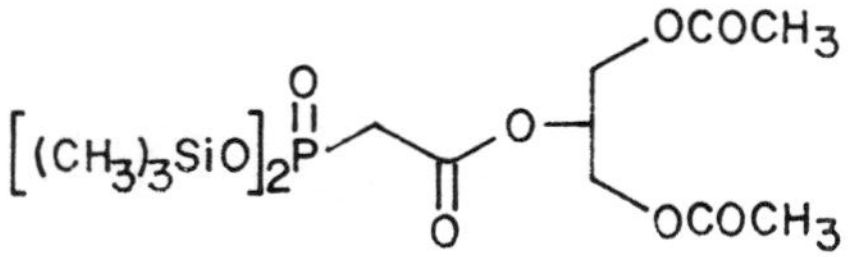 $[(CH_3)_3SiO]_2P$ structure with $OCOCH_3$ groups	$[(CH_3)_3SiO]_3P$	RX, 165°C, 1 hr	93	261
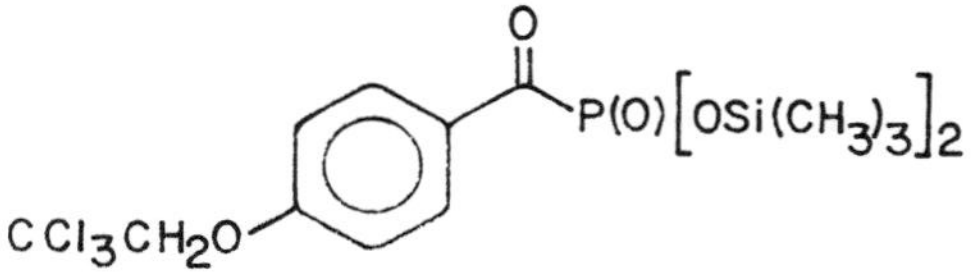 CCl_3CH_2O–$P(O)[OSi(CH_3)_3]_2$ (benzoyl)	$[(CH_3)_3SiO]_3P$	AcX	91	218
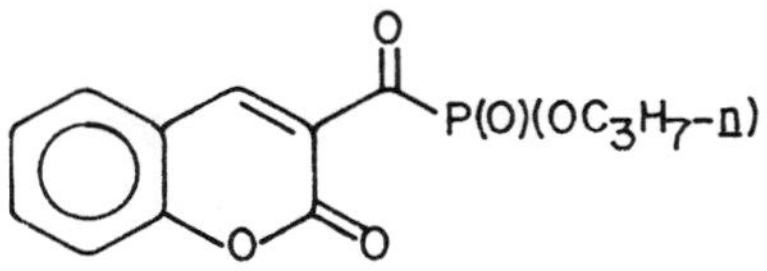 coumarin-3-$P(O)[OC_3H_7\text{-}n]$	$(n\text{-}C_3H_7O)_3P$	AcX	70	210
$[(C_2H_5)_3SiO]_2P(O)COCH(CF_3)_2$	$[(C_2H_5)_3SiO]_3P$	AcX	52	226
$C_6H_5COP(O)[OCH(C_2H_5)_2]_2$	$[(C_2H_5)_2CHO]_3P$	AcX	62	199
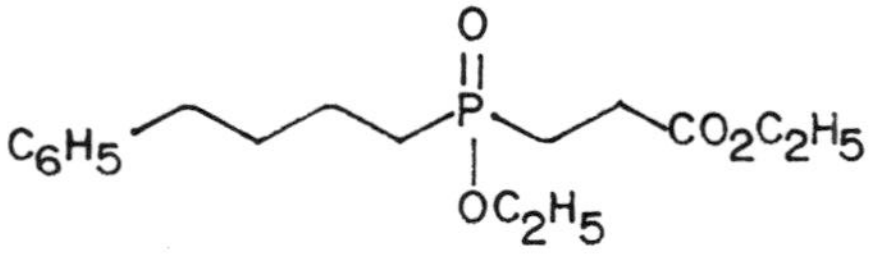 C_6H_5 chain, P, $OSi(CH_3)_3$, OC_2H_5, $CO_2C_2H_5$ / OC_2H_5 structure	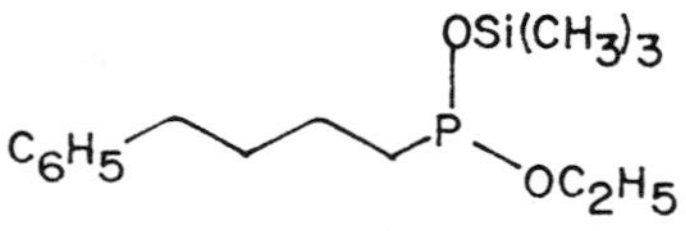	RX, r.t., 5 hr	96	267

Table 2 (continued)
FORMATION OF CARBON TO PHOSPHORUS BONDS VIA THE MICHAELIS-ARBUZOV REACTION

Product	Phosphorus reagent	Method	Yield (%)	Refs.
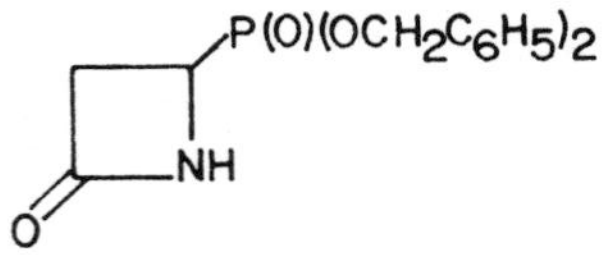	$(C_6H_5CH_2O)_3P$	ROAc, 120°C, 7 hr	46	245
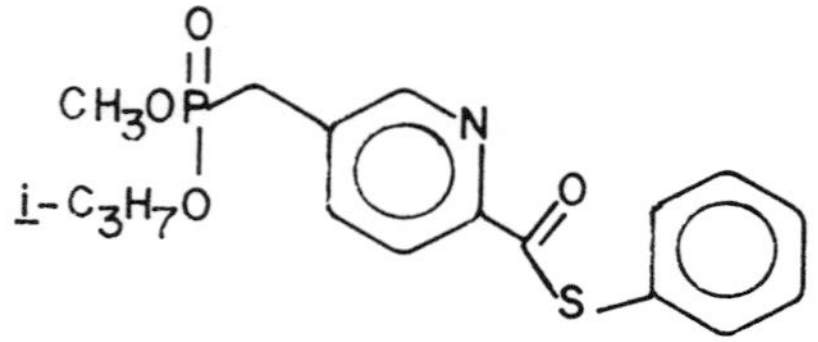	$(CH_3O)_2POCH(CH_3)_2$	RX	45	143
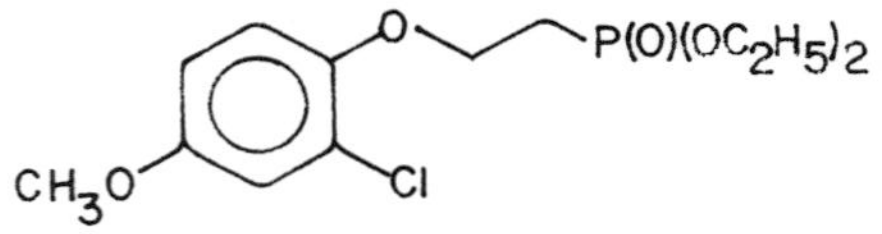	$(C_2H_5O)_3P$	RX, 185°C, 2 hr	30	117
$(n\text{-}C_4H_9)_3SnCH_2P(O)(OC_2H_5)_2$	$(C_2H_5O)_3P$	RX	60	119
$C_6H_5P(O)(OC_6H_5)_2$	$(C_2H_5O)_3P$	ArX, Ni, 320°C, 8.5 hr	91	273
		ArX, NiCl$_2$, 250°C, 37 hr	90	272
$(C_6H_5)_2P(O)COC(CH_3)_2OCOCH_3$	$(C_6H_5)_2POCH_3$	AcX, 50°C, 1 hr	90	232
$[(n\text{-}C_4H_9O)_2P(O)]_2CHOCH_3$	$(n\text{-}C_4H_9O)_3P$	RX, ZnCl$_2$, 200°C, 4 hr	38	276

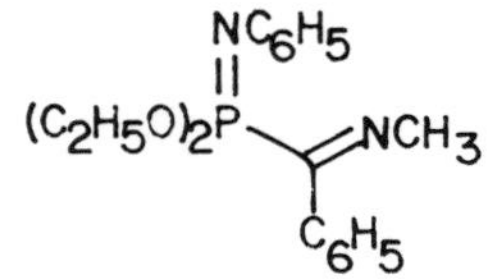

Product	Reagent	Conditions	Yield (%)	Ref.
(C2H5O)2P(=NC6H5)(=NCH3)C6H5	(C2H5O)2PNHC6H5	(C2H5)3N	90	303
(C6H5O)2P(O)CO2C6H5	(C6H5O)2POC2H5	AcX, 110°C, 10 hr / AcX	98 / 95	215 / 213
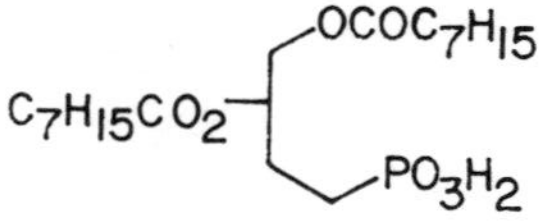 C7H15CO2 (OCOC7H15) PO3H2	[(CH3)3SiO]3P	RX, 125°C, 16 hr	71	258
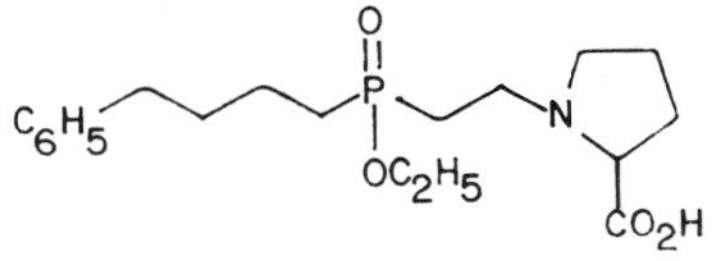 pyrrolidine-CO2H, P(=O)(OC2H5)C6H5	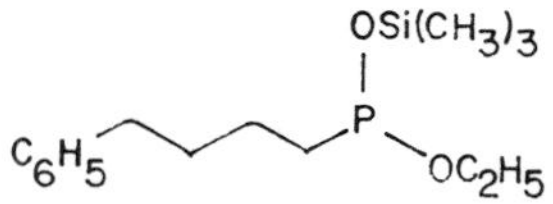 (CH3)3SiO / OC2H5 / C6H5 phosphine	RX, r.t., 5 hr	57	267
[(C2H5)3SiO]2P(O)C7H15-n	[(C2H5)3SiO]3P	RX, heat	80	251
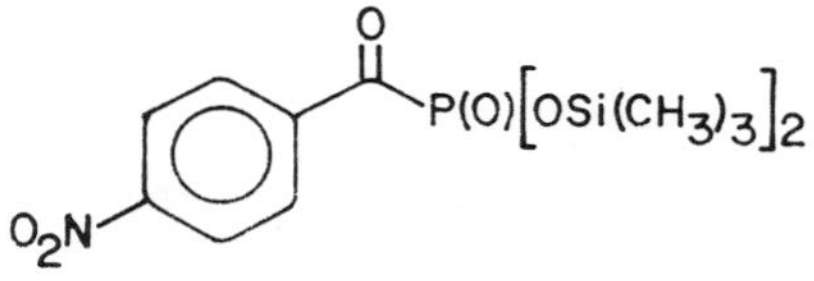 p-O2N-C6H4-C(O)-P(O)[OSi(CH3)3]2	[(C2H5)3SiO]3P	AcX	83	225
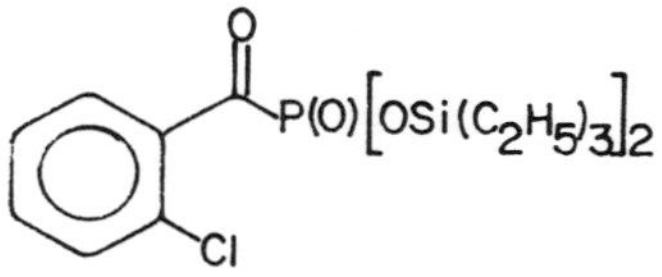 o-Cl-C6H4-C(O)-P(O)[OSi(C2H5)3]2	[(C2H5)3SiO]3P	AcX	70	225

Table 2 (continued)
FORMATION OF CARBON TO PHOSPHORUS BONDS VIA THE MICHAELIS-ARBUZOV REACTION

Product	Phosphorus reagent	Method	Yield (%)	Refs.
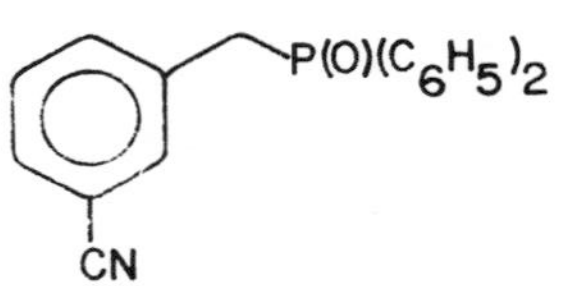	$(C_6H_5)_2POC_2H_5$	RX	87	149
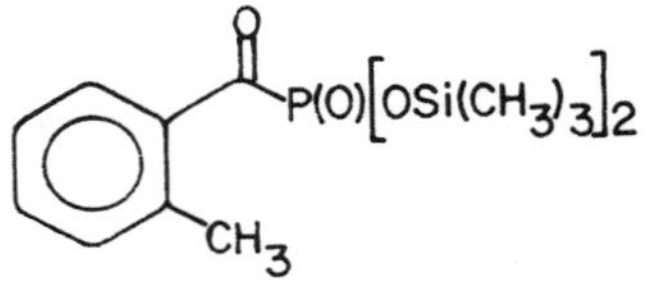	$[(C_2H_5)_3SiO]_3P$	AcX	90	225
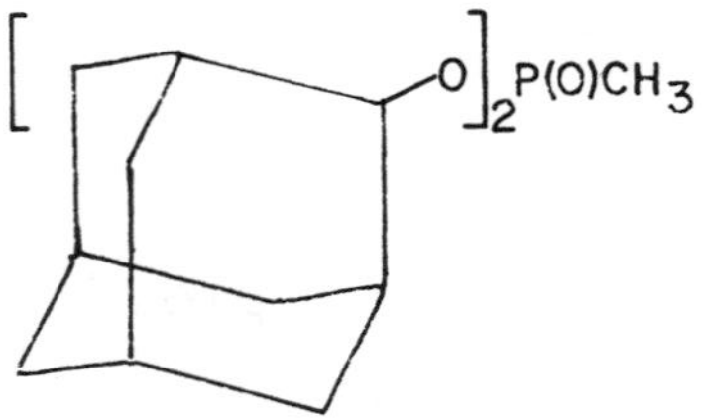	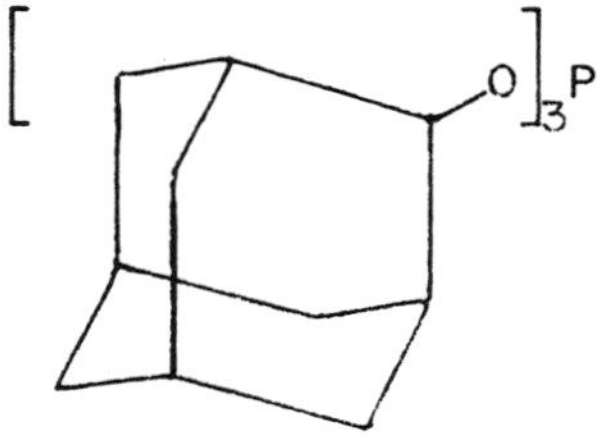	RX	72	95,96
$C_6H_5CH_2OCONHCH_2P(O)(OC_6H_5)_2$	$(C_6H_5O)_3P$	ROAc, HOAc, 120°C, 2 hr	48	243
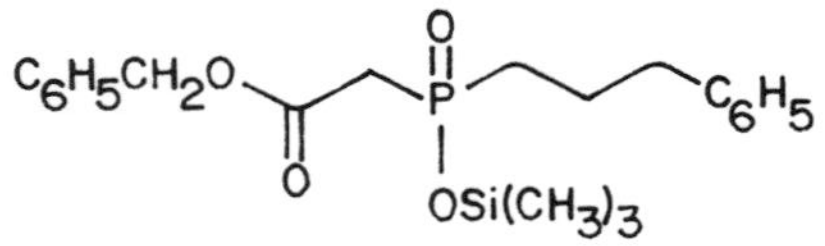	$C_6H_5CH_2CH_2CH_2CH_2P[OSi(CH_3)_3]_2$	RX	71	265

Product	Reagent	Conditions	Yield (%)	Ref.
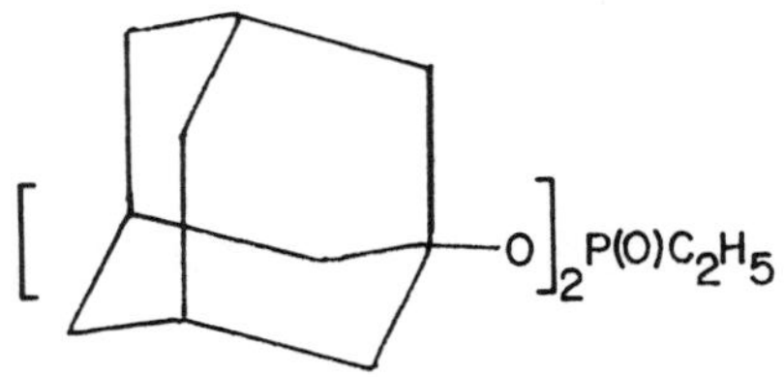 [adamantyl-O]$_2$P(O)C$_2$H$_5$ / [adamantyl-O]$_3$P	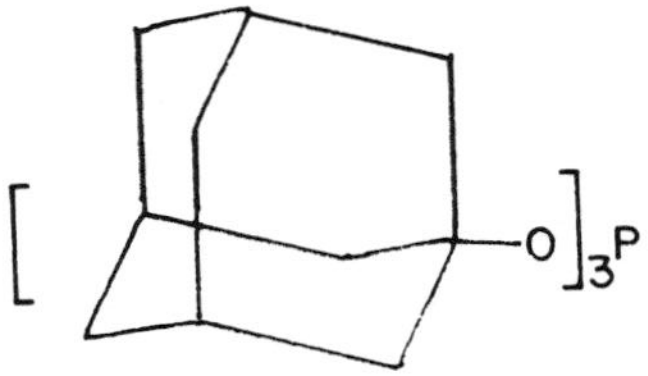 (C$_2$H$_5$O)$_3$P	RX, heat, 3 hr	75	95
(C$_2$H$_5$O)$_2$P(O)CH$_2$C(O)NH–CH(C$_3$H$_7$)P(O)(OC$_6$H$_5$)$_2$	(C$_2$H$_5$O)$_3$P	RX	72	300
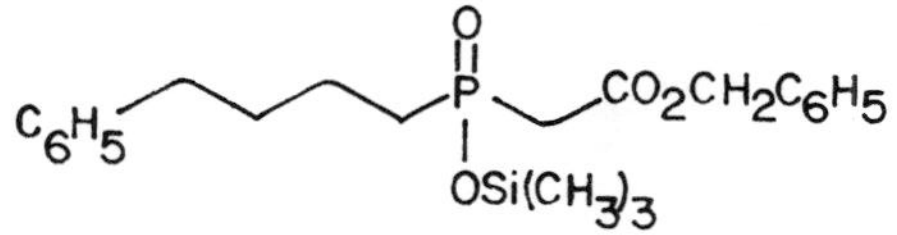 C$_6$H$_5$–(CH$_2$)$_3$–P(O)[OSi(CH$_3$)$_3$]CO$_2$CH$_2$C$_6$H$_5$	C$_6$H$_5$CH$_2$CH$_2$CH$_2$P[OSi(CH$_3$)$_3$]$_2$	RX, r.t., 5 hr	71	267
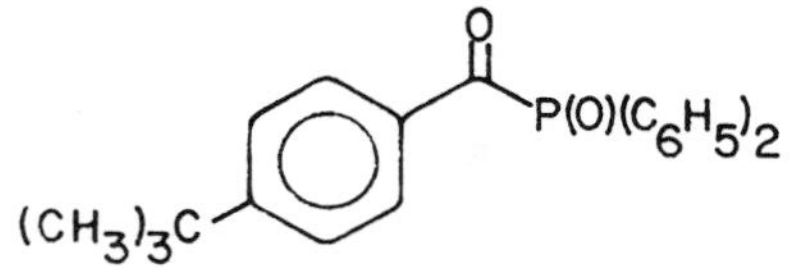 (CH$_3$)$_3$C–C$_6$H$_4$–C(O)–P(O)(OC$_6$H$_5$)$_2$	(C$_6$H$_5$)$_2$POCH$_3$	AcX, 50°C, 1 hr	83	232
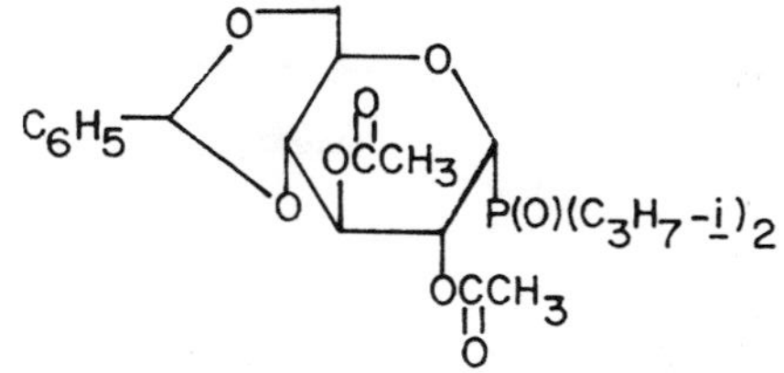 C$_6$H$_5$ sugar, OCCH$_3$, OCH$_3$, P(O)(C$_3$H$_7$–i)$_2$, OCCH$_3$	(i-C$_3$H$_7$O)$_3$P	RX, 180°C, 24 hr	43	105

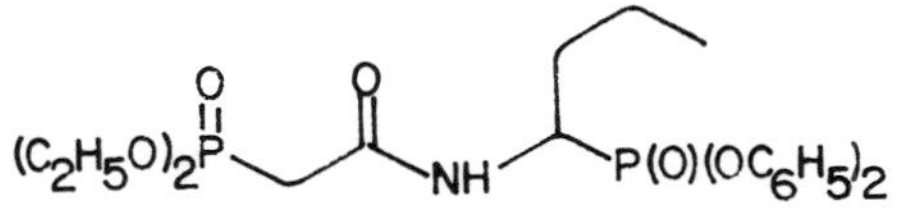

Table 2 (continued)
FORMATION OF CARBON TO PHOSPHORUS BONDS VIA THE MICHAELIS-ARBUZOV REACTION

Product	Phosphorus reagent	Method	Yield (%)	Refs.
$(C_2H_5O)_2P(O)CH_2COCH = P(C_6H_5)_3$	$(C_2H_5O)_3P$	RX, heat	80	170
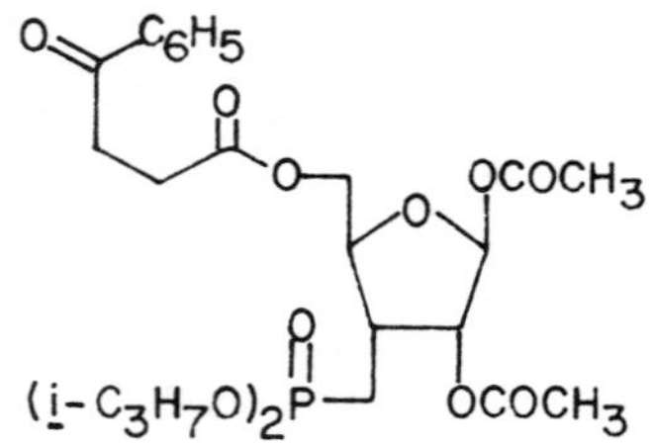	$(i\text{-}C_3H_7O)_3P$	RX, 180°C, 3 d	68	106
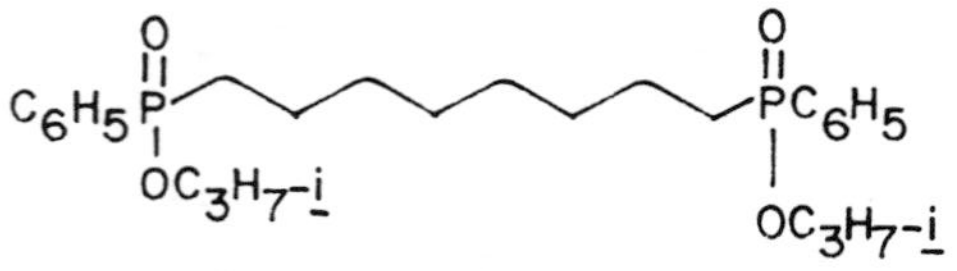	$C_6H_5P(OC_3H_7\text{-}i)_2$	RX, 190°C, 3.5 hr	94	134
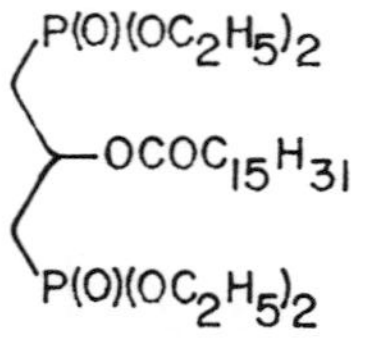	$(C_2H_5O)_3P$	RX, 145°C, 48 hr	45	114

Product	Reagent	Conditions	Yield	No.
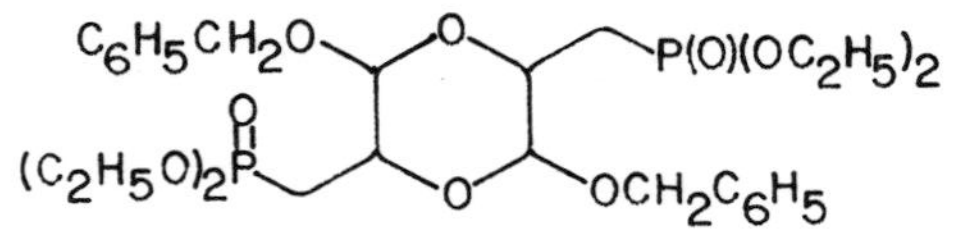 $C_6H_5CH_2O$, $(C_2H_5O)_2P(O)$... CH_2–$P(O)(OC_2H_5)_2$, $OCH_2C_6H_5$ (pyranose)	$(C_2H_5O)_3P$	RX, 145°C	93	111
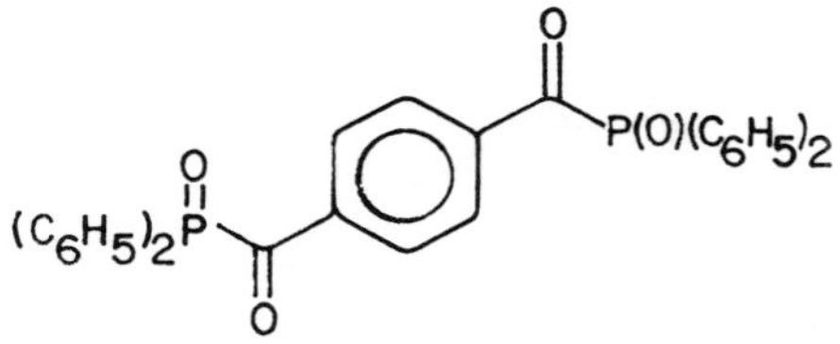 $(C_6H_5)_2P(O)$–C_6H_4–$C(O)$... $P(O)(C_6H_5)_2$	$(C_6H_5)_2POCH_3$	AcX, 50°C, 1 hr	35	232
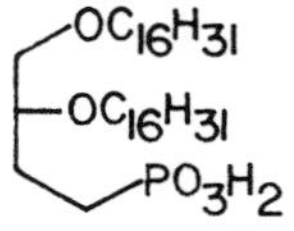 $OC_{16}H_{31}$, $OC_{16}H_{31}$, PO_3H_2	$[(CH_3)_3SiO]_3P$	1. RX; 2. H_2O	75	260
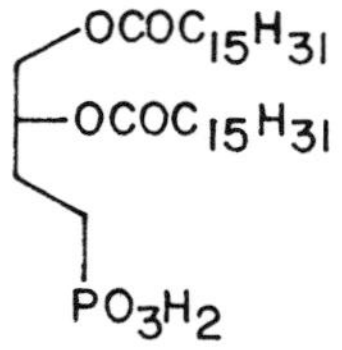 $OCOC_{15}H_{31}$, $OCOC_{15}H_{31}$, PO_3H_2	$[(CH_3)_3SiO]_3P$	1. RX; 2. H_2O	66	259
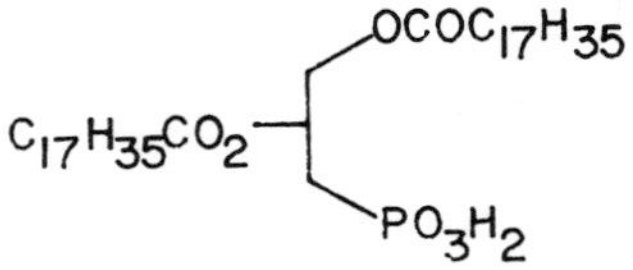 $OCOC_{17}H_{35}$, $C_{17}H_{35}CO_2$, PO_3H_2	$[(CH_3)_3SiO]_3P$	RX, 125°C, 16 hr	89	257

Table 2 (continued)

FORMATION OF CARBON TO PHOSPHORUS BONDS VIA THE MICHAELIS-ARBUZOV REACTION

Product	Phosphorus reagent	Method	Yield (%)	Refs.
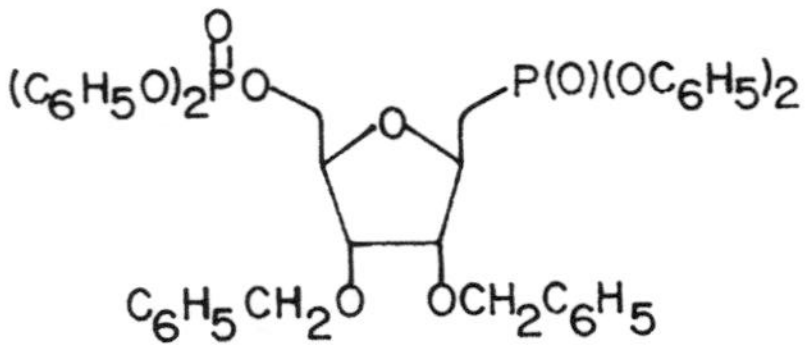	$(C_6H_5O)_2POC_2H_5$	RX, 160°C	40	102
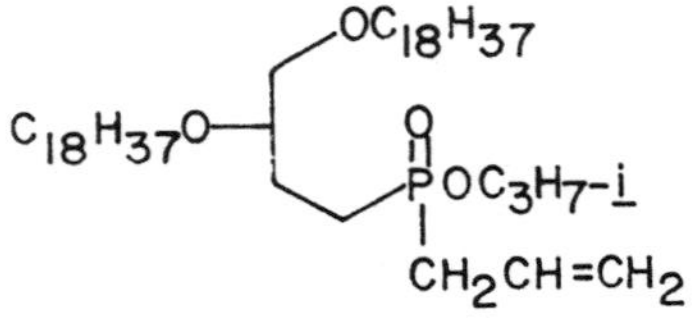	$(i\text{-}C_3H_7O)_2PCH_2CH=CH_2$	RX, heat	90	128, 129
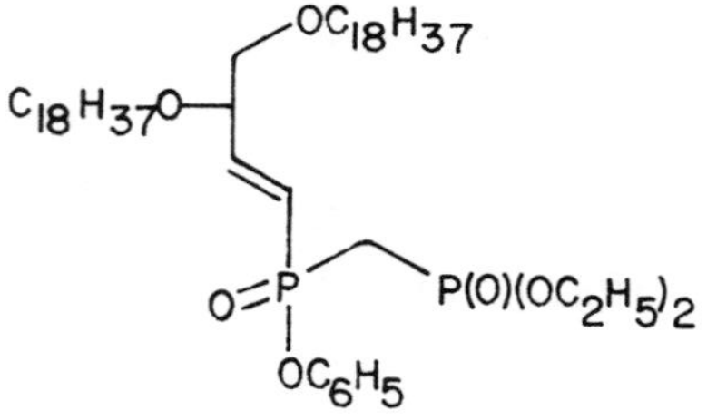	$(C_2H_5O)_3P$	RX, 150°C, 48 hr	54	194

REFERENCES

1. **Allen, D. W.**, Phosphines and phosphonium salts, *Organophosphorus Chem.*, 8, 1, 1977.
2. **Allen, D. W.**, Phosphines and phosphonium salts, *Organophosphorus Chem.*, 9, 1, 1978.
3. **Smith, D. J. H.**, Phosphines, phosphonium salts, and halogenophosphines, *Compr. Org. Chem.*, 2, 1127, 1979.
4. **Allen, D. W.**, Phosphines and phosphonium salts, *Organophosphorus Chem.*, 11, 1, 1980.
5. **Hudson, H. R.**, Quasi-phosphonium intermediates and compounds, in *Topics in Phosphorus Chemistry,* Vol. 11, Grayson, M. and Griffith, E. J., Eds., John Wiley & Sons, New York, 1983, 339.
6. **Michaelis, A., and Becker, T.**, Uber die Constitution der phosphorigen Saure, *Chem. Ber.*, 30, 1003, 1897.
7. **Crofts, P. C.**, Compounds containing carbon-phosphorus bonds, *Quart. Rev.*, 12, 341, 1958.
8. **Harvey. R. G. and deSombre, E. R.**, The Michaelis-Arbuzov and related reactions, in *Topics in Phosphorus Chemistry*, Vol. 1, Grayson, M. and Griffith, E. J., Eds., John Wiley & Sons, New York, 1964, 57.
9. **Roudier, L.**, The rate of reaction of organic halides with diesters of sodium phosphite, *Compt. Rend.*, 245, 2296, 1957.
10. **Farnham, W. B., Murray, R. K., and Mislow, K.**, Stereospecific alkylation of menthyl phenylphosphinate, *J. Am. Chem. Soc.*, 92, 5809, 1970.
11. **Malenko, D. M., Repina, L. A., and Sinitsa, A. D.**, Synthesis and rearrangement of benzimidoyl phosphites (trans), *J. Gen. Chem. U.S.S.R.*, 54, 1925, 1984.
12. **Reiff, L. P. and Aaron, H. S.**, Stereospecific synthesis and reactions of optically active isopropyl methylphosphinate, *J. Am. Chem. Soc.*, 92, 5275, 1970.
13. **Aaron, H. S., Braun, J., Shryne, T. M., Frack, H. F., Smith, G. E., Uyeda, R. T., and Miller, J. I.**, The stereochemistry of asymmetric phosphorus compounds. III. The resolution of a series of *O*-alkyl alkylphosphonothioic acids, *J. Am. Chem. Soc.*, 82, 596, 1960.
14. **Emmick, T. L. and Letsinger, R. L.**, Unsymmetrical secondary phosphine oxides. Synthetic, isotopic, and sterochemical studies, *J. Am. Chem. Soc.*, 90, 3459, 1968.
15. **Benschop, H. P. and Van den Berg, G. R.**, Stereospecific inclusion in cycloamyloses: partial resolution of isopropyl methylphosphinate and related compounds, *Chem. Commun.*, p. 1431, 1970.
16. **Cramer, F. and Dietsche, W.**, Occlusion compounds. XV. Resolution of racemates with cyclodextrins, *Chem. Ber.*, 92, 378, 1959.
17. **Van Hooidonk, C. and Breebaart-Hansen, J. C. A. E.**, Stereospecific reaction of isopropyl methylphosphonofluoridate (Sarin) with α-cyclodextrin. Model for enzyme inhibition, *Rec. Trav. Chim.*, 89, 289, 1970.
18. **Van den Berg, G. R., Platenberg, D. H. J. M., and Benschop, H. P.**, Stereochemistry of a Michaelis-Arbusov reaction: alkylation of optically active ethyl trimethylsilyl phenylphosphonite with retention of configuration, *Chem. Commun.*, p. 606, 1971.
19. **Kunuyants, I. L., Bykhovskaya, E. G., and Sizov, Y. A.**, Acylation of phosphites and phosphonites by fluorocarboxylic acid chlorides and bis(trifluoromethyl)ketene, *Zh. Vses. Khim. Va.*, 17, 354, 1972.
20. **Issleib, K., Koetz, J., Balszuweit, A., Lettau, H., Thust, U., and Pallas, M.**, Phosphonoformates, East German Patent 215,085, 1984.
21. **Issleib, K., Balszuweit, A., Moegelin, W., and Bertram, D.**, Phosphonoformates and thiophosphonoformates by conversion of bistrimethylsilyl hypophosphite, East German Patent 219,198, 1985.
22. **Petrillo, E. W. and Spitzmiller, E. R.**, Synthesis of 2-phosphonopyrrolidine and its substitution for proline in an inhibitor of angiotensin-converting enzyme, *Tetrahedron Lett.*, 4929, 1979.
23. **Ratcliffe, R. W. and Christensen, B. G.**, Total synthesis of β-lactam antibiotics. I. α-thioformamidodiethylphosphonoacetates, *Tetrahedron Lett.*, 4645, 1973.
24. **Petersen, H. and Reuther, W.**, α-Ureidoalkylierung von Phosphor (III)-Verbindungen, *Liebig's Ann. Chem.*, 766, 58, 1972.
25. **Ivanov, B. E., Krokhina, S. S., Ryzhkina, I. S., Gaidai, V. I., and Smirnov, V. N.**, Interaction of diethyl thiophosphite with phenolic mannich bases (transl.), *Bull. Acad. Sci. U.S.S.R.*, 569, 1979.
26. **Kong, A. and Engel, R.**, A mechanistic investigation of the Todd reaction, *Bull. Chem. Soc. Jpn.*, 58, 3671, 1985.
27. **Minowa, N., Fukatsu, S., Niida, T., and Mase, S.**, Phosphinic Acid Esters and Process for Preparing the Same, U.S. Patent 4,510,102, 1985.
28. Nippon Kayaku Co., Ltd., Alkyl and Aryl Phosphonites, Japanese Patent 82 46,993, 1982.
29. **Kem, K. M., Nguyen, N. V., and Cross, D.**, Phase-transfer catalyzed Michaelis-Becker reaction, *J. Org. Chem.*, 46, 5188, 1981.
30. **Makosza, M. and Wojciehowski, K.**, Phosphonic Acid Esters, Polish Patent 105,428, January 1980.
31. **Khachatryan, R. A., Sayadayan, S. V., Torgomyan, A. M., and Indzhikyan, M. G.**, Synthesis of allylphosphonates by interphase catalysis, *Arm. Khim. Zh.*, 34, 889, 1981.

32. **Makosza, M. and Wojciechowski, K.,** Synthesis of phosphonic acid esters in a solid-liquid catalytic two-phase system, *Bull. Pol. Acad. Sci. Chem.,* 32, 175, 1984.
33. **Paulsen, H., Thiem, J., and Moner, M.,** Phosphorus containing carbohydrates. I. Preparation of 3,4,6-tri-*O*-acetyl-1,2-*O*-[1'-(dialkylphosphono)ethylidene]- α -D-hexopyranoses, *Tetrahedron Lett.,* p. 2105, 1971.
34. **Paulsen, H. and Thiem, J.,** Reaktion von Glucopyranosylhalogeniden mit Dialkylphosphit Salzen und Trialkylphosphit, *Chem. Ber.,* 106, 115, 1973.
35. **Paulsen, H., Thiem, J., and Moner, M.,** Preparation of 3,4,6-tri-*O*-acetyl-1,2-*O*–[1'-(dialkylphosphono)ethylidene]-α-D-hexopyranoses, *Tetrahedron Lett.,* p. 2105, 1971.
36. **Paulsen, H. and Thiem, J.,** Darstellung von 1,2-*O*-[1-(Dialkylphosphono)athyliden]hexosen und -pentosen, *Chem. Ber.,* 106, 132, 1973.
37. **Schaub, F.,** Phosphonic Acid Ester Derivatives, West German Patent Application 2,944,598, 1980.
38. **Godovikov, N. N., Polyakova, L. A., Kireeva, E. G., and Kabachnik, M. I.,** Synthesis and properties of derivatives of heptamethylcyclotetrasiloxanomethylphosphonic acid (transl.), *Bull. Acad. Sci. U.S.S.R.,* 365, 1982.
39. **Hoffman, J. A.,** Cyclic Diphosphonates, U.S. Patent 4,268,459, 1981.
40. **Jagodic, V., Bozic, B., Tusek-Bozic, L., and Herak, M. J.,** Synthesis and spectroscopic studies of some quinolylmethylphosphonates, *J. Heterocycl. Chem.,* 17, 685, 1980.
41. **Kamiya, T., Hemmi, K., Takeno, H., and Hashimoto, M.,** Studies on phosphonic acid antibiotics. I. Structure and synthesis of 3-(*N*-acetyl-*N*-hydroxyamino)propylphosphonic acid (FR-900098) and its *N*-formyl analogue (FR-31564), *Tetrahedron Lett.,* 95, 1980.
42. **Tang, K.-C., Tropp, B. E., and Engel, R.,** The synthesis of phosphonic acid and phosphate analogues of glycerol 3-phosphate and related metabolites, *Tetrahedron,* 34, 2873, 1978.
43. **Schaub, F.,** Benzyl Phosphonic Acid Ester Derivatives, Australian Patent 52,640/79, 1979.
44. **Pevzner, L. M., Terekhova, M. I., Ignatev, V. M., Petrov, E. S., and Ionin, B. I.,** Synthesis and CH acidities of some diethyl (furylmethyl)phosphonates (transl.), *J. Gen. Chem. U.S.S.R.,* 54, 1775, 1984.
45. **Novikova, Z. S., Zdorova, S. N., and Lutsenko, I. F.,** Silicon-substituted benzylphosphonic acids (transl.), *J. Gen. Chem. U.S.S.R.,* 42, 108, 1972.
46. **Rudinskas, A. J. and Hullar, T. L.,** Pyridoxal phosphate. 5. 2-Formylethynylphosphonic acid and 2-formylethylphosphonic acid, potent inhibitors of pyridoxal phosphate binding and probes of enzymes topography, *J. Med. Chem.,* 19, 1367, 1976.
47. **Flynn, G. A., Beight, D. W., Bohme, E. H. W., and Metcalf, B. W.,** The synthesis of fluorinated aminophosphonic acid inhibitors of alanine racemase, *Tetrahedron Lett.,* 285, 1985.
48. **Sturtz, G.,** Action des phosphites sodes sur les cetones-halogenees, *Bull. Soc. Chim. Fr.,* 2333, 1964.
49. **Sturtz, G.,** Action des phosphites sodes, des phosphonites sodes et des phosphinites sur les cetones-halogenees prises song forme de cetals ou d'ethers enoliques, *Bull. Soc. Chim. Fr.,* 2340, 1964.
50. **Russell, G. A. and Ros, F.,** Reactions of α-Haloketones with Nucleophiles, *J. Am. Chem. Soc.,* 107, 2506, 1985.
51. **Bodalski, R., Pietrusiewicz, K. M., Monkiewicz, J., and Koszuk, J.,** A new efficient synthesis of substituted Nazarov reagents. A Wittig-Horner-Emmons approach, *Tetrahedron Lett.,* 2287, 1980.
52. **Brittelli, D. R.,** Preparation of Dialkyl- and Diarylphosphonoalkanoic Acids and Substituted Acrylic Acids, U.S. Patent 4,307,232, 1981.
53. **Burton, D. J. and Flynn, R. M.,** Method for Preparing Fluorine-containing Phosphonates, U.S. Patent 4,478,761, 1984.
54. **Myers, T. C., Harvey, R. G., and Jensen, E. V.,** Phosphonic acids. II. Synthesis of γ-ketophosphonic acids from methyl ketones via mannich bases, *J. Am. Chem. Soc.,* 77, 3101, 1955.
55. **Fiaud, J.-C.,** The palladium-catalysed reaction of lithium diphenylthiophosphides with allylic carboxylates. A stereoselective synthesis of phosphine sulfides, *Chem. Commun.,* p. 1055, 1983.
56. **Tsivunin, V. S., Zaripova, V. G., Zykova, T. V., Salakhutdinov, R. A., and Khazeeva, G. G.,** Reaction of sodium diethyl phosphite with 3-chloro-1,4-dioxan-2-ol acetic ester (transl.), *J. Gen. Chem. U.S.S.R.,* 49, 418, 1979.
57. **Baboulene, M. and Sturtz, G.,** Reactive de phosphites sodes, en Milieu alcoolique, vis-a-vis d'oxiranes. Synthese de phosphonates funtionnels, *Phosphorus and Sulfur,* 7, 101, 1979.
58. **Stamm, H., and Gerster, G.,** Reactions with aziridines. XXI. The (Michaelis-)Arbuzov reaction with *N*-acyl aziridines and other amidoethylations at phosphorus, *Tetrahedron Lett.,* 1623, 1980.
59. **Michaelis, A., and Kaehne, R.,** Ueber des Verhalten der Jodalkyle gegen die Sogen. Phosphorigsaureester oder *O*-Phosphine, *Chem. Ber.,* 31, 1048, 1898.
60. **Kosolapoff, G. M.,** *Organophosphorus Compounds,* John Wiley & Sons, New York, 1950.
61. **Kosolapoff, G. M.,** The Synthesis of phosphonic and phosphinic acids, in *Organic Reactions,* Vol. 6, Adams, R., Ed., John Wiley & Sons, New York, 1951, 273.
62. **Harvey, R. G., and de Sombre, E. R.,** The Michaelis-Arbuzov and related reactions, in *Topics in Phosphorus Chemistry,* Vol. 1, Grayson, M. and Griffith, E. J., Eds., John Wiley & Sons, New York, 1964, 57.

63. **Bhattacharya, A. K. and Thyagarajan, G.,** The Michaelis-Arbuzov rearrangement, *Chem. Rev.,* 81, 415, 1981.
64. **Shokol, V. A. and Kozhushko, B. N.,** Phosphorohalidites and phosphoropseudohalidites in the Arbuzov reaction with halogen-containing compounds (transl.), *Russ. Chem. Rev.,* 54, 98, 1983.
65. **Pudovik, A. N.,** Mechanism of the Arbuzov rearrangement *Dokl. Akad. Nauk S.S.S.R.,* 84, 519, 1952.
66. **Razumov, A. I.,** Mechanism of the Arbuzov Rearrangement, *Zh. Obshch. Khim.,* 29, 1609, 1959.
67. **Razumov, A. I. and Bankovskaya, N. N.,** Preparation and some properties of intermediate products of the Arbuzov rearrangement, *Izv. Akad. Nauk S.S.S.R.,* 863, 1957.
68. **Razumov, A. I., Mukhacheva, O. A., and Sim-Do-Khen,** Some esters of alkanethiophosphonic, alkeneselenophosphonic, dialkylphosphinic, and alkanephosphonous acids and the mechanism of the addition reactions of these esters, *Izv. Akad. Nauk S.S.S.R.,* 894, 1952.
69. **Smith, B. E. and Burger, A.,** Dialkylaminoalkyl phosphonates and phosphinates, *J. Am. Chem. Soc.,* 75, 5891, 1953.
70. **Dimroth, K. and Nurrenbach, A.,** Zur Einwirkung von Carbonium-Ionen auf Phosphorigsaure-triester und Bildung von Phosphonsaureestern, *Angew. Chem.,* 70, 26, 1958.
71. **Dimroth, K. and Nurrenbach, A.,** Reaktionen von Phosphorigsaure-triesters und -triamiden mit Carbonium-Ionen, *Chem. Ber.,* 93, 1649, 1960.
72. **Garabadzhin, A. V., Shibaev, V. I., Laurentev, A. N., and Sochilin, E. G.,** Reaction of perfluoralkyliodides with trialkyl phosphites (transl.), *J. Gen. Chem. U.S.S.R.,* 49, 1309, 1979.
73. **Brown, C., Hudson, R. F., Rice, V. T., and Thompson, A. R.,** Reduced nucleophilic reactivity in five-membered cyclic phosphites: correlation with transition state structure, *Chem. Commun.,* p. 1255 1971.
74. **Streitwieser, A., and Schaeffer, W. D.,** Stereochemistry of the primary carbon. III. Optically active 1-aminobutane-1-d, *J. Am. Chem. Soc.,* 78, 5597, 1956.
75. **Berry, R. S.,** Correlation of rates of intramolecular tunneling processes, with application to some group V compounds, *J. Chem. Phys.,* 32, 933, 1960.
76. **Bodkin, C. L. and Simpson, P.,** The stereochemical course of the Arbusov reaction with reference to configuration at phosphorus, *Chem. Commun.,* p. 1579, 1970.
77. **Bodkin, C. L. and Simpson, P.,** The role of pentaco-ordinate species in the mechanism of the Arbusov reaction, *J. Chem. Soc., Perkin II,* p. 2049, 1972.
78. **Adamcik, R. D., Chang, L. L., and Denney, D. B.,** Stereochemistries of the reactions of *cis* and *trans*-2-methoxy-4-methyl-1,3,2-dioxaphosphorinan with methyl iodide, *Chem. Commun.,* p. 986, 1974.
79. **Bentrude, W. G. and Hargis, J. H.,** Conformations of six-membered ring phosphorus heterocycles. I. The ring conformations and phosphorus configurations of isomeric six-membered ring phosphites, *J. Am. Chem. Soc.,* 92, 7136, 1970.
80. **Haque, M., Caughlan, C. N., Hargis, J. H., and Bentrude, W. G.,** Crystal and molecular structure of 5-*t*-butyl-2-methyl-2-oxo-1,3,2-dioxaphosphorinan, *J. Chem. Soc. (A),* p. 1786, 1970.
81. **Mikolajczyk, M.,** Optically active trivalent phosphorus acid esters: synthesis, chirality at phosphorus and some transformations, *Pure Appl. Chem.,* 52, 959, 1980.
82. **Razumov, A. I., Liorber, B. G., Zykova, T. V., and Bambushek, I. Y.,** Investigations in the series of phosphinous and phosphinic acid derivatives. LXXIV. Intermediate products in Arbuzov rearrangement of monoalkylphosphinous esters (transl.), *J. Gen. Chem. U.S.S.R.,* 40, 1996, 1970.
83. **Maslinnekov, I. G., Laurentev, A. N., Prokofeva, G. N., and Alekseeva, T. B.,** Intermediate compounds in Arbuzov reactions of fluorine-containing phosphonites (transl.), *J. Gen. Chem. U.S.S.R.,* 52, 464, 1982.
84. **Fluck, E. and Lorenz, J.,** Nuclear magnetic resonance of phosphorus compounds. XIV. Chemical shifts of phosphines, phosphonium salts, and diphosphinonickel(II) chlorides, *Z. Naturforsch.,* 22B, 1095, 1967.
85. **Verheyden, J. P. H. and Moffatt, J. G.,** Halo sugar nucleosides. I. Iodination of the primary hydroxyl groups of nucleosides with methyltriphenoxyphosphonium iodide, *J. Org. Chem.,* 35, 2319, 1970.
86. **Nesterov. L. V., Kessel, A. Y., Samitov, Y. Y., and Musina, A. A.,** Structure and reactivity of quasiphosphonium salts, *Dokl. Akad. Nauk S.S.S.R.,* 180, 116, 1968.
87. **Crutchfield, M. M., Dungan, C. H., Letcher, J. H., Mark, V., and van Wazer, J. R.,** The measurement and interpretation of high resolution ^{31}P nuclear magnetic resonance spectra, in *Topics in Phosphorus Chemistry,* Vol. 5, Grayson, M. and Griffith, E. J., Eds., John Wiley & Sons, New York, 1967, 227.
88. **Hudson, H. R., Rees, R. G., and Weekes, J. E.,** Preparation, structure, and nuclear magnetic resonance spectroscopy of triphenyl and trineopentyl phosphite — alkyl halide adducts, *J. Chem. Soc., Perkin I,* p. 982, 1974.
89. **Hudson, H. R., Rees, R. G., and Weekes, J. E.,** Methyltrineopentyloxyphosphonium iodide: a crystalline Michaelis-Arbuzov intermediate and its mode of decomposition, *Chem. Commun.,* p. 1297, 1971.
90. **Colle, K. S. and Lewis, E. S.,** Methoxyphosphonium ions: intermediates in the Arbuzov reaction, *J. Org. Chem.,* 43, 571, 1978.
91. **Henrick, K., Hudson, H. R., and Kow, A.,** Michaelis-Arbuzov intermediates: X-Ray crystal structures of the methyl bromide adducts of neopentyl diphenylphosphinite and dineopentyl phenylphosphonite, *Chem. Commun.,* p. 226, 1980.

92. **Gerrard, W. and Green, W. J.,** Mechanism of the formation of dialkyl alkylphosphonates, *J. Chem. Soc.,* p. 2550, 1951.

93. **Ford-Moore, A. H. and Williams, J. H.,** The reaction between trialkyl phosphites and alkyl halides, *J. Chem. Soc.,* p. 1465, 1947.

94. **Nesterov, L. V. and Aleksandrova, N. A.,** Decomposition of ethyldiisobutoxymethylphosphonium iodide in acetonitrile and methylene chloride (transl.), *J. Gen. Chem. U.S.S.R.,* 50, 29, 1980.

95. **Yurchenko, R. I., Klepa, T. I., Bobrova, O. B., Yurchenko, A. G., and Pinchuk, A. M.,** Phosphorylated adamantanes. II. Adamantyl phosphites in Arbuzov reaction (transl.), *J. Gen. Chem. U.S.S.R.,* 51, 647, 1981.

96. **Clark, R. D., Kozar, L. G., and Heathcock, C. H.,** Cyclopentenones from β,ε -diketophosphonates. Synthesis of *cis*-Jasmone, *Syn. Commun.,* 5, 1, 1975.

97. **Rocca, J. R., Tumlinson, J. H., Glancey, B. M., and Lofgren, C. S.,** Synthesis and stereochemistry of tetrahydro-3,5-dimethyl-6-(1-methoxybutyl)-2*H*-pyran-2-one, a component of the queen recognition pheromone of *Solenopsis Invicta, Tetrahedron Lett.,* p. 1893, 1983.

98. **Seo, K. and Inokawa, S.,** Sugars containing a carbon-phosphorus bond. VI. 5-(Alkylphosphinyl)-5-deoxy-D-xylopyranose, *Bull. Chem. Soc. Jpn.,* 48, 1247, 1975.

99. **Seo, K. and Inokawa, S.,** Sugars containing a carbon-phosphorus bond. IV. 5-(Alkylphosphonyl)-5-deoxy-O-methyl-D-xylopyranose, *Bull. Chem. Soc. Jpn.,* 46, 3301, 1973.

100. **Inokawa, S., Tsuchiya, Y., Seo, K., Yishida, H., and Ogata, T.,** 3-O-benzyl-5-deoxy-5-(ethylphosphinyl)-D-xylopyranose, *Bull. Chem. Soc. Jpn.,* 44, 2279, 1971.

101. **Inokawa, S., Kitagawa, H., Seo, K., Yoshida, H., and Ogata, T.,** Sugars containing a carbon-phosphorus bond. Part V. 5-Deoxy-5-(ethylphosphinyl)-D-ribopyranose, *Carbohydr. Res.,* 30, 127, 1973.

102. **Reitz, A. B., Nortey, S. O., and Maryanoff, B. E.,** Stereoselectivity in the electrophile-promoted cyclizations of a hydroxyolefin derived from arabinose, *Tetrahedron Lett.,* p. 3915, 1985.

103. **Zhdanov, Y. A., Uzlova, L. A., Glebova, Z. I., and Kistyan, G. K.,** Borodine-Hunsdiecker reaction in sugar chemistry. Synthesis of a phosphorus analog of a 2-deoxy aldonic acid (transl.), *J. Gen. Chem. U.S.S.R.,* 45, 1579, 1975.

104. **Chmielewski, M., BeMiller, J. N., and Cerretti, D. P.,** Synthesis of phosphonate analogs of α-D-glucopyranosyl and α-D-galactopyranosyl phosphate, *Carbohydr. Res.,* 97, C1, 1981.

105. **Cerretti, P.,** Synthesis of phosphonate analogs of α-D-glucopyranosyl and α-D-galactopyranosyl phosphate, *Carbohydr. Res.,* 94, C10, 1981.

106. **Mazur, A., Tropp, B. E., and Engel. R.,** Synthesis of a phosphonic acid analogue of an oligonucleotide, *Tetrahedron,* 40, 3949, 1984.

107. **Brigot, D., Collignon, N., and Savignac, P.,** Preparation d'acides aminoalkyl phosphoniques a l'aide d' ω-halogenoalkyl amines phosphorylees, *Tetrahedron,* 35, 1345, 1979.

108. **Wasielewski, C. and Antczak, K.,** A new and facile synthesis of phosphinothricin and 2-amino-4-phosphonobutyric acid, *Synthesis,* 540, 1981.

109. **Kabak, J., DeFilippe, L., Engel, R., and Tropp, B.,** Synthesis of the phosphonic acid isostere of glycerol 3-phosphate, *J. Med. Chem.,* 15, 1074, 1972.

110. **Pfeiffer, F. R., Mier, J. D., and Weisbach, J. A.,** Synthesis of phosphonic acid isosteres of 2-phospho-, 3-phospho-, and 2,3-diphosphoglyceric acid, *J. Med. Chem.,* 17, 112, 1974.

111. **Baer, E. and Robinson, R.,** Phosphonic acid analogs of carbohydrate metabolites. IV. Synthesis of DL-glyceraldehyde-3-phosphonic acid, *Can. J. Chem.,* 51, 104, 1973.

112. **Sarin, V., Tropp, B. E., and Engel, R.,** Isosteres of natural phosphates. 7. The preparation of 5-carboxy-4-hydroxy-4-methylpentyl-1-phosphonic acid, *Tetrahedron Lett.,* 351, 1977.

113. **Sturtz, G., Belbeoc'h, A., Damin, B. Clement, J.-C., and Lecolier, S.,** Synthese d'halohydrines phosphonates et leur utilisation a l'ignifugation des mousses de polyurethenne, French Patent 2,332,282, 1977.

114. **Robinson, R. and Baer, E.,** Synthesis of glycerol-1,3-diphosphonic acid, *Can. J. Biochem.,* 51, 1203, 1973.

115. **Bonsen, P. P. M., Burbach-Westerhuis, G. J., DeHaas, G. H., and VanDeenen, L. L. M.,** Chemical synthesis of some lecithin analogues. Potential inhibitors of phospholipase A, *Chem. Phys. Lipids,* 8, 199, 1972.

116. **Moschides, M. C.,** Synthesis of 1,2-dipalmitoyloxypropyl-3-(2-trimethylammoniummethyl)phosphinate, *Chem. Phys. Lipids,* 36, 343, 1985.

117. **Diana, G. D.,** Arylalkyl and Aryloxyalkyl Phosphonates and Use as Antiviral Agents, U.S. Patent 4,182,759, 1980.

118. **Diana, G. D.,** Aralkyl- and Aryloxyalkylphosphonates, West German Patent Application 2,922,054, 1979.

119. **Tzschach, A., Weichmann, H., Klepl, M., and Ochsler, B.,** Biologically Active Organophosphorus Functional Tetraorganotin Compounds, German (East) Patent 142,887, 16 July 1980.

120. **Abramson, A. and Weil, E. D.**, Arbuzov Reactions Employing an Aliphatic Solvent, U.S. Patent 4,311,652, 1982.

121. **Nikonova, L. Z. and Nuretdinova, A. A.**, Synthesis and isomerization of 2-(3′-chloroalkoxy)-1,3,2-azathia-, 2-(3′-chloroalkoxy)1,3,2-diaza-, 2-(3′-chloroalkoxy)-1,3,2-oxathia-, and 2-(3′-chloroalkoxy)-1,3,2-dithiaphospholanes (transl.), *Bull. Acad. Sci. U.S.S.R.*, 29, 663, 1980.

122. **Tandara, M. and Pilipovic, I.**, Arbuzov rearrangement of tris(β-chloroethyl) 2-chloroethylphosphonate in a film reactor, *Kem. Ind.*, 30, 673, 1981.

123. Nissan Chemical Industries, β-Haloethyl Methyl(β-haloethyl)phosphinates, JP. Patent 80 62,096, 1980.

124. **Hirao, T., Hagihara, M., Ohshiro, Y., and Agawa, T.**, Versatile synthesis of diethyl cyclopropane-phosphonates, *Synthesis*, 60, 1984.

125. **Bergesen, K. and Berge, A.**, The conformation of 5-methyl-2-alkoxy-2-oxo-1,2-oxaphosphorinanes, *Acta Chem. Scand.*, 26, 2975, 1972.

126. **Bayer, E., Gugel, K. H., Hagenmaier, H., Jessipow, S., Konig, W. A., and Zahner, H.**, Phosphin-othricin und Phosphinothricyl-Alanyl-Alanin, *Helv. Chim. Acta*, 55, 224, 1972.

127. **Gruszecka, E., Mastalerz, P., and Soroka, M.**, New synthesis of phosphinothricin and analogues, *Rocz. Chem.*, 49, 2127, 1975.

128. **Rosenthal, A. F. and Chodsky, S. V.**, New synthetic phosphinate analogs of lecithin, *J. Lipid Res.*, 12, 277, 1971.

129. **Rosenthal, A. F., Vargas, L., and Han, S. C. H.**, Synthesis of optically active diether phosphinate analogs of lecithin, *Biochim. Biophys. Acta*, 260, 369, 1972.

130. **Ogawa, Y., Inoue, S., and Niida, T.**, Fungicidal Compound, Japanese Patent 73 91,019, 1973.

131. **Walz, K., Nolte, W., and Muller, F.**, Surface-active Phosphonic Acid Esters, U.S. Patent 4,385,000, 1983.

132. **Horner, L. and Hoffmann, H.**, Wege zur Darstellung primarer, sekundarer und tertiarer Phosphine, *Chem. Ber.*, 91, 1583, 1958.

133. **Wetzel, R. B. and Kenyon, G. L.**, A convenient procedure for the synthesis of phosphorus-carbon bonds using sodium bis(2-methoxyethoxy)aluminum hydride, *J. Am. Chem. Soc.*, 94, 1774, 1972.

134. **Chan, T. H. and Ong, B. S.**, Macrocyclic diphosphines synthesis and stereochemistry, *J. Org. Chem.*, 39, 1748, 1974.

135. **Pascik, I., Arend, G., Reich, F., Schaefer, K., and Ludke, H.**, Halogenhaltige Phosphonate, West German Patent Application 2,712,175, 1978.

136. **Ionin, B. I. and Petrov, A. A.**, Prototropic isomerization of esters of alkenylphosphonic acids (transl.), *J. Gen. Chem. U.S.S.R.*, 33, 426, 1963.

137. **Bodesheim, F., Velker, E., Bentz, F., and Guenter, N.**, Reactions of 2-methylenepropane-1,3-diol derivatives with organic phosphorus compounds, *Chem.-Ztg.*, 96, 581, 1972.

138. **Akermark, B., Nystrom, J.-E., Rein, T., Badevall, J.-E., Helquist, P., and Aslanian, R.**, Facile route to 1-phosphonyl- and 1-sulfonyl-1,3-dienes *via* palladium-catalyzed elimination of allylic acetates, *Tetrahedron Lett.*, p. 5719, 1984.

139. **Stubbe, J. A. and Kenyon, G. L.**, Analogs of phosphoenolpyruvate. Substrate specificities of enolase and pyruvate kinase from rabbit muscle, *Biochemistry*, 11, 338, 1972.

140. **Crenshaw, M. D. and Zimmer, H.**, Synthesis of trisubstituted vinyl chlorides, *J. Org. Chem.*, 48, 2782, 1983.

141. **Liaw, B. R. and Guo, W. J.**, Synthesis of some dialkyl bromo-substituted benzyl phosphonates, *J. Chin. Chem. Soc. (Taipei)*, 31, 311, 1984.

142. **Okada, I., Nakazawa, M., Suzuki, S., and Kabe, H.**, Phosphonomethylphenyltetrahydrophthalimides and Their Use in Herbicides, Japanese Patent 60,246,392, 1985.

143. Kanebo, Ltd., Phosphonic Acid Esters, Japanese Patent 80,108,886, 1980.

144. **Belokrinitskii, M. A. and Orlov, N. F.**, Reaction of alkyl halides with triorganosilyl derivatives of phosphorous acid, *Kremniiorg. Mater.*, 145, 1971.

145. **Rakov, A. P., Rudnitskaya, G. F., and Andreev, G. F.**, Dialkyl (2,5-dimethoxybenzyl)phosphonates: their synthesis and some of their properties (transl.), *J. Gen. Chem. U.S.S.R.*, 46, 1450, 1976.

146. **Novikova, Z. S., Zdorova, S. N., and Lutsenko, I. F.**, Silicon-substituted benzylphosphonic acids (transl.), *J. Gen. Chem. U.S.S.R.*, 42, 108, 1972.

147. **Agranat, I., Rabinowitz, M., and Shaw, W.-C.**, Multiple Horner-Emmons cyclizations as a route to nonbenzenoid aromatics. Synthesis of polycyclic dodecanes, *J. Org. Chem.*, 44, 1936, 1979.

148. **Schellhammer, C. W. and Klanke, E.**, [(Fluorosulfonyl)benzyl]phosphonic Acid Esters, West German Patent Application 3,001,896, 1981.

149. **Yafarova, R. L., Ismagilov, R. K., Zabusova, N. G., and Baranova, N. Y.**, Synthesis of α-(dialkyl-phosphinyl)- and α-(diphenylphosphinyl)-*m*-tolunitriles (transl.), *J. Gen. Chem. U.S.S.R.*, 52, 193, 1982.

150. **Maier, L. and Lea, P. J.**, Organic phosphorus compounds. 76. Synthesis and properties of phosphinothricin derivatives, *Phosphorus and Sulfur*, 17, 1, 1983.

151. **Garibina, V. A., Dogadina, A. V., Zakharov, V. I., Ionin, B. I., and Petrov, A. A.,** (Haloethynyl)phosphonates. Synthesis and electrophilic reactions of (chloroethynyl)phosphonic esters (transl.), *J. Gen. Chem. U.S.S.R.,* 49, 1728, 1979.

152. **Kruglov, S. V., Ignatev, V. M., Ionin, B. I., and Petrov, A. A.,** Synthesis of symmetrical and mixed diphosphonic esters (transl.), *J. Gen. Chem. U.S.S.R.,* 43, 1470, 1973.

153. **Kluge, A. F.,** Phosphonate reagents for the synthesis of enol ethers and aldehyde homologation, *Tetrahedron Lett.,* p. 3629, 1978.

154. **Kluge, A. F. and Cloudsdale, I. S.,** Phosphonate reagents for the synthesis of enol ethers and one-carbon homologation to aldehydes, *J. Org. Chem.,* 44, 4847, 1979.

155. **Binder, J. and Zbiral, E.,** Diethyl[trimethylsilyl-ethoxymethyl]phosphonat — ein neuartiges Reagens für variable Aufbaustrategien von Carbonyl-, α-Hydroxycarbonylverbindungen und Vinylphosphonaten, *Tetrahedron Lett.,* p. 4213, 1984.

156. **Meleki, M., Miller, J. A., and Lever, O. W.,** Preparation and properties of α-alkoxyallyl phosphorus ylides, in *Phosphorus Chemistry. Proc. 1981 Int. Conf.* Quin, L. D. and Verkade, J. G., Eds., American Chemical Society, Washington, D.C., 1981, 145.

157. **Schaumann, E. and Grabley, F.-F.,** Preparation and synthetic reactions of α-alkoxyallyl phosphorus ylides, *Justus Liebigs Ann. Chem.,* 88, 1977.

158. **Earnshaw, C., Wallis, C. J., and Warren, S.,** Stereo- and Regio-specific vinyl ether synthesis by the Horner-Wittig reaction, *Chem. Commun.,* p. 314, 1977.

159. **Gross, H., Medved, T. Y., Nowak, S., Belski, F. I., Keitl, I., and Kabachnik, M. I.,** Synthesis and complex-formation properties of [(o-phenylenedioxy)methylene]bisphosphonic acid (transl.), *J. Gen. Chem. U.S.S.R.,* 55, 654, 1985.

160. **Tsivunin, V. S., and Zaripova, V. G.,** Reactions of trialkyl phosphites with 2,3-dichloro-2-methyl-p-dioxane (transl.), *J. Gen. Chem. USSR,* 54, 2277, 1984.

161. **Mikolajczyk, M. and Zatorski, A.,** α-Phosphorylsulphoxides. I. Synthesis, *Synthesis,* 669, 1973.

162. **Farrington, G. K., Kumar, A., and Wedler, F. C.,** Design and synthesis of new transition-state analogue inhibitors of aspartate transcarbamylase, *J. Med. Chem.,* 28, 1668, 1985.

163. **Kozlova, T. F., Grapov, A. F., and Melnikov, A. E.,** Reactions of methyl phosphorodichloridite with alkyl chloromethyl ethers catalyzed by the boron trifluoride ether complex (transl.), *J. Gen. Chem. U.S.S.R.,* 42, 1277, 1972.

164. **Kozhushko, B. N., Paliichuk, Y. A., Negrebetskii, V. V., and Shokol, V. A.,** (1-Alkoxyalkyl)phosphonic dihalides (transl.), *J. Gen. Chem. U.S.S.R.,* 50, 795, 1980.

165. **Stukalo, E. A., Yureva, E. M., and Markovskii, L. N.,** 1-(Dihalophosphinyl)vinyl isocyanates (transl.), *J. Gen. Chem. U.S.S.R.,* 50, 278, 1980.

166. **Herzig, C. and Gasteiger, J.,** Reaction of 2-chlorooxiranes with phosphites and phosphanes: A new route to β-carbonylphosphonic esters and -phosphonium salts, *Chem. Ber.,* 115, 601, 1982.

167. **Pfliegal, T., Seres, J., Gajary, A., Davoczy-Csuka, K., and Nagy, L. T.,** N- Phosphonomethylglycine, Swiss Patent 620,223, 1980.

168. **Felix, R. A.,** N-Acylaminomethyl-N-cyanomethyl Phosphonates, U.S. Patent 4,476,063, 1984.

169. **Guervich, P. A., Kiselev, V. V., Moskva, V. V., and Moskva, N. A.,** Two paths of reaction of N-(ω-Chloroalkyl)-2-pyrrolidones (transl.), *J. Gen. Chem. U.S.S.R.,* 54, 421, 1984.

170. **Ogawa, Y., Inoue, S., and Niida, T.,** Fungicidal Compound, Japanese Patent 73 91,019, 1973.

171. **Diana, G. D.,** Arylalkyl- and Aryloxyalkylphosphonates as Antiviral Agents, U.S. Patent 4,217,346, 1980.

172. **Southgate, C. C. B. and Dixon, H. B. F.,** Phosphonate analogues of aminoacyl adenylates, *Biochem. J.,* 175, 461, 1978.

173. **Varlet, J.-M., Collignon, N., and Savignac, P.,** A new route to 2-aminoalkanephosphonic acids, *Synthetic Commun.,* 8, 335, 1978.

174. **Collignon, N., Fabre, G., Varlet, J.-M., and Savignac, P.,** Amination reductrice d'aldehydes phosphoniques, un reexamen, *Phosphorus and Sulfur,* 10, 81, 1981.

175. **Cates, L. A., Jones, G. S., Good, D. J., Tsai, H. Y.-L., Li, V.-S., Caron, N., Tu, S.-C., and Kimball, A. P.,** Cyclophosphamide potentiation and aldehyde oxidase inhibition by phosphorylated aldehydes and acetals, *J. Med. Chem.,* 23, 300, 1980.

176. **Razumov, A. I., Krasilnikova, E. A., Zykova, T. V., and Nevzorova, O. L.,** Reactivity and structure of phosphorylated carbonyl compounds. XVIII. Synthesis and investigation of the properties of [alkoxy(2-thienyl)phosphinyl]acetaldehydes and their derivatives (transl.), *J. Gen. Chem. U.S.S.R.,* 49, 469, 1979.

177. **Malone, G. R. and Myers, A. I.,** The chemistry of 2-chloromethyloxazines. Formation of phosphoranes and phosphonates. The use of α,β-unsaturated oxazines as a common intermediate for the synthesis of aldehydes, ketones, and acids *J. Org. Chem.,* 39, 623, 1974.

178. **Burri, K. F., Cardone, R. A., Chen, W. Y., and Rosen, P.,** Preparation of macrolides *via* the Wittig reaction. A total synthesis of (−)-vermiculine, *J. Am. Chem. Soc.,* 100, 7069, 1978.

179. **Herrin, T. R., Fairgrieve, J. S., Bower, R. R., Shipkowitz, N. L., and Mao, J. C.-H.,** Synthesis and anti-Herpes Simplex activity of analogues of phosphonoacetic acid, *J. Med. Chem.*, 20, 660, 1977.

180. **Berry, J. P., Isbell, A. F., and Hunt, G. E.,** Aminoalkylphosphonic acids, *J. Org. Chem.*, 37, 4396, 1972.

181. **Isbell, A. F., Berry, J. P., and Tansby, L. W.,** The synthesis and properties of 2-aminoethylphosphonic acid and 3-aminopropylphosphonic acids, *J. Org. Chem.*, 37, 4399, 1972.

182. **Kafarski, P. and Soroka, M.,** An improved synthesis of *N*-(phosphonoacetyl)-amino acids, *Synthesis*, 219, 1982.

183. **Salbeck, G.,** Arbuzow Reaktion mit 2-Chlorimidchloriden, *Phosphorus and Sulfur*, 22, 353, 1985.

184. **Kukhar, V. P. and Sagina, E. I.,** Reactions of trialkyl phosphites and triphenylphosphine with trichloroacetic derivatives. II (transl.), *J. Gen. Chem. U.S.S.R.*, 49, 889, 1979.

185. **Kukhar, V. P. and Sagina, E. I.,** Reactions of trialkyl phosphites with trichloroacetic derivatives (transl.), *J. Gen. Chem. U.S.S.R.*, (Engl.) 49, 50, 1979.

186. **Yaouanc, J. J., Sturtz, G., Kraus, J. L., Chastel, C., and Colin, J.,** Synthese d'analogues Tetrazoles de l'acide phosphonoacetique, *Tetrahedron Lett.*, p. 2689, 1980.

187. **Kutyrev, A. A., Moskva, V. V., and Alparova, M. V.,** Reactions of phosphites with trichloroacetonitrile (transl.), *J. Gen. Chem. U.S.S.R.*, 54, 1332, 1984.

188. **Sauers, R. F.,** Herbicidal Phosphonates, U. S. Patent 4,225,521, 1980.

189. **Dahl, O.,** Alkyl di-*t*-butylphosphinites. Exceptionally halogenophilic phosphinites in Arbuzow reactions, *J. Chem. Soc. Perkin I*, 947, 1978.

190. **Francis, M. D. and Martodam, R. R.,** Chemical, biochemical, and medicinal properties of the diphosphonates, in *The Role of Phosphonates in Living Systems*, Hilderbrand, R. L., Ed., CRC Press, Boca Raton, Fla., 1983, 55.

191. **Kukhar, V. P. and Sagina, E. I.,** Reactions of trialkyl phosphites with polyhalomethanes (transl.), *J. Gen. Chem. U.S.S.R.*, 49, 1284, 1979.

192. **Menge, M., Muenzenberg, K. J., and Reimann, E.,** 2,2-Propanediphosphonic acid, *Arch. Pharm.*, 314, 218, 1981.

193. **Rosenthal, A. F., and Vargas, L. A.,** A synthetic phosphinate-phosphonate liponucleotide analogue, *Chem. Commun.*, p. 976, 1981.

194. **Vargas, L. and Rosenthal, A. F.,** A reagent for introducing the phosphinic acid isostere of phosphodiesters, *J. Org. Chem.*, 48, 4775, 1983.

195. **Kabachnik, M. I. and Rossiiskaya, P. A.,** Esters of α-Ketophosphonic Acids, *Izv. Akad. Nauk S.S.S.R. Otd. Khim. Nauk*, p. 365, 1945.

196. **Scherer, H., Hartmann, A., Regitz, M., Tungyal, B. D., and Gunther, H.,** Carbene. V. 7-Phosphono-7-aryl-norcaradiene, *Chem. Ber.*, 105, 3357, 1972.

197. **Kugler, R., Pike, D. C., and Chin, J.,** Kinetics and mechanism of the reaction of dimethyl acetylphosphonate with water. Expulsion of a phosphonate from a carbonyl hydrate, *Can. J. Chem.*, 56, 1792, 1978.

198. **Kojima, M., Yamashita, M., Yoshida, H., Ogata, T., and Inokawa, S.,** Useful methods for the preparation of unsaturated phosphonates (1-methylenealkanephosphonates), *Synthesis*, 147, 1979.

199. **Sekine, M., Satoh, M., Yamagata, H., and Hata, T.,** Acylphosphonates: P-C bond cleavage of dialkyl acylphosphonates by means of amines. Substituent and solvent effects for acylation of amines, *J. Org. Chem.*, 45, 4162, 1980.

200. **Zon, J.,** A simple preparation of Diethyl 1-acylamino-1-ethenephosphonates, *Synthesis*, p. 324, 1981.

201. **Zon, J.,** "Synthesis of diisopropyl 1-Nitroalkenephosphonates from diisopropyl 1-oxoalkanephosphonates," *Synthesis*, p. 661, 1984.

202. **Ohler, E., El-Badawi, M., and Zbiral, E.,** Ein einfacher Weg zu α-substititierten (*E*)-3-Oxo-1-alkenylphosphonsaureestern, *Monatsh. Chem.*, 116, 77, 1985.

203. **Subotkowski, W., Kowalik, J., Tyka, R., and Mastalerz, P.,** The phosphonic analog of tryptophan, *Pol. J. Chem.*, 55, 853, 1981.

204. **Asano, S., Kitahara, T., Ogawa, T., and Matsui, M.,** The synthesis of α-amino phosphonic acids, *Agr. Biol. Chem.*, 37, 1193, 1973.

205. **Gandurina, I. A., Zhukov, Y. N., Osipova, T. I., and Khomutov, R. M.,** α-Amino(alkylamino)phosphonic Acids, U.S.S.R. Patent 697,519, 1979.

206. **Kowalik, J., Zygmunt, J., and Mastalerz, P.,** 1-Amino-2-mercaptoethanephosphonic acid, the phosphonic analog of cysteine, *Pol. J. Chem.*, 55, 713, 1981.

207. **Subotkowski, W., Tyka, R., and Mastalerz, P.,** Large scale preparation of dialkyl 2-pyrrolidinephosphonates, *Pol. J. Chem.*, 57, 1389, 1983.

208. **Pfister, T., Eue, L., and Schmidt, R. R.,** Phenoxypropionylphosphonic Acid Esters, West German Patent 3,337,540, 1985.

209. **Pfister, T., Eue, L., Satel, H. J., Schmidt, R. R., and Henssler, G.,** Chlorinated Phosphonylmethylcarbonylpyrazole Derivatives, West German Patent 3,409,081, 1985.

210. **Vishnyakova, G. M., Smirnova, T. V., and Tarakanova, L. A.,** Synthesis of Coumarin-containing α-ketophosphonates (transl.), *J. Gen. Chem. U.S.S.R.,* 55, 1076, 1985.

211. **Arbuzov, A. E. and Dunin, A. A.,** Action of halogen derivatives of aliphatic esters on alkyl phosphites, *J. Russ. Phys. Chem. Soc.,* 46, 291, 1914.

212. **Warren, S. and Williams, M. R.,** The acid-catalysed decarboxylation of phosphonoformic acid, *J. Chem. Soc. B,* 618, 1971.

213. **Helgstrand, A. J. E., Johansson, K. N. G., Misiorny, A., Noren, J. O., and Stening, G. B.,** Aromatic Esters of Phosphonoformic Acid, European Patent 3,275, 1979.

214. **Helgstrand, A. J. E., Johansson, K. N. G., Misiorny, A., Noren, J. O., and Stening, G. B.,** Aliphatic Derivatives of Phosphonoformic Acid, Pharmaceutical Compositions and Methods for Combating Virus Infections, European Patent 3,007, 1979.

215. **Helgstrand, A. J. E., Johansson, K. N. G., Misiorny, A., Noren, J. O. and Stening, G. B.,** Phosphonoformic Acid Esters, U.S. Patent 4,372,894, 1983.

216. **Helgstrand, A. J. E., Johansson, K. N. G., Misiorny, A., Noren, J. O., and Stening, G. B.,** Phosphonoformic Acid Esters and Pharmaceutical Compositions Containing Same, U.S. Patent 4,386,081, 1983.

217. **Engel, R.,** Thermolysis of phosphonates: phosphonoformate esters, *Phosphorus and Sulfur,* 29, 369, 1987.

218. **Sekine, M. and Hata, T.,** Convenient synthesis of unesterified acylphosphonic acids, *Chem. Commun.,* p. 285, 1978.

219. **Sekine, M., Kume, A., Nakajima, M., and Hata, T.,** A new method for acylation of enolates by means of dialkyl acylphosphonates as acylating agents, *Chem. Lett.,* p. 1087, 1981.

220. **Sekine, M., Yamagata, H., and Hata, T.,** Silyl phosphite equivalents: 2,2,2-trichloroethoxycarbonylphosphonates, *Chem. Commun.,* p. 970, 1981.

221. **Alfonsov, V. A., Zamaletdinova, G. U., Batyeva, E. S., and Pudovik, A. N.,** New path in reaction of silyl phosphites with acetyl chloride (transl.), *J. Gen. Chem. U.S.S.R.,* 54, 414, 1984.

222. **Welter, W., Hartmann, A., and Regitz, H.,** Isomerization reactions of phosphoryl-vinyl-carbenes to phosphorylated cyclopropenes, allenes, acetylenes, indenes, and 1,3-butadienes, *Chem. Ber.,* 111, 3068, 1978.

223. **Horner, L. and Roder, H.,** Notiz uber die reduktive Umwandlung von Carbonsaure in ihre Aldehyde, *Chem. Ber.,* 103, 2984, 1970.

224. **Terauchi, K. and Sakurai, H.,** Photochemical studies of the esters of aroylphosphonic acids, *Bull. Chem. Soc. Jpn.,* 43, 883, 1970.

225. **Dyakov, V. M., Orlov, N. F., Gusakova, G. S., and Zakharova, N. M.,** Triorganosilyl derivatives of *o*- and *p*-substitutedbenzoylphosphonic acids, *Kremniiorg. Mater.,* p. 139, 1971.

226. **Dyakov, V. M. and Voronkov, M. G.,** Trialkylsilyl esters of polyfluoroacylphosphonic acids (transl.), *Bull. Acad. Sci. U.S.S.R.,* p. 379, 1973.

227. **Winkler, T. and Bencze, W. L.,** Perkow, reaction induced C,C-bond formation, *Helv. Chim. Acta,* 63, 402, 1980.

228. **Hoffmann, H., Maurer, F., Priesnitz, U., and Riebel, H. J.,** Alkenes, West German Patent Application 2,917,620, 1980.

229. **Tam, C. C., Mattocks, K. L., and Tishler, M.,** Enol-keto tautomerism of α-ketophosphonates, *Proc. Natl. Acad. Sci. U.S.A.,* 78, 3301, 1981.

230. **Szpala, A., Tebby, J. C., and Griffiths, D. V.,** Reaction of phosphites with unsaturated acid chlorides: synthesis and reactions of dimethyl but-2-enoylphosphonate, *J. Chem. Soc., Perkin I,* p. 1363, 1981.

231. **Griffiths, D. V., Jamali, H. A. R., and Tebby, J. C.,** Reactions of phosphites with acid chlorides. Phosphite attack at the carbonyl oxygen of α-ketophosphonates, *Phosphorus and Sulfur,* 11, 95, 1981.

232. **Lechtken, P., Buethe, I., and Hesse, A.,** Acylphosphine oxide compounds, U.S. Patent 4,324,744, 1982.

233. **Ohler, E. and Zbiral, E.,** Cyclisierungsreaktionen von Diazoalkenyl-phosphonsaureestern Synthese von Pyrazolyl- und 2,3-Benzodiazepinylphosphonsaureestern, *Monatsh. Chem.,* 115, 629, 1984.

234. **Cann, P. F., Warren, S., and Williams, M. R.,** Electrophilic substitutions at phosphorus: reactions of diphenylphosphinyl systems with carbonyl compounds, *J. Chem. Soc., Perkin I.,* p. 2377, 1972.

235. **Gazizov, T. K., Belyalov, R. U., and Pudovik, A. N.,** Mechanism of reactions of phenyl esters of phosphorus (III) acids with carboxylic acid chlorides (transl.), *J. Gen. Chem. U.S.S.R.,* 50, 1355, 1980.

236. **Taira, K. and Gorenstein, D. C.,** Experimental tests of the stereoelectronic effect at phosphorus: Michaelis-Arbusov reactivity of phosphite esters, *Tetrahedron,* 40, 3215, 1984.

237. **Schollkopf, U., Schroeder, R., and Stafforst, D.,** Reactions of α-metalated diethyl isocyanomethyl- and α-isocyanobenzylphosphonates with carbonyl compounds, *Justus Liebigs Ann. Chem.,* p. 44, 1974.

238. **Rachon, J., Schollkopf, U., and Wintel, T.,** Synthesis of 1-aminoalkylphosphonic acids by alkylation of α-metalated diethyl isocyanomethylphosphonates, *Justus Liebigs Ann. Chem.,* p. 709, 1981.

239. **Schollkopf, U. and Schroeder, R.,** Umsetzungen von α-Metalliertem Isocyanmethyl-phosphonsaure Diathylester mit Carbonylverbindungen, *Tetrahedron Lett.,* p. 633, 1973.

240. **Fresneda, P. M. and Molina, P.,** Synthesis of diethyl arylmethanephosphonates from arylmethanamines, *Synthesis,* p. 222, 1981.

241. **Akiba, K.-Y., Negishi, Y., and Inamoto, N.,** Preparation of 13-substituted 8*H*-dibenzo[*a,g*]quinolizin-8-ones by intramolecular Wittig-Horner reaction of dialkyl 2-(*o*Acylbenzoyl)-1,2-dihydro-1-isoquinolylphosphonates, *Bull. Chem. Soc. Jpn.,* 57, 2188, 1984.

242. **Akiba, K.-Y., Negishi, Y., and Inamoto, N.,** Preparation of tetraalkyl 2,2′-phthaloylbis[1,2-dihydro-1-isoquinolylphosphonate]s and their reactions with base, *Bull. Chem. Soc. Jpn.,* 57, 2333, 1984.

243. **Oleksyszyn, J. and Subotkowska, L.,** Aminomethanephosphonic acid and its diphenyl ester, *Synthesis,* 906, 1980.

244. **Clauss, K., Grimm, D., and Prossel, G.,** β-Lactams bearing substituents bonded *via* hetero atoms, *Justus Liebigs Ann. Chem.,* 539, 1974.

245. **Campbell, M. M. and Carruthers, N.,** Synthesis of α-aminophosphonic and α-aminophosphinic acids and derived dipeptides from 4-acetoxyazetidin-2-ones, *Chem. Commun.,* p. 730, 1980.

246. **Vaultier, M., Ouali, M. S., and Carrie, R.,** Synthese d'esters α-aminophosphoniques fonctionnels a partir d'aziridines, *Bull. Soc. Chim. Fr.,* II, 343, 1979.

247. **Burkhouse, D. and Zimmer, H.,** Novel synthesis of 1-alkoxy-1-arylmethanephosphonic acid esters, *Synthesis,* p. 330, 1984.

248. **Alfonsov, V. A., Girfanova, Y. N., Batyeva, E. S., and Pudovik, A. N.,** Isomerization of P (III) esters under the action of organic acids (transl.), *J. Gen. Chem. U.S.S.R.,* 51, 2290, 1981.

249. **Block, H.-D. and Ernst, M.,** Verfahren zur Herstellung von Alkanphosphonsaurediarylestern, West German Patent Application 2,747,554, 1979.

250. **Hata, T. and Sekine, M.,** Silyl- and stannyl-esters of phosphorus oxyacids — intermediates for the synthesis of phosphate derivatives of biological interest, in *Phosphorus Chemistry Directed Toward Biology,* Stec, W. J., Ed., Pergamon, New York, 1980, 197.

251. **Belokrinitsky, M. A. and Orlov, N. F.,** Reaction of alkyl halides with triorganosilyl derivatives of phosphorous acid, *Kremniiorg. Mater.,* p. 145, 1971.

252. **Orlov, N. F., Kaufman, B. L., Sukhi, L., Selserand, L. N., and Sudakova, E. V.,** Synthesis of triorganosilyl derivatives of phosphorous acid, *Khim. Prim. Soedin.* p. 111, 1966.

253. **Sekine, M., Yamagata, H., and Hata, T.,** A general and convenient method for the synthesis of unesterified carbamoyl- and thiocarbamoyl-phosphonic acids, *Tetrahedron Lett.,* p. 3013, 1979.

254. **Voronkov, M. S. and Skorik, Y. I.,** Reaction of phosphorus trihalides with trialkyloxysilanes and hexaalkyldisiloxanes, *Zh. Obshch. Khim. S.S.S.R.,* 35, 106, 1965.

255. **Sekine, M., Mori, H., and Hata, T.,** Protection of phosphonate function by means of ethoxycarbonyl group. A new method for generation of reactive silyl phosphite intermediates, *Bull. Chem. Soc. Jpn.,* 55, 239, 1982.

256. **Hata, T., Sekine, M., and Kagawa, N.,** Reactions of tris(Trimethylsilyl) phosphite with alkyl halides, *Chem. Lett.,* p. 635, 1975.

257. **Rosenthal, A. F., Vargas, L. A., Isaacson, Y. A., and Bittman, R.,** A simple synthesis of phosphonate-containing lipids. Introduction of the phosphonic acid moiety into hydrolytically-labile compounds, *Tetrahedron Lett.,* p. 977, 1975.

258. **Deroo, P. W., Rosenthal, A. F., Isaacson, Y. A., Vargas, L. A., and Bittman, R.,** Synthesis of DL-2,3-diacyloxypropylphosphonylchlolines from DL-2,3-diacyloxyiodopropanes, *Chem. Phys. Lipids,* 16, 60, 1976.

259. **Tang, J.-C., Tropp, B. E., Engel, R., and Rosenthal, A. F.,** Isosteres of natural phosphates. 4. The synthesis of phosphonic acid analogues of phosphatidic acid and acyldihydroxyacetone phosphate, *Chem. Phys. Lipids,* 17, 169, 1976.

260. **Doerr, I. L., Tang, J.-C., Rosenthal, A. F., Engel, R., and Tropp, B. E.,** Synthesis of phosphonate and ether analogs of *rac*-phosphatidyl-L-serine, *Chem. Phys. Lipids,* 19, 185, 1977.

261. **Herrin, T. R., and Fairgrieve, J. S.,** Triglyceride Ester of Phosphonoacetic Acid Having Antiviral Activity, U.S. Patent 4,150,125, 1979.

262. **Sekine, M., Mori, H., and Hata, T.,** Preparation of phosphonate function by means of ethoxycarbonyl group. A new method for generation of reactive silyl phosphite intermediates, *Bull. Chem. Soc. Jpn.,* 55, 239, 1982.

263. **Sekine, M., Tetsuaki, T., Yamada, K., and Hata, T.,** A facile synthesis of phosphoenolpyruvate and its analogue utilizing *in situ* generated trimethylsilyl bromide, *J. Chem. Soc., Perkin I,* 2509, 1982.

264. **Issleib, K. and Walther, B.,** Silyl, germanyl, and stannyl esters of phosphinous acids of the (R₂PO)ₙ ER′₄₋ₙ type (E = silicon, germanium, tin), *J. Organomet. Chem.,* 22, 375, 1970.

265. **Thottathil, J. K. and Moniot, J. L.,** Phosphinic Acid Intermediates, European Patent 138,403, 1985.

266. **Rosenthal, A. F., Gringauz, A., and Vargas, L. A.,** Bis(trimethylsilyl) trimethylsiloxymethylphosphonite: A useful reagent for the introduction of the hydroxymethylphosphinate group, *Chem. Commun.,* p. 384, 1976.

267. **Thottathil, J. K., Przybyla, C. A., and Moniot, J. L.,** Mild Arbuzov reactions of phosphonous acids, *Tetrahedron Lett.,* p. 4737, 1984.

268. **Siegfried, B.,** Calcium salt of hydroxymethylphosphinic acid, Swiss Patent 311,982, 1956.

269. **Lutsenko, I. F. and Kraits, Z. S.,** Arbuzov rearrangement of vinyl esters of phosphorous and phenyl-phosphonous Acids, *Dokl. Akad. Nauk S.S.S.R.,* 132, 612, 1960.

270. **Nesmeyanov, A. N., Lutsenko, I. F., Kraits, Z. S., and Bokovoi, A. P.,** Vinyl Esters of Phosphorous Acid, *Dokl. Akad. Nauk S.S.S.R.,* 124, 1251, 1959.

271. **Honig, M. L. and Weil, E. D.,** Process for Preparing Diaryl Methylphosphonate and Derivatives Thereof, U.S. Patent 4,152,373, 1979.

272. **Hechenbleikner, I. and Enlow, W. P.,** Arbuzov Rearrangement of Triphenyl Phosphite, U.S. Patent 4,133,807, 1978.

273. **Block, H.,** Aryl Phosphonyl Compounds, European Patent 43,482, 1982.

274. **Block, H.-D. and Frohlen, H.-G.,** Preparation of Alkane Phosphonic and Phosphinic Acid Aryl Esters, U.S. Patent 4,377,537, 1983.

275. **Sturtz, G., Damin, B., and Clement, J.-C.,** Propene-1,3- and -2,3-diyldiphosphonates. Synthesis of dienephosphonates by the Wittig-Horner reaction, *J. Chem. Res.* (M), p. 1209, 1978.

276. **Petrov, K. A., Chauzov, V. A., Agafonov, S. V., and Pazhitnova, N. V.,** (methoxymethy-lene)bisphosphonic esters (transl.), *J. Gen. Chem. U.S.S.R.,* 50, 1225, 1980.

277. **Arend, G., Schaffner, H., and Schramm, J.,** Process for the Production of (Meth)allyl Phosphonic Acid Dialkyl Esters, U.S. Patent 4,017,564, 1977.

278. **Martin, D. J.,** Catalyzed Process for Producing Pentavalent Phosphorus Derivatives, U.S. Patent 3,705,214, 1972.

279. **Hampton, A., Perini, F., and Harper, P. J.,** Derivatives of phosphonate and vinyl phosphate analogs of D-ribose 5-phosphate, *Carbohydr. Res.,* 37, 359, 1974.

280. **Baboulene, M. and Sturtz, G.,** Aminomethyl-1 benzoyl-2 cyclopropanes. I. Synthese, *Bull. Soc. Chim. Fr.,* p. 1585 1974.

281. **Bianchini, J.-P. and Gaydou, E. M.,** Effect of alkyl substituents of phosphites in the synthesis of ketophosphonates and enol phosphates, *C. R.,* 280(C), 1521, 1975.

282. **Gaydou, E. M. and Bianchini, J.-P.,** Kinetics and mechanism of ketophosphonate formation from triethyl phosphite and aryl substituted α-bromoacetophenones, *Chem. Commun.,* p. 541, 1975.

283. **Borowitz, I. J., Firstenberg, S., Borowitz, G. B., and Schuessler, D.,** On the kinetics and mechanism of the Perkow reaction, *J. Am. Chem. Soc.,* 94, 1623, 1972.

284. **Gaydou, E. M. and Bianchini, J.-P.,** Etude du mechanisme de la formation de phosphates d'enols a partir de composes carbonyles α-halogenes et de triacoylphosphites, *Can. J. Chem.,* 54, 3626, 1976.

285. **Chopard, P. A., Clark, V. M., Hudson, R. F., and Kirby, A. J.,** A short synthesis of several gambir alkaloids, *Tetrahedron,* 26, p. 1961, 1970.

286. **Brown, C. and Hudson, R. F.,** Stereoelectronic suppression of an Arbuzov-type reaction, *Tetrahedron Lett.* p. 3191, 1971.

287. **Costisella, B. and Gross, H.,** Zur Reaktion von P(III)Amiden mit Orthoameisensaureesterchloriden, *Phosphorus and Sulfur,* 8, 99, 1980.

288. **Tsivunin, V. S., Zaripova, V. G., Zykova, T. V., and Salakhutdinov, R. A.,** Reaction of triethyl phosphite with 3-chloro-1, 4-dioxan-2-ol acetic ester (transl.), *J. Gen. Chem. U.S.S.R.,* 49, 1679, 1979.

289. **McEwen, G. K. and Verkade, J. G.,** An intermediate in the bromination of the trialkyl phosphite P(OCH$_2$)$_3$CMe, *Chem. Commun.,* p. 668, 1971.

290. **Michalski, J., Pakulski, M., and Skowronska, A.,** Arbuzov reaction of alkyl and silyl phosphites with halogens involving four- and five-co-ordinate intermediates, *J. Chem. Soc., Perkin I,* p. 833, 1980.

291. **Efimova, V. D., Kharrasova, F. M., Zykova, T. V., and Salakhutdinov, R. S.,** Reaction of carbon tetrachloride and chloral with alkyl esters of several alkylphosphonous acids (transl.), *J. Gen. Chem. U.S.S.R.,* 44, 51, 1974.

292. **Kharrasova, F. M., Efimova, V. D., Nesterov, L. V., Martynov, A. A., and Salakhutdinov, R. A.,** Action of carbon tetrachloride on alkyl phosphinites (transl.), *J. Gen. Chem. U.S.S.R.,* 48, 1301, 1978.

293. **Kharrasova, F. M., Efimova, V. D., Bikeev, S. S., and Salakhutdinov, R. A.,** Reactions of carbon tetrachloride with alkyl esters of phosphorus (III) acids in presence of benzaldehyde (transl.), *J. Gen. Chem. U.S.S.R.,* 49, 2156, 1979.

294. **Kolodyazhnyi, O. I., Yakovlev, V. N., and Kukhar, V. P.,** Reactions of [(alkoxycar-bonyl)methyl]phosphonites with carbon tetrabromide. II (transl.), *J. Gen. Chem. U.S.S.R.,* 49, 2170, 1979.

295. **Krutikov, V. I., Laurentev, A. N., Maslennikov, I. G., Blinova, G. G., and Sochilin, E. G.,** Reactions of carbon tetrachloride with ethylphosphonous esters (transl.), *J. Gen. Chem. U.S.S.R.,* 50, 1796, 1980.

296. **Bittman, R., Rosenthal, A. F., and Vargas, L. A.,** Synthesis of phospholipids *via* dimethylphosphoryl chloride, *Chem. Phys. Lipids,* 34, 201, 1984.

297. **Gupta, A., Sacks, K., Khan, S., Tropp, B. E., and Engel, R.,** An improved synthesis of vinylic phosphonates from ketones, *Synthetic Commun.,* 10, 299, 1980.

298. **Bobkova, R. G., Ignatova, N. P., Shvetsov-Shilovskii, N. I., Negrebetskii, V. V., and Vasilev, A. F.,** Preparation and properties of substituted 1,2,3-diazaphospholines (transl.), *J. Gen. Chem. U.S.S.R.,* 46, 588, 1974.

299. **Garabadzhin, A. V., Rodin, A. A., Lavrentev, A. N., and Sochilin, E. G.,** Reactions of perfluoroalkyl iodides with phosphorus (III) esters. IV. Synthesis and NMR and mass spectra of alkyl(fluoroalkyl)phosphinates (transl.), *J. Gen. Chem. U.S.S.R.,* 51, 34, 1981.

300. **Kafarski, P., Lejczak, B., Mastalerz, P., Dus, D., and Radzikowski, C.,** *N*-(Phosphonomethyl)amino phosphonates. Phosphonate analogues of *N*-(phosphonoacetyl)-L-aspartic acid (PALA), *J. Med. Chem.,* 28, 1555, 1985.

301. **Ohler, E., Haslinger, E., and Zbiral, E.,** Synthese und ^{1}H-NMR-Spektren von (3-Acylbicyclo[2.2.1]hept-5-en-2-yl)phosphonsaureestern, *Chem. Ber.,* 115, 1028, 1982.

302. **Pfister, T., Eue, L., Schmidt, R. R., and Hanssler, G.,** Chlorinated Phosphorylmethylcarbonyl Derivative Plant Protection Agents, U.S. Patent 4,541,858, 1985.

303. **Tupchienko, S. K., Dudchenko, T. N., and Sinitsa, A. D.,** Reactions of phosphoramidous esters with imidoyl chlorides (transl.), *Gen. Chem. U.S.S.R.* 55, 691, 1985.

Chapter 3

TRIVALENT PHOSPHORUS ADDITION TO CARBONYL CARBON — THE ABRAMOV AND PUDOVIK REACTIONS

I. INTRODUCTION

In addition to the nucleophilic substitution reactions which were discussed in the previous chapter, fully esterified trivalent phosphorus acids and anionic forms of monobasic trivalent phosphorus acids are capable of undergoing facile addition reactions at unsaturated electrophilic centers. A hint of this reactivity was seen in the prior discussion of the "abnormal" Perkow reaction [1-3] and reactions involving substrates bearing a carbonyl group distant from the halide to be displaced. [4,5] In this section we will be concerned with recent developments in the utilization of this reactivity for the formation of carbon to phosphorus bonds.

Several general developments in chemical science have led to a particular interest in these reactions in recent years. One is the growth of attention being given to the use of *umpolung* reagents for chemical syntheses. [6,7] The products of trivalent phosphorus addition to carbonyl (or carbonyl-related) groups lead directly to quinquevalent phosphorus compounds bearing a hydroxyl group (or related polar functionality) at the carbon attached to phosphorus. Such structural elements facilitate both further functionalization of that carbon and the ultimate cleavage of phosphorus from carbon. Overall, this approach generates a masked carbonyl function capable of reacting with a polarity inverted from the normal pattern and provides for regeneration of the normal carbonyl group.

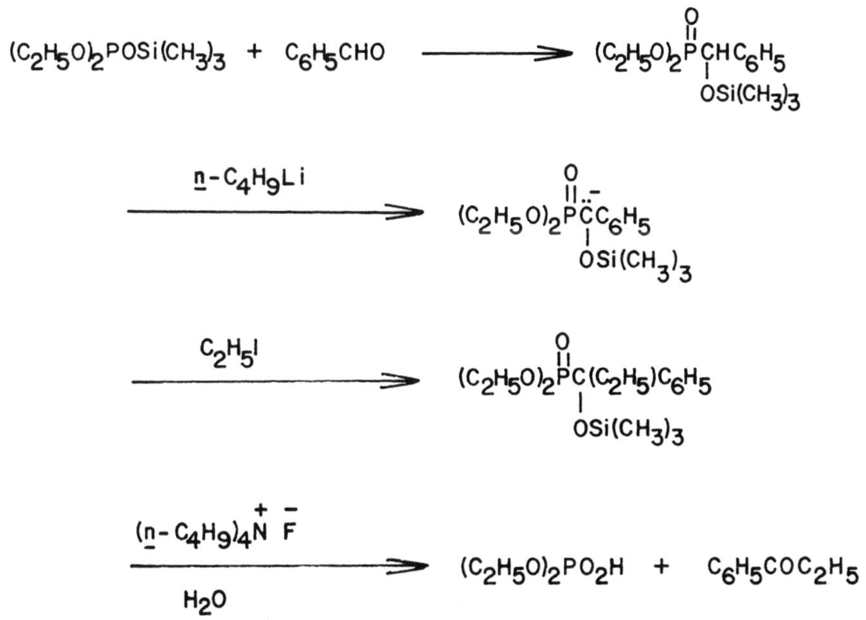

$$(C_2H_5O)_2POSi(CH_3)_3 + C_6H_5CHO \longrightarrow (C_2H_5O)_2\overset{O}{\overset{\|}{P}}\underset{OSi(CH_3)_3}{\overset{|}{C}}HC_6H_5$$

$$\xrightarrow{\underline{n}-C_4H_9Li} (C_2H_5O)_2\overset{O}{\overset{\|}{P}}\underset{OSi(CH_3)_3}{\overset{|}{\overset{..-}{C}}}C_6H_5$$

$$\xrightarrow{C_2H_5I} (C_2H_5O)_2\overset{O}{\overset{\|}{P}}\underset{OSi(CH_3)_3}{\overset{|}{C}}(C_2H_5)C_6H_5$$

$$\xrightarrow[H_2O]{(\underline{n}-C_4H_9)_4\overset{+}{N}\,\overset{-}{F}} (C_2H_5O)_2PO_2H + C_6H_5COC_2H_5$$

A second area of research activity stimulating recent interest in carbonyl addition reactions of trivalent phosphorus is the use of analogues of natural metabolites for the regulation and probing of biological systems. For example, addition of trivalent phosphorus reagents to imine carbon can be used to generate analogues of the natural amino acids.

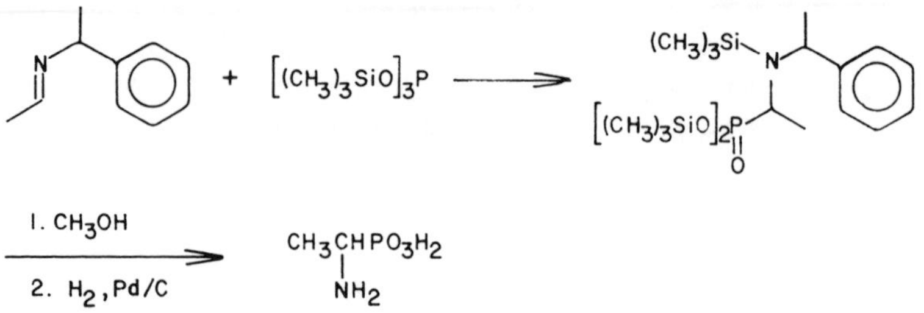

In addition, other α-functionalized phosphonic acids have found utility for a variety of biological applications. These applications include their use as probes of metabolic processes,[8-13] antibacterial agents,[14-16] herbicides,[17-24] pharmaceuticals,[25-29] insecticides,[30] fungicides,[31] and as analogues of vitamins.[32] Use of these types of materials for nonbiological applications such as flame retardants,[33,34] ion-exchange resins,[35] and scale inhibitors[36] has also been noted.

With this breadth of utility for α-functionalized quinquevalent organophosphorus compounds, it is quite understandable that the methods for their synthesis should receive intensive investigation.

II. THE ABRAMOV REACTION

A. Reactions and Mechanism

The original work of Abramov[37] involved the heating of an aldehyde with a trialkyl phosphite at 70 to 100°C for several hours in a sealed tube. Under these rather stringent conditions there could be isolated in variable yield the dialkyl α-alkoxyalkylphosphonates.

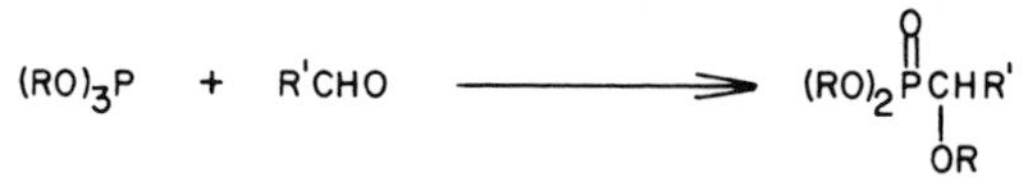

Later investigations demonstrated that reaction between trialkyl phosphites and carbonyl compounds occurred under much milder conditions, even at 20°.[38,39] In either the presence or absence of solvent, trialkyl phosphites add readily to the carbonyl carbon of aliphatic aldehydes. Under these mild conditions the initial adduct does not undergo alkyl group transfer, but rather adds to a second molecule of aldehyde, subsequently undergoing cyclization to a 1,4,2-dioxaphospholane.

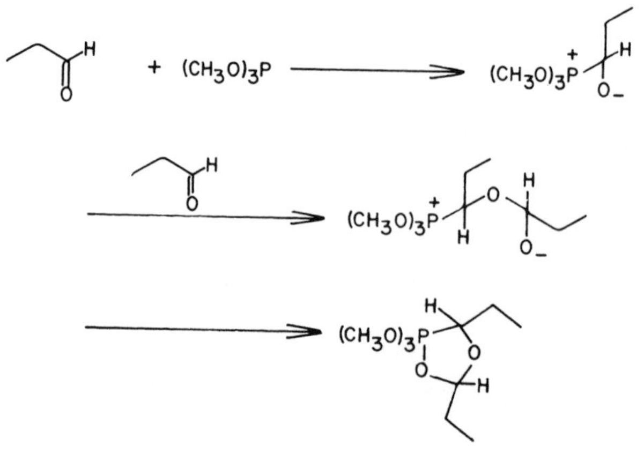

60 %

With ketones and aromatic aldehydes bearing strongly electron-withdrawing groups, reaction proceeds via initial attack at oxygen to generate 1,3,2-dioxaphospholanes.[39,40] No intermediates could be detected in this reaction process.

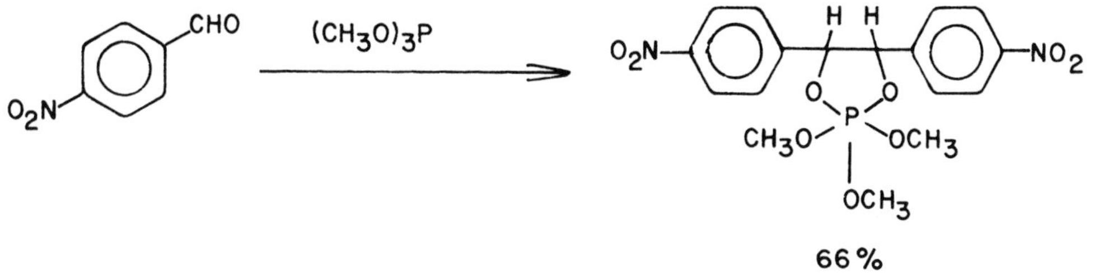

66%

Further investigations with chloral and fluoral found similar results.[41,42] Reaction of chloral with a silyl-alkyl phosphorous acid ester resulted in the formation of a simple Abramov-type product at low temperature, but gave rearranged Perkow-type product at higher temperatures.[43,44] Excellent results have also been obtained with chloral when phosphorus trichloride is used as the trivalent phosphorus reagent for the Abramov-type addition reaction.[45,46] Use of methanol in the reaction system generates dimethyl 1-hydroxy-2,2,2-trichloroethylphosphonate in extremely high yield. (In a similar vein, another phosphorus-halide, phenyldichlorophosphine has been found to give Abramov-type adducts with aldimines.[47]

A fundamental difficulty arises in attempting to use the simple Abramov reaction with alkyl esters of trivalent phosphorus acids for synthetic purposes. Although nucleophilic addition to the carbonyl carbon occurs readily, the intermediate species is unfavorably suited for intramolecular dealkylation. Standard "rear-side attack" on carbon (with a trajectory near 180°) by the α-oxyanion function is prohibited sterically, and "front-side attack" is prohibited electronically. When reaction does occur, the addition of phosphorus to carbon would seen to involve a simple aldol-type process.[48]

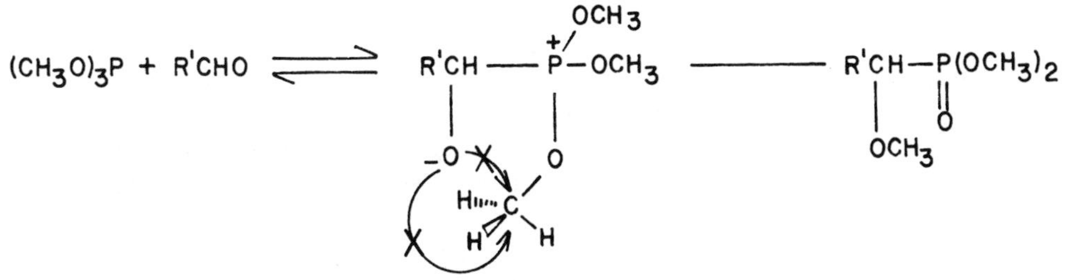

Moreover, when the reaction does occur using alkyl esters, stringent conditions are used under which relatively high concentrations of the initial adduct are generated and dealkylation proceeds via an intermolecular route.

Thus, it is not surprising that poor results have been obtained when the simple Abramov reaction using full alkyl esters has been attempted for synthetic purposes. The yields of target material have generally been extremely low.[49-51] The best yields noted with carbohydrate derivatives are still only fair, with significant amounts of Perkow-type side-product being formed as well.

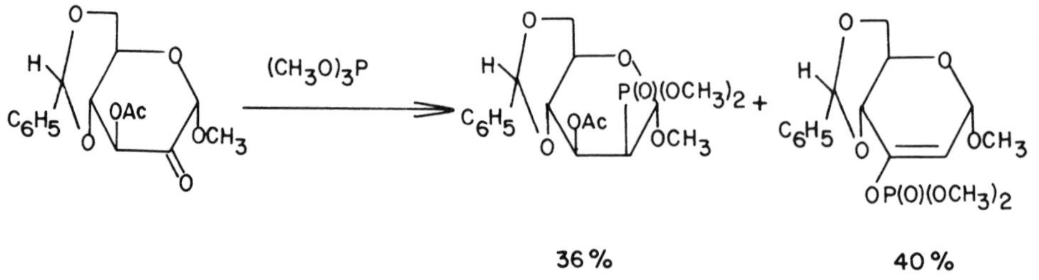

36%   40%

Under quite stringent conditions of heating, fair yields of carbonyl addition product have been obtained as well in the cephalosporin series.[15,16]

The use in Abramov reactions of carboxylate mixed anhydrides of the phosphorus acids have been reported in several instances,[52-55] including one in which the species is presumably generated *in situ*.[53] With such reagents, the steric requirements for "deacylation" after phosphorus addition to the carbonyl carbon are met much more easily.

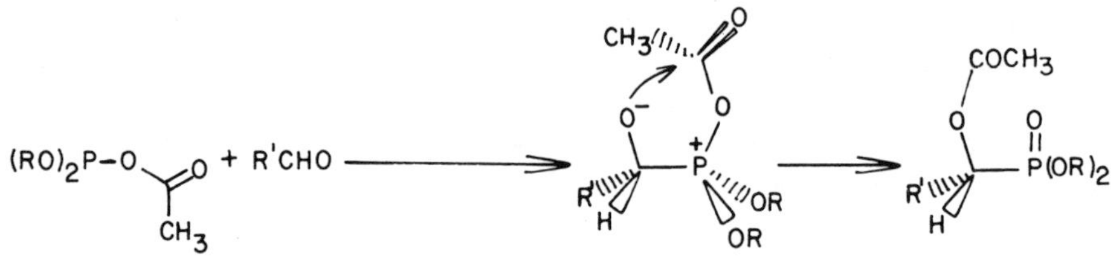

The fundamental problem in making the Abramov reaction be of synthetic utility may thereby be seen to be the facilitation of either or both: (1) the formation of the P=O linkage by "dealkylation", and (2) the neutralization of charge at the α-oxide site by the formation of a new covalent bond to the oxygen. Several approaches to accomplishing these ends have been made.

One basic approach to accomplish the second of these two goals involves the addition to the reaction mixture of a "trapping-agent". The most convenient and successful of such agents, a silyl halide, was reported in 1963 by Birum and Richardson.[56] An equivalent amount of the silyl halide is added to the initial reaction mixture to serve as a substrate for the nucleophilic α-oxyanion generated at low temperatures. In this manner excellent yields of the α-siloxyphosphonates may be generated with minimum difficulties in purification.

$$CH_3CHO + (CH_3)_3SiCl + (C_2H_5O)_3P \xrightarrow{-35°} CH_3CH\left[OSi(CH_3)_3\right]P(O)(OC_2H_5)_2$$

97%

Since the original effort, this approach has been used quite successfully by others for the preparation of a variety of the target α-siloxyalkylphosphonates.[57-60] This reaction system is particularly convenient for the preparation of carbonyl-related *umpolung* reagents. There is generated in a "one-pot" procedure a reagent which is suitably protected and activated for alkylation and subsequent regeneration of the carbonyl function.[59]

$$(C_2H_5O)_3P + (CH_3)_3SiCl + C_6H_5CHO \longrightarrow (C_2H_5O)_2\overset{\overset{\displaystyle O}{\|}}{P}CH\left[OSi(CH_3)_3\right]C_6H_5$$

94 %

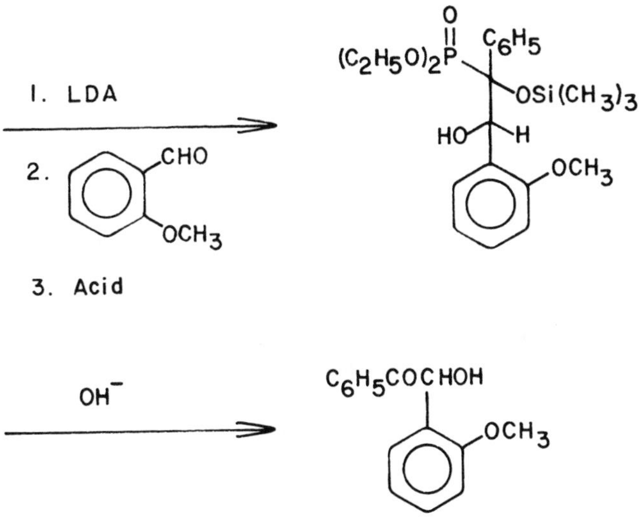

1. LDA

2. ⟨benzene ring⟩ CHO, OCH_3

3. Acid

$$(C_2H_5O)_2\overset{\overset{\displaystyle O}{\|}}{P}\underset{HO}{\overset{C_6H_5}{\diagdown}}C\overset{OSi(CH_3)_3}{\diagup}\underset{H}{\diagup}$$ ⟨benzene ring⟩ OCH_3

OH⁻

$$C_6H_5COCHOH$$ ⟨benzene ring⟩ OCH_3

90 %

It should be noted that the reaction has been demonstrated to proceed via trapping of the adduct of the carbonyl compound and trialkyl phosphite, rather than via initial attack of the silyl halide on the phosphorus ester.[58] Either route might be anticipated to lead to the observed products.

Other electrophilic reagents may be used to trap the intermediate α-oxyanions which are initially formed. Phosphorus trichloride has been reported to trap such species, yielding a product bearing a trivalent phosphorus monoester linkage at the α-position.[61]

$$PCl_3 + CH_3CHO + (C_2H_5O)_3P \longrightarrow (C_2H_5O)_2\overset{\overset{\displaystyle O}{\|}}{P}CH(CH_3)OPCl_2 + C_2H_5Cl$$

61 %

Both charge neutralization and dealkylation are accomplished intramolecularly through the use of mixed carboxylate-phosphorus anhydride reagents as mentioned previously.[52-55] A similar result has been obtained using tetraethyl pyrophosphite and other trivalent phosphorus anhydrides.[42]

$(C_2H_5O)_2POP(OC_2H_5)_2$ + CF_3CHO $\longrightarrow$ $CF_3CHP(O)(OC_2H_5)_2$
$\qquad\qquad\qquad\qquad\qquad\qquad\qquad\qquad\qquad\qquad\qquad$ $OP(OC_2H_5)_2$

$\qquad\qquad\qquad\qquad\qquad\qquad\qquad\qquad\qquad\qquad\qquad$ **56 %**

A favorable steric array for intramolecular dealkylation is also obtained by enlarging the ring involved. This may be seen in the reaction of triethyl phosphite with oximes, wherein the oxygen of the oxime performs the dealkylation step.[62]

$(C_2H_5O)_3P$ + $\longrightarrow$

$\qquad\qquad\qquad\qquad\qquad\qquad\qquad\qquad\qquad\qquad\qquad$ **39 %**

An alternative approach which facilitates both dealkylation and charge neutralization involves the use of silyl esters of trivalent phosphorus acids. If a silyl ester linkage is used, intramolecular desilylation and resultant charge neutralization can occur readily without steric restraint. Displacement is possible at silicon from the same side as the leaving group (incipient phosphoryl oxygen) departs.[63] Thus, there would be anticipated facile reaction with transfer specifically of the silyl ester linkage in preference to other ester linkages which might be present.

With the preparation of mixed silyl-alkyl esters of phosphorous acid,[54,64] the fulfillment of this possibility could be realized. The reaction of these mixed silyl-alkyl esters of phosphorous acid with both aldehydes and ketones proceeds in excellent yield under extremely mild conditions with the exclusive transfer of the silyl-ester function to the α-oxyanion site.[58,64,65]

$(CH_3O)_2POSi(CH_3)_3$ + C_6H_5CHO $\longrightarrow$ $C_6H_5CH\left[OSi(CH_3)_3\right]P(O)(OCH_3)_2$

$\qquad\qquad\qquad\qquad\qquad\qquad\qquad\qquad\qquad\qquad\qquad$ **97 %**

In addition to those addition reactions involving simple aldehydes and ketones,[55,55-71] mixed silyl-alkyl phosphorous acid esters have been observed to undergo facile reaction with ketene,[72] α-ketophosphonates,[73] ketoesters,[74,75] and α,β-unsaturated carbonyl compounds.[57,65,76,77] The products of these reactions have found particular utility for *umpolung* processes.[78,79]

With the synthesis of tris(trimethylsilyl) phosphite,[8,80,81] a direct approach for the preparation of the unesterified form of the α-hydroxyphosphonic acids became available. Removal of the residual silyl ester linkages at the phosphonate center is accomplished with water or alcohol under mild conditions.[82]

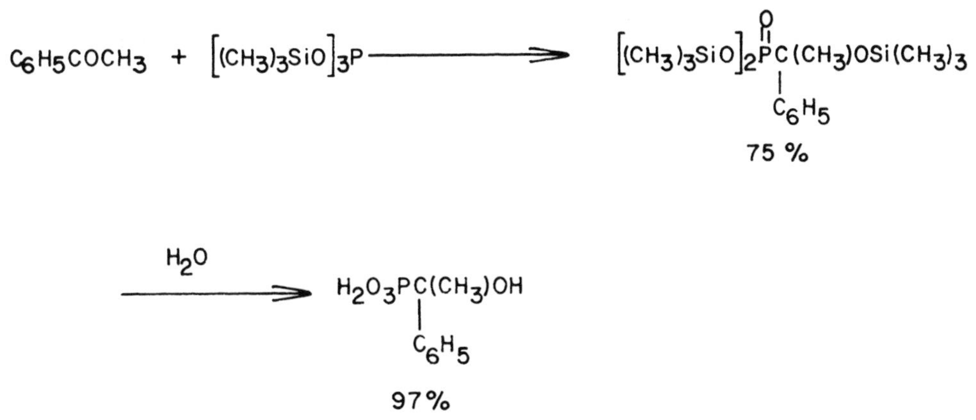

$$C_6H_5COCH_3 \ + \ [(CH_3)_3SiO]_3P \longrightarrow [(CH_3)_3SiO]_2\overset{\overset{O}{\|}}{P}\underset{\underset{C_6H_5}{|}}{C}(CH_3)OSi(CH_3)_3$$

$$75\%$$

$$\overset{H_2O}{\longrightarrow} \ H_2O_3P\underset{\underset{C_6H_5}{|}}{C}(CH_3)OH$$

$$97\%$$

In addition to reactions with simple aldehydes and ketones,[8,82] tris(trimethylsilyl)phosphite has been investigated in carbonyl addition reaction with α-haloketones,[83-85] ketoesters,[84,85] isocyanates,[86] α-ketophosphonates,[87] and α,β-unsaturated carbonyl compounds.[82,88]

In two instances, mixtures of products may be anticipated. Reaction of α-halocarbonyl compounds generally leads to the formation of carbonyl-carbon addition and Perkow-type reaction products.[85] With chloral, the carbonyl-carbon adduct is obtained at low temperatures with the Perkow-type product being formed at elevated temperatures.[43] Recent efforts have determined that the initial attack by phosphorus is at the carbonyl carbon with the intermediate then undergoing a rearrangement to give the phosphate.[44]

The second instance where multiple product formation might be anticipated is in reaction with α,β-unsaturated carbonyl compounds. Michael-type addition might be expected to compete with the simple reaction at the carbonyl carbon site. Both types of products have been observed using tris(trimethylsilyl) phosphite, aldehydes yielding the simple carbonyl-carbon adducts, and ketones giving exclusively the conjugate adducts.[82] The conjugate addition reactions will be considered in detail in the following chapter (Chapter 4 — Hydrophosphinylation Reactions).

The use of silyl esters of trivalent phosphorus acids in these types of reactions for the preparation of biologically interesting molecules has recently been reviewed.[89]

A report has been made of the reaction of tris(trimethylsilyl) phosphite with chiral aldimines.[90] Aldimines, which were prepared from aldehydes and optically active 2-phenylethylamine, reacted with the phosphite to produce adducts exhibiting asymmetric induction. The extent of asymmetric induction found in the α-aminoalkylphosphonic acids produced upon deprotection was dependent upon the structure of the original aldehyde and the specific reaction conditions employed, but could be as high as 80%. This direct addition route provides an alternative to the multistep route to optically active α-aminoalkylphosphonic acids which involves Michaelis-Arbuzov reaction of an acyl halide, imine formation, reduction, and resolution.[91,92]

$$CH_3CHO \ + \ \underset{CH_3}{\overset{H_{\prime\prime\prime\prime}}{\diagup}}\overset{NH_2}{\underset{C_6H_5}{\diagdown}} \longrightarrow CH_3CH=N_{\prime\prime\prime\prime}\underset{C_6H_5}{\overset{CH_3}{\diagup}}H$$

$$\xrightarrow{P\left[OSi(CH_3)_3\right]_3} \quad CH_3CHN\overset{Si(CH_3)_3}{\diagup}\underset{\underset{P(O)\left[OSi(CH_3)_3\right]_2}{\overset{|}{H}}}{\overset{\prime\prime\prime\prime\prime C_6H_5}{\diagdown CH_3}}$$

$$\xrightarrow[\text{2. } H_2, Pd(OH)_2/C]{\text{1. } CH_3OH} \quad \underset{CH_3}{\overset{H_{\prime\prime\prime\prime}}{\diagup}}\overset{NH_2}{\underset{PO_3H_2}{\diagdown}}$$

40 % yield

40 % optical purity

Another effort involving asymmetric induction utilized triethyl phosphite in reaction with a Schiff base formed *in situ* from an aldehyde and a chiral *N*-substituted urea.[93] The initial reaction was catalyzed by boron trifluoride etherate, and the target chiral free α-aminoalkylphosphonic acid was isolated in moderate yield upon hydrolytic work-up. Unfortunately, the optical purity of the material isolated *via* this route is low. The determination of the absolute configurations of these species has been reported elsewhere.[91,94]

$$\underset{CH_3}{\overset{H_{\prime\prime\prime\prime}}{\diagup}}\overset{NHCONH_2}{\underset{C_6H_5}{\diagdown}} + C_6H_5CHO + (C_2H_5O)_3P \xrightarrow[\text{2. aq. acid}]{\text{1. } BF_3} \underset{H}{\overset{C_6H_{5\prime\prime\prime\prime}}{\diagup}}\overset{NH_2}{\underset{PO_3H_2}{\diagdown}}$$

47 % yield

17 % optical purity

The primary interest in these α-aminoalkylphosphonic acids has, of course, been as analogues of the natural α-amino acids. Numerous other reports have been made of the use of Abramov-type reactions in approaches to their synthesis. While one other report involved the use of trialkyl phosphites in these reactions (in which the target materials were to be used as flame retardants[34]), most of the efforts have utilized either triphenyl phosphite or unesterified phosphorous acid.

The leading effort in the use of triphenyl phosphite for this purpose involved an *in situ* reaction with mono- and dicondensation products of a series of aldehydes with urea or thiourea.[95] In contrast to the Michaelis-Arbuzov reaction, triphenyl phosphite is found to provide a significantly more facile reaction than are trialkyl phosphites. This finding is rationalized to relate to the acid-base characteristics of the phosphites involved. As the initial condensation reaction is acid catalyzed, the more basic trialkyl phosphites provide an inhibitory effect not present with triphenyl phosphite. Moreover, the loss of the aromatic ester as phenol is presumed to result with proton transfer to the esteric oxygen and subsequent fission of the O-P bond, rather than by displacement at carbon.[95] Yields of α-ureidophosphonates in the range of 60 to 90% are found with this method.

$$2 \ (C_6H_5O)_2P \ + \ H_2NCONH_2 \ + \ 2 \ C_6H_5CHO \ \longrightarrow \ (C_6H_5O)_2\overset{O}{\overset{\|}{P}}\underset{\underset{C_6H_5}{|}}{CH}NH\overset{O}{\underset{\|}{C}}NH\underset{\underset{C_6H_5}{|}}{CH}P(OC_6H_5)_2$$

71 %

Numerous reports have followed which are concerned with syntheses of diphenyl α-aminoalkylphosphonates, or the free acids, which utilize the same fundamental approach. The variety of nitrogenous reagents used in these efforts include thiourea,[96-99] *N*-substituted thioureas,[98,100,101] ethyl carbamate,[102] 2-imidazolidinone,[33] 2-imidazolidinethione,[33] and benzyl carbamate.[103-105] In most instances yields in the good to excellent range could be obtained using acetic acid in the reaction medium.

$$C_6H_5NHCSNH_2 \ + \ CH_3CHO \ + \ (C_6H_5O)_3P \ \xrightarrow{\ CH_3CO_2H\ } \ C_6H_5NHCSNHCH(CH_3)P(O)(OC_6H_5)_2$$

85 %

The direct use of phosphorous acid for the formation of carbon-phosphorus bonds with Schiff bases was accomplished by Moedritzer and Irani in the development of a Mannich-type procedure for phosphorus-centered systems.[106] While both primary and secondary amines were found to react in extremely high yield with formaldehyde and phosphorous acid, other carbonyl compounds were not investigated. This approach was later used to bind a phosphonic acid function to a polymer for the generation of a cation exachange resin.[35]

$$\underline{n}\text{-}C_4H_9NH_2 \ + \ 2 \ CH_2O \ + \ 2 \ H_3PO_3 \ \longrightarrow \ \underline{n}\text{-}C_4H_9N(CH_2PO_3H_2)_2$$

81 %

A report[107] from another laboratory indicating that other carbonyl compounds could be used for preparation of substituted α-aminophosphonic acids under the simple "one-pot" reaction procedure was later found to be in error.[108,109]

The α-substituted-α-aminoalkylphosphonic acids can be prepared from phosphorous acid, but separate generation of the Schiff base appears to be required.[108] In most instances reaction of the imine with phosphorous acid is complete within a few minutes at 110°C and isolated; yields are in the good to excellent range. A variety of carbonyl compounds have subsequently been utilized in this way.[36,110]

$$C_6H_5CH_2N=CHC_6H_5 \ + \ H_3PO_3 \ \longrightarrow \ C_6H_5CH_2NHCH(C_6H_5)PO_3H_2$$

98 %

The "one-pot" procedure involving phosphorous acid has been successful using Schiff bases derived from aldehydes with amides rather than amines for the preparation of α-aminoalkylphosphonic acids.[111] The α-amidoalkylphosphonic acid formed directly is hydrolyzed to give the target analogues of natural α-amino acids in moderate yield. One report has been made of the addition of phosphorous acid to Schiff bases derived from carboxylic acids and urea, formed *in situ* with phosphorus pentoxide.[112] Hypophosphorous acid has also been used in similar carbon-phosphorus bond-forming reactions with Schiff bases.[113]

Interesting reports have also appeared concerning the use of nitriles as substrates for Abramov-type reaction of phosphorous acid.[114,115] High yields of the 1-aminoalkyl-1,1-diphosphonic acids can be obtained in this way.

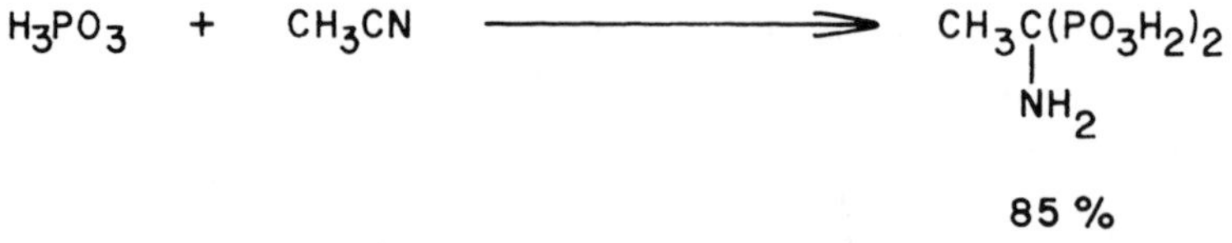

$$H_3PO_3 \quad + \quad CH_3CN \quad \longrightarrow \quad CH_3\underset{\underset{NH_2}{|}}{C}(PO_3H_2)_2$$

85 %

There has recently been reported the use of phosphorous acid (along with phosphorus pentoxide) in addition reactions with aldehydes. The reactions produce the α-hydroxyalkyl-phosphonic acids in excellent yield. [116]

Abramov-type addition reactions have also been investigated at thiocarbonyl centers. A series of geminal dithiols have been studied in thermal reaction with trialkyl phosphites.[117,118] On heating, the geminal dithiol releases hydrogen sulfide leaving the thioketone for reaction. Addition to carbon by the phosphorus center then occurs readily. The dealkylation step is found to proceed via alternative routes, depending on the nature of the ester group. Methyl and allylic ester linkages are transferred to sulfur by displacement, whereas isopropyl esters undergo an elimination process generating propene and the phosphonate bearing a free α-thiol functionality. Ethyl esters give a mixture of products.

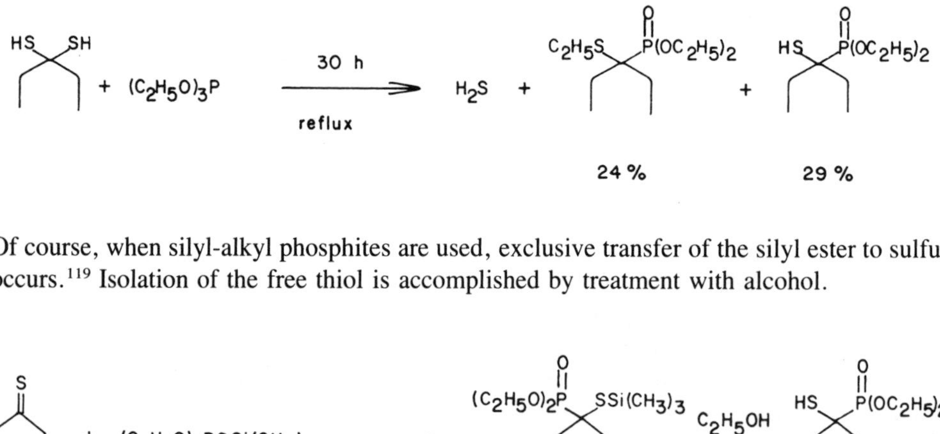

Of course, when silyl-alkyl phosphites are used, exclusive transfer of the silyl ester to sulfur occurs.[119] Isolation of the free thiol is accomplished by treatment with alcohol.

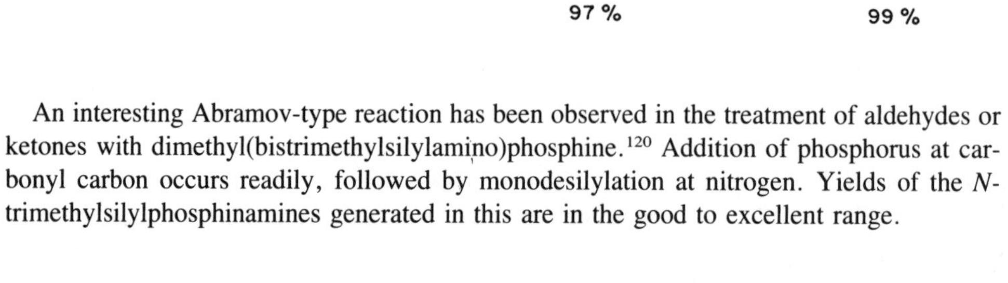

An interesting Abramov-type reaction has been observed in the treatment of aldehydes or ketones with dimethyl(bistrimethylsilylamino)phosphine.[120] Addition of phosphorus at carbonyl carbon occurs readily, followed by monodesilylation at nitrogen. Yields of the *N*-trimethylsilylphosphinamines generated in this are in the good to excellent range.

$$[(CH_3)_3Si]_2NP(CH_3)_2 + CH_3COCH_3 \longrightarrow (CH_3)_3SiN=P(CH_3)_2C(CH_3)_2OSi(CH_3)_3$$

88 %

B. Comparison of Methods

As noted previously, the direct Abramov reaction of trialkyl phosphites with simple carbonyl compounds presents significant difficulties if it is to be used for synthetic purposes. However, the use of silyl esters of trivalent phosphorus acids, or silyl trapping agents, allows these difficulties to be overcome with relative ease.

If the synthetic target is the free quinquevalent phosphorus acid, the use of tris(trimethylsilyl) phosphite is highly preferred to other approaches which would require more drastic conditions for ester cleavage. Silyl esters at phosphorus may be removed upon treatment with water or alcohol, liberating the free acid.

For further use as carbonyl *umpolung* reagents, the phosphorus acid sites must be blocked as either esters or amides. Similarly, the α-hydroxyl function must be masked with a reversible blocking agent. The trivalent phosphorus reagents bearing a single silyl ester function have been demonstrated to be ideal for this purpose as they provide immediately upon reaction the desired functionality. Procedures involving the *in situ* generation of the phosphorus reagent (*e.g.*, from a dialkyl phosphite and trimethylchlorosilane) in the presence of the reacting aldehyde, or the trapping of the initial adduct with trimethylchlorosilane, provide particularly convenient approaches. Yields by these methods are as good as when a preformed and purified reagent is used, with fewer laboratory operations.

For the preparation of α-aminophosphonates, several methods have been noted. In those instances where the phosphorus center of the target molecule is to remain esterified, the use of triphenyl phosphite with acetic acid and the imine generated *in situ* would seem to be the favored method. In those instances when the free α-aminophosphonic acid is desired, the several approaches using phosphorous acid should be given serious consideration.

In some instances the use of the monobasic form of the trivalent phosphorus reagent should be considered. This approach (the Pudovik reaction) is discussed in the next section of this chapter.

C. Experimental Procedures

[1-(Diethoxyphosphinyl)ethoxy]dimethylsilane — Reaction of an aldehyde with a trialkyl phosphite in the presence of a silyl halide.[56]

To a mixture of dimethyldichlorosilane (64.5 g, 0.5 mol) and triethyl phosphite (166 g, 1.0 mol) in a 500-mℓ flask there was added freshly distilled acetaldehyde (47 g, 1.07 mol) over a period of 18 min while the temperature was maintained at 15 to 22°C. The mixture was then warmed to 100°C and then vacuum distilled (148 to 154°C/0.2 to 0.4 torr) to give the desired [1-(diethoxyphosphinyl)ethoxy]dimethylsilane (165 g, 79%) as a colorless liquid which exhibited analytical data in accord with the proposed formulation.

***O,O*-Diphenyl 2-Methylthio-1-(*N*-phenylthioureido)ethylphosphonate** — Reaction of triphenyl phosphite with an imide generated *in situ* in the presence of acetic acid.[101]

To a solution of triphenyl phosphite (6.2 g, 0.02 mol) and thiomethoxyacetaldehyde (2.25 g, 0.025 mol) in glacial acetic acid (18 mℓ), powdered *N*-phenylthiourea was added in a single portion. The reaction mixture was stirred at room temperature for 30 min and then for 30 min at 80°C. After the mixture was cooled to room temperature, water (5 mℓ) was added and the solution was maintained at room temperature for 10 hr. The precipitate was removed by suction filtration, washed with 1:1 acetic acid:water (2 × 10 mℓ), dried over potassium hydroxide in an evacuated dessicator, and recrystallized from chloroform/methanol. In this manner there was isolated pure O,O-diphenyl 2-methylthio-1-(N-phenylthioureido)ethylphosphonate (8.61 g, 94%) of mp 136 to 138°C which exhibited spectra and analytical data in accord with the proposed structure.

1-Aminoethane-1,1-diphosphonic Acid — Reaction of phosphorous acid with a nitrile.

Into a 250-mℓ three-necked round-bottomed flask suspended in an oil bath, equipped with a three-way adapter, carrying a pressure equalizing addition funnel, a reflux condenser with a drying tube, a mechanical stirrer, and a thermometer was placed phosphorous acid (260 g, 3.17 mol). To the reaction flask was introduced acetonitrile (33 g, 0.80 mol) dropwise over a two hr period while the phosphorous acid was agitated and maintained at a temperature of 138 to 142°C. After completion of the addition, the reaction mixture was maintained at that temperature for an additional 12 hr. Methanol was then added to precipitate the pure 1-aminoethane-1,1-diphosphonic acid (13.9 g, 85%), which exhibited spectral and analytical data in accord with the proposed structure.

N-Benzyl α-aminobenzylphosphonic Acid — Reaction of phosphorous acid with an imine.[110]

The imine from benzaldehyde and benzyl amine (65 g, 0.33 mol) was added to phosphorous acid (27.3 g, 0.33 mol) and the mixture was stirred with heating. As the temperature reached 95 to 100°C the entire mixture became a homogeneous liquid which reacted vigorously as the temperature reached 115 to 120°C. After the reaction mixture became very viscous, it was allowed to cool, whereupon it condensed to a glass. This material was dissolved in aqueous sodium carbonate, which upon acidification precipitated pure N-benzyl α-aminobenzylphosphonic acid (90 g, 98%) of mp 233 to 234°C. The material exhibited analyses in accord with the proposed structure.

Benzyliminodimethylenediphosphonic Acid — Reaction of phosphorous acid with a formaldimine generated *in situ*.[106]

Benzylamine (53.5 g, 0.5 mol) and crystalline phosphorous acid (82 g, 1.0 mol) were dissolved in water (100 mℓ) and there was added concentrated hydrochloric acid (100 mℓ), and the mixture was heated to reflux. Over the course of 1 hr, 37% aqueous formaldehyde (160 mℓ, 2.0 mol) was added dropwise and the reaction mixture was maintained at reflux for another hour. After cooling, the solvent was evaporated and the residual syrup was dissolved in hot ethanol. Upon cooling, the crude benzyliminodimethylenediphosphonic acid precipitated and was recrystallized from hot dilute hydrochloric acid to give the pure material (127 g, 85.7%) of mp 248°C which exhibited analytical data in accord with the proposed structure.

Diethyl 1-(trimethylsiloxy)octylphosphonate — Reaction of a mixed silyl-alkyl phosphite with an aldehyde.[70]

To a solution of diethyl trimethylsilyl phosphite (18.3 g, 87.1 mmol) in dry benzene (10 mℓ) was added dropwise octanal (9.67 g, 75.4 mmol) at room temperature. After stirring for 3 hr the solvent was removed and distillation of the residual oil (104 to 105°C/0.04 torr) gave pure diethyl 1-(trimethylsiloxy)octylphosphonate (23.1 g, 91%) which exhibited spectra and analytical data in accord with the proposed structure.

Table 1
ABRAMOV REACTION OF CARBONYL COMPOUNDS

Product	Phosphorus reagent	Method	Yield (%)	Refs.
$(CH_3O)_2P(O)CH(OH)CCl_3$	PCl_3	A	84	46
$n\text{-}C_4H_9NHCOPO_3H_2$	$P[OSi(CH_3)_3]_3$	B	93	86
$(C_2H_5O)_2P(O)CH(CH_3)OPCl_2$	$(C_2H_5O)_3P$	C	61	61
$C_6H_5NHCOPO_3H_2$	$P[OSi(CH_3)_3]_3$	B	98	86
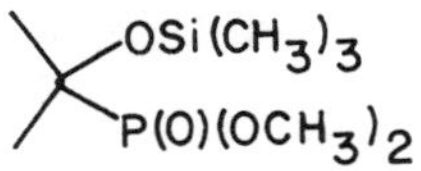 (CH₃)₃C group bearing OSi(CH₃)₃ and P(O)(OCH₃)₂	$(CH_3O)_2POSi(CH_3)_3$	D	74	58
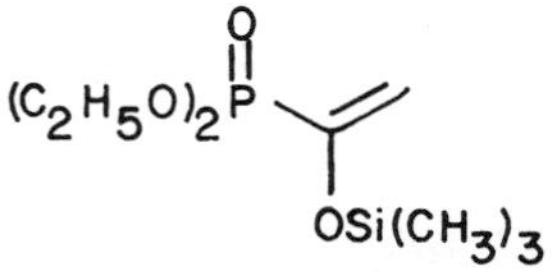 $(C_2H_5O)_2P(O)$ with =CH₂ and OSi(CH₃)₃	$(C_2H_5O)_2POSi(CH_3)_3$	D	87	64
$(C_2H_5O)_2P(O)CH(CH_3)OSi(CH_3)_3$	$(C_2H_5O)_2POSi(CH_3)_3$	D	72 / 70 / 88	54 / 64 / 70
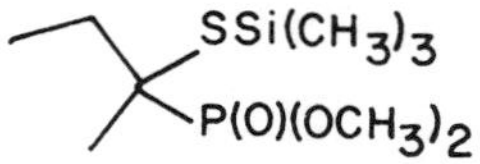 group bearing SSi(CH₃)₃ and P(O)(OCH₃)₂	$(CH_3O)_2POSi(CH_3)_3$	E	88	119
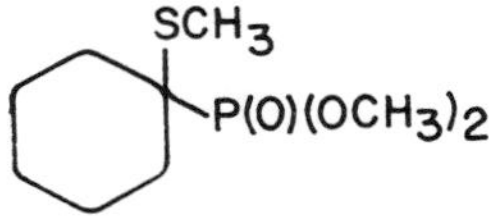 cyclohexane bearing SCH₃ and P(O)(OCH₃)₂	$(CH_3O)_3P$	E	59	117

Table 1 (continued)
ABRAMOV REACTION OF CARBONYL COMPOUNDS

Product	Phosphorus reagent	Method	Yield (%)	Refs.
$(C_2H_5O)_2P(O)CH(CCl_3)OSi(CH_3)_3$	$(C_2H_5O)_2POSi(CH_3)_3$	D	92	43
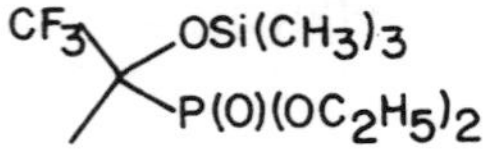 $CF_3\!-\!C(OSi(CH_3)_3)\!-\!P(O)(OC_2H_5)_2$	$(C_2H_5O)_2POSi(CH_3)_3$	D	66	55
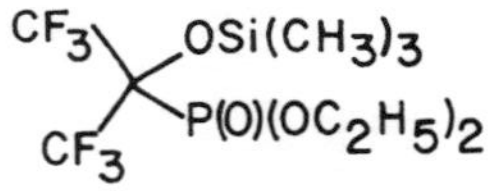 $(CF_3)_2C(OSi(CH_3)_3)P(O)(OC_2H_5)_2$	$(C_2H_5O)_2POSi(CH_3)_3$	D	96	55
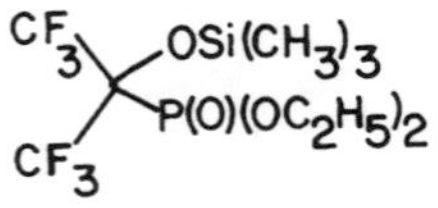 $(CF_3)_2C(OSi(CH_3)_3)P(O)(OC_2H_5)_2$	$(C_2H_5O)_2POSi(CH_3)_3$	D	95	66
$(C_2H_5O)_2P(O)CH(CF_3)OP(OC_2H_5)_2$	$(C_2H_5O)_2POP(OC_2H_5)_2$	D	56	42
$(C_2H_5O)_2P(O)CH[CH(CH_3)_2]OSi(CH_3)_2$	$(C_2H_5O)_2POSi(CH_3)_3$	D	83	70
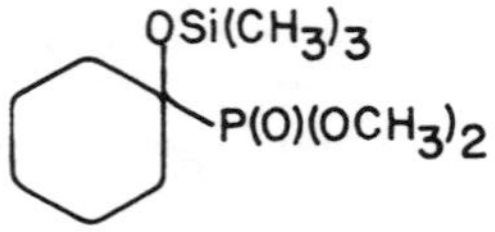 cyclohexane with $OSi(CH_3)_3$ and $P(O)(OCH_3)_2$	$(CH_3O)_2POSi(CH_3)_3$	D	86	58
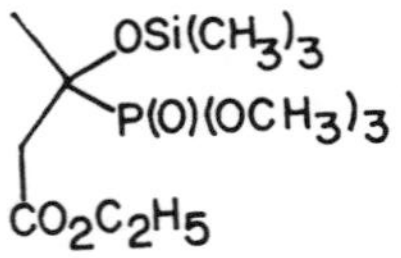 $C(OSi(CH_3)_3)(P(O)(OCH_3)_3)CH_2CO_2C_2H_5$	$(CH_3O)_2POSi(CH_3)_3$	D	60	75

Structure	Reagent	Method		
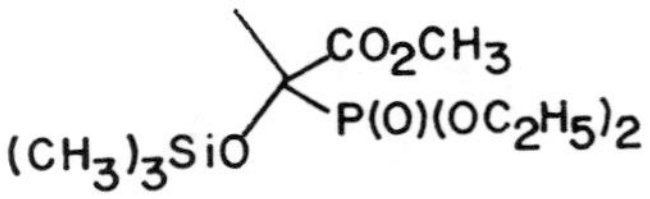 $(CH_3)_3SiO$ ... CO_2CH_3 / $P(O)(OC_2H_5)_2$	$(C_2H_5O)_2POSi(CH_3)_3$	D	36	74
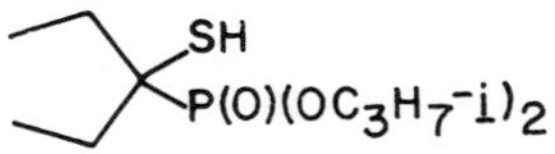 SH cyclopentane $P(O)(OC_3H_7-i)_2$	$(i\text{-}C_3H_7O)_3P$	E	80	118
$(CH_3O)_2P(O)CH(C_6H_5)OSi(CH_3)_3$	$(CH_3O)_2POSi(CH_3)_3$	D	78	54
	$(CH_3O)_2POSi(CH_3)_3$	D	62	58
$OSi(CH_3)_3$ / $\underline{n}\text{-}C_5H_9$ $P(O)(OCH_3)_2$				
$[(CH_3)_3SiO]_2P(O)CH[OSi(CH_3)_3]CH=CH_2$	$P[OSi(CH_3)_3]_3$	D	94	82
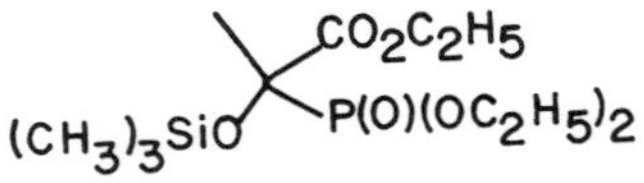 $(CH_3)_3SiO$ ... $CO_2C_2H_5$ / $P(O)(OC_2H_5)_2$	$(C_2H_5O)_2POSi(CH_3)_3$	D	40	74
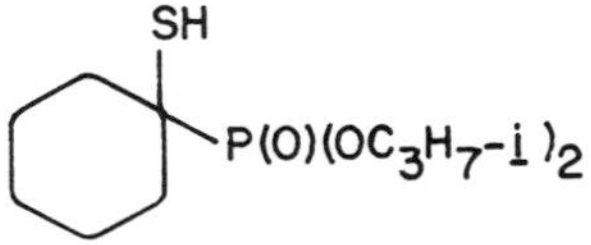 SH cyclohexane $P(O)(OC_3H_7-i)_2$	$(i\text{-}C_3H_7O)_3P$	E	89	117
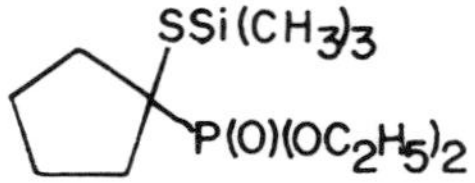 $SSi(CH_3)_3$ cyclopentane $P(O)(OC_2H_5)_2$	$(C_2H_5O)_2POSi(CH_3)_3$	E	97	117

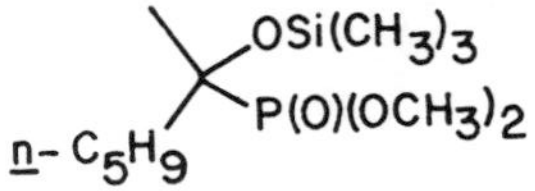

Table 1 (continued)
ABRAMOV REACTION OF CARBONYL COMPOUNDS

Product	Phosphorus reagent	Method	Yield (%)	Refs.
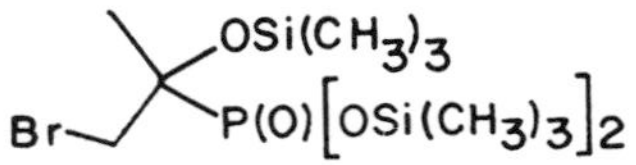 Br—...—C(OSi(CH_3)_3)—P(O)[OSi(CH_3)_3]_2	P[OSi(CH_3)_3]_3	D	89	84
[(CH_3)_2N]_2P(O)CH[OSi(C_2H_5)_3]CH=CH_2	[(CH_3)_2N]_2POSi(C_2H_5)_3	D	90	65
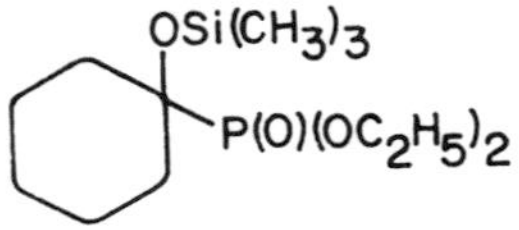 cyclohexane with OSi(CH_3)_3 and P(O)(OC_2H_5)_2	(C_2H_5O)_2POSi(CH_3)_3	D	52	64
[(CH_3)_3SiO]_2P(O)CH[OSi(CH_3)_3]CH=CHCH_3	P[OSi(CH_3)_3]_3	D	89	82
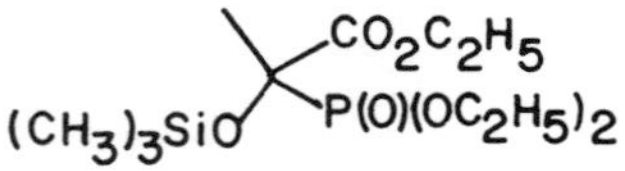 (CH_3)_3SiO—C(CO_2C_2H_5)—P(O)(OC_2H_5)_2	(C_2H_5O)_2POSi(CH_3)_3	D	62	64
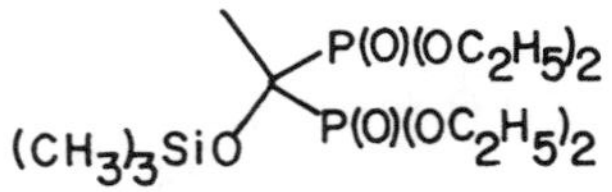 (CH_3)_3SiO—C[P(O)(OC_2H_5)_2][P(O)(OC_2H_5)_2]	(C_2H_5O)_2POSi(CH_3)_3	F	62	73
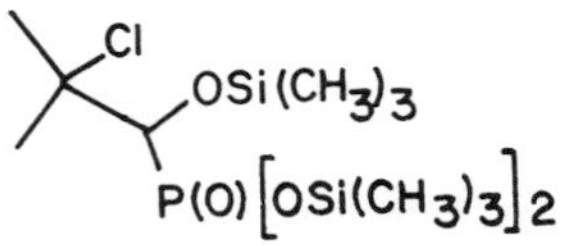 Cl—C(OSi(CH_3)_3)—P(O)[OSi(CH_3)_3]_2	P[OSi(CH_3)_3]_3	D	92	83

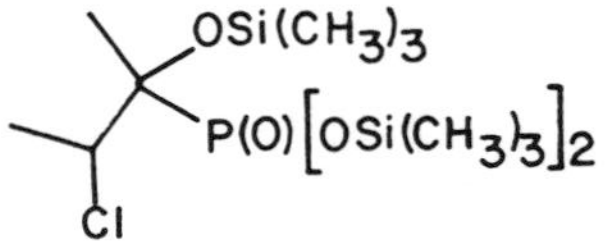	$P[OSi(CH_3)_3]_3$	D	92	84
$(C_2H_5O)_2P(O)CH(C_6H_5)OSi(CH_3)_3$	$(C_2H_5O)_2POSi(CH_3)_3$	D	66	54
		D	55	64
			91	70
$(C_2H_5O)_2P(O)CH(C_6H_5)OSi(CH_3)_3$	$(C_2H_5O)_3P$	G	94	60
$[(CH_3)_2N]_2P(O)CH[OSi(C_2H_5)_3]CH=CHCH_3$	$[(CH_3)_2N]_2POSi(C_2H_5)_3$	D	95	65
$[(CH_3)_2N]_2P(O)CH[CH(CH_3)_2]OSi(C_2H_5)_3$	$[(CH_3)_2N]_2POSi(C_2H_5)_3$	D	90	65
		B	84	8
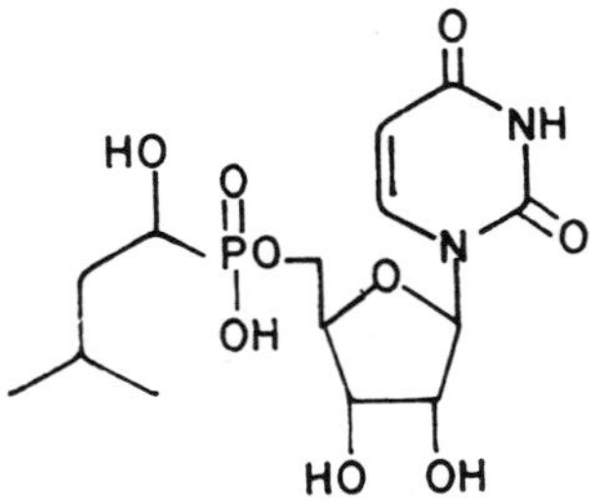	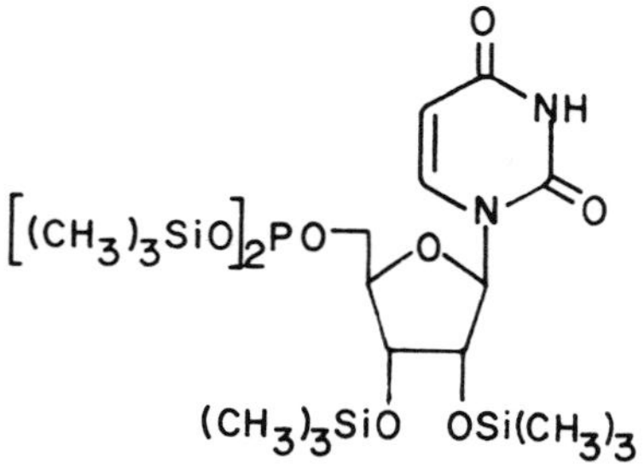			
$(CH_3O)_2P(O)CH(C_5H_{11}\text{-}n)OSi(CH_3)_2C(CH_3)_3$	$(CH_3O)_2POSi(CH_3)_2C(CH_3)_3$	D	81	58
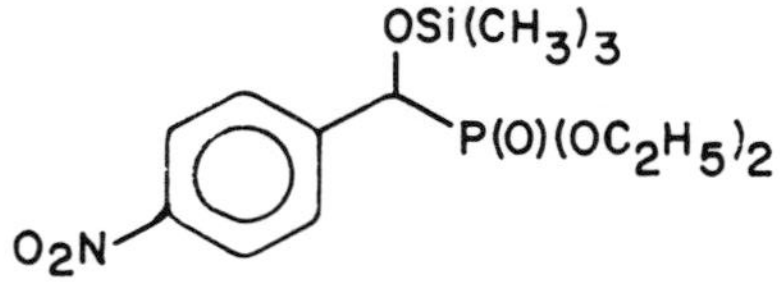	$(C_2H_5O)_2POSi(CH_3)_3$	D	73	71

Table 1 (continued)
ABRAMOV REACTION OF CARBONYL COMPOUNDS

Product	Phosphorus reagent	Method	Yield (%)	Refs.
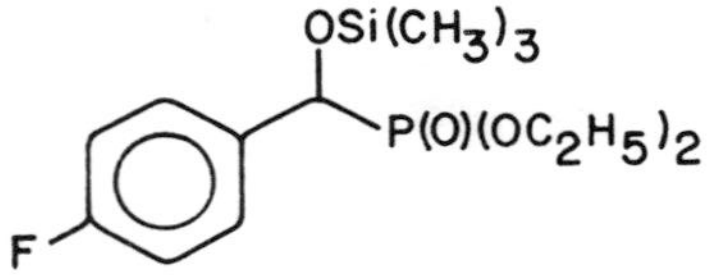 $OSi(CH_3)_3$ / $P(O)(OC_2H_5)_2$ (4-fluorophenyl)	$(C_2H_5O)_2POSi(CH_3)_3$	D	66	71
$[(C_2H_5O)_2P(O)CH(CH_3)O]_2Si(CH_3)_2$	$(C_2H_5O)_3P$	G	79	56
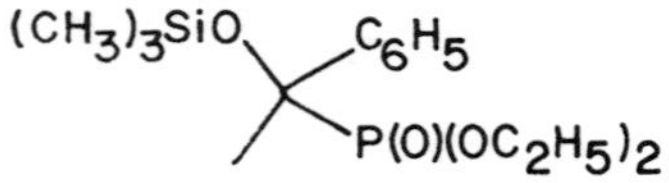 $(CH_3)_3SiO$, C_6H_5, $P(O)(OC_2H_5)_2$	$(C_2H_5O)_2POSi(CH_3)_3$	D	78	68
$(C_2H_5O)_2P(O)CH(C_7H_{15}\text{-}n)OSi(CH_3)_3$	$(C_2H_5O)_2POSi(CH_3)_3$	D	91	70
$[(CH_3)_3SiO]_2P(O)CH(C_6H_5)OSi(CH_3)_3$	$P[OSi(CH_3)_3]_3$	D	88	82
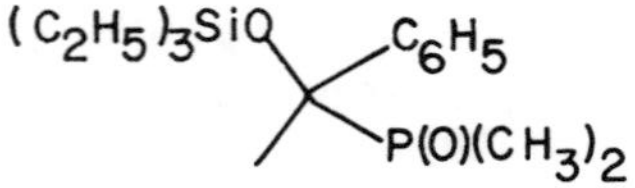 $(C_2H_5)_3SiO$, C_6H_5, $P(O)(CH_3)_2$	$(CH_3O)_2POSi(C_2H_5)_3$	D	27	54
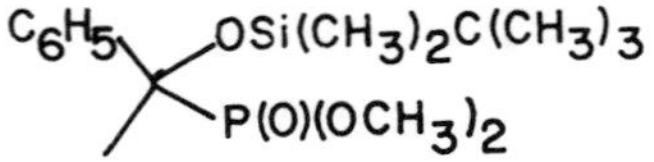 C_6H_5, $OSi(CH_3)_2C(CH_3)_3$, $P(O)(OCH_3)_2$	$(CH_3O)_2POSi(CH_3)_2C(CH_3)_3$	D	34	58

Product	Reagent			
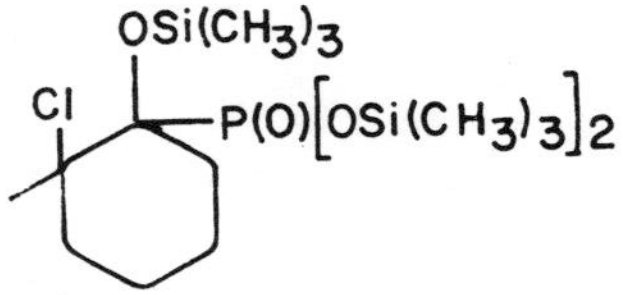 ![structure] OSi(CH$_3$)$_3$, Cl, cyclohexane-P(O)[OSi(CH$_3$)$_3$]$_2$	P[OSi(CH$_3$)$_3$]$_3$	D	82	83
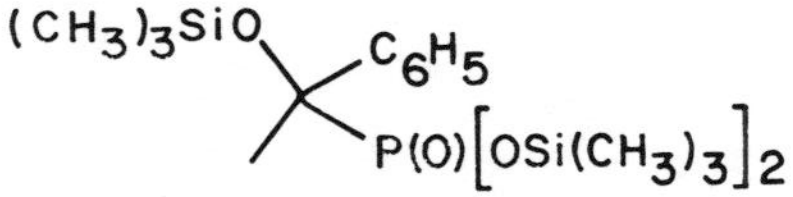 (CH$_3$)$_3$SiO—C(CH$_3$)$_2$—C$_6$H$_5$—P(O)[OSi(CH$_3$)$_3$]$_2$	P[OSi(CH$_3$)$_3$]$_3$	D	75	82
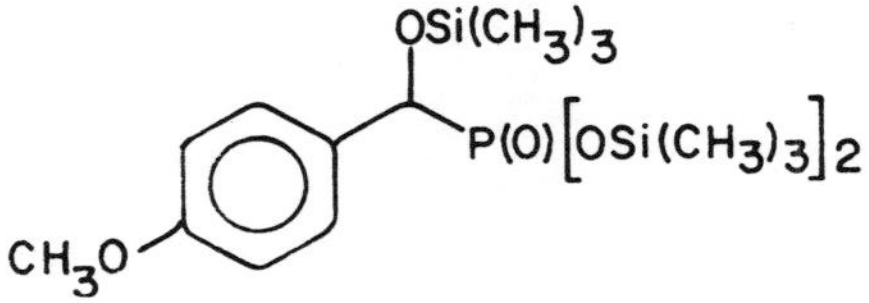 OSi(CH$_3$)$_3$ / CH$_3$O—C$_6$H$_4$—CH—P(O)[OSi(CH$_3$)$_3$]$_2$	P[OSi(CH$_3$)$_3$]$_3$	D	83	82
[(CH$_3$)$_2$N]$_2$P(O)CH(C$_6$H$_5$)OSi(C$_2$H$_5$)$_3$	[(CH$_3$)$_2$N]$_2$POSi(C$_2$H$_5$)$_3$	D	92	65
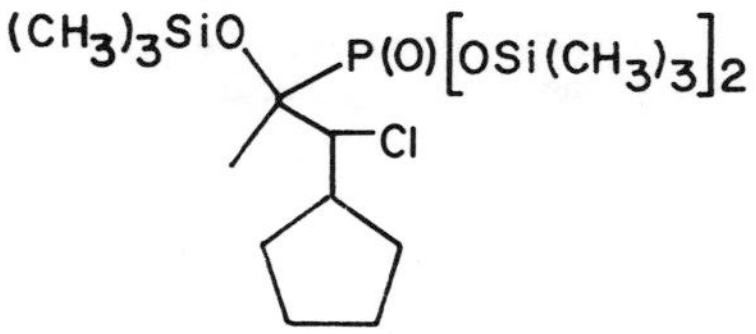 (CH$_3$)$_3$SiO—C(CH$_3$)(Cl)(cyclopentyl)—P(O)[OSi(CH$_3$)$_3$]$_2$	P[OSi(CH$_3$)$_3$]$_3$	D	82	84
[(C$_2$H$_5$)$_2$N]$_2$P(O)CH(C$_6$H$_5$)OSi(CH$_3$)$_3$	[(C$_2$H$_5$)$_2$N]$_2$POSi(CH$_3$)$_3$	D	71	67
[(CH$_3$)$_2$N]$_2$P(O)CH[OSi(C$_2$H$_5$)$_3$]CH=CHC$_6$H$_5$	[(CH$_3$)$_2$N]$_2$POSi(C$_2$H$_5$)$_3$	D	93	57

Table 1 (continued)

ABRAMOV REACTION OF CARBONYL COMPOUNDS

Product	Phosphorus reagent	Method	Yield (%)	Refs.
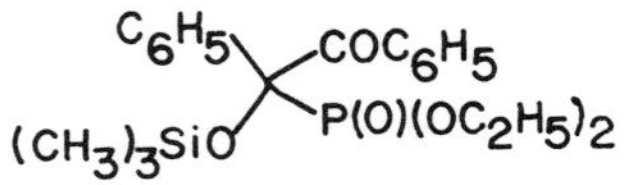 C6H5, COC6H5 / (CH3)3SiO — P(O)(OC2H5)2	$(C_2H_5O)_2POSi(CH_3)_3$	D	95	71

Note: A, carbonyl compound added to phosphorus reagent in alcohol solution; B, hydrolytic work-up of reaction; C, carbonyl compound with phosphorus trichloride added to phosphorus reagent; D, carbonyl compound used alone with phosphorus reagent; E, thiocarbonyl compound used alone with phosphorus reagent; F, 1-ketophosphonate added to phosphorus reagent; G, carbonyl compound with silyl halide added to phosphorus reagent.

Table 2

ABRAMOV REACTION OF IMINES

Product	Phosphorus reagent	Method	Yield (%)	Refs.
$H_2O_3PCH(NH_2)CH_3$ $[H_2O_3P]_2C(NH_2)CH_3$	$P[OSi(CH_3)_3]_3$ $(C_6H_5O)_3P$ H_3PO_3	A B C	50 41 85	90 102 114
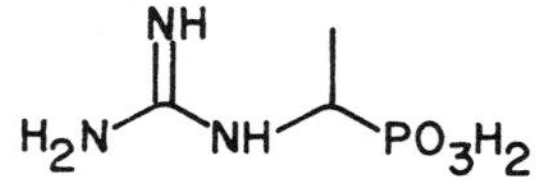	$(C_6H_5O)_3P$	B	13	96
$H_2O_3PCH(NH_2)CH_2CH_2CH(CH_3)_2$ $[H_2O_3PCH_2]_2CH_2CH=CH_2$ $[H_2O_3PCH_2]_2CH_2CH_2CH_3$	$P[OSi(CH_3)_3]_3$ H_3PO_3 H_3PO_3	A D D	70 49 81	90 106 106
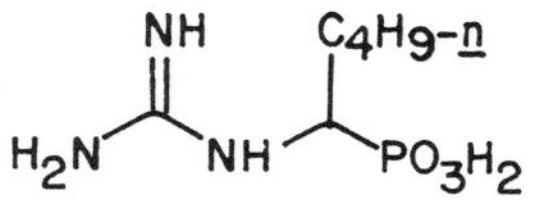	$(C_6H_5O)_3P$	B	22	96
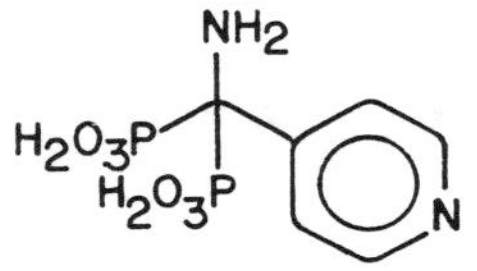	H_3PO_3	C	86	114
$H_2O_3PCH(NH_2)C_6H_5$	H_3PO_3	B	75	111

Table 2 (continued)
ABRAMOV REACTION OF IMINES

Product	Phosphorus reagent	Method	Yield (%)	Refs.
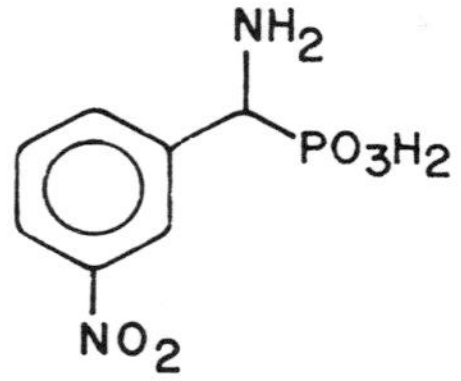 (3-nitrophenyl)CH(NH$_2$)PO$_3$H$_2$	P[OSi(CH$_3$)$_3$]$_3$	A	88	90
	(C$_2$H$_5$O)$_3$P	B	47	93
	H$_3$PO$_3$	B	59	111
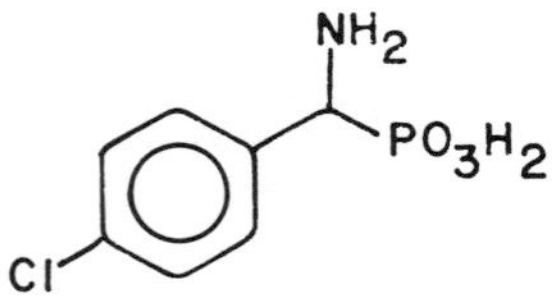 (4-chlorophenyl)CH(NH$_2$)PO$_3$H$_2$	H$_3$PO$_3$	B	67	111
H$_2$O$_3$PCH(C$_6$H$_5$)NHC$_2$H$_5$	H$_3$PO$_3$	E	68	108
[H$_2$O$_3$PCH$_2$]$_2$NCH$_2$C$_6$H$_5$	H$_3$PO$_3$	D	86	106
[H$_2$O$_3$PCH$_2$]$_2$NC$_7$H$_{15}$-n	H$_3$PO$_3$	D	90	106
H$_2$O$_3$PCH[CH(CH$_3$)$_2$]NHCH$_2$C$_6$H$_5$	H$_3$PO$_3$	E	40	108
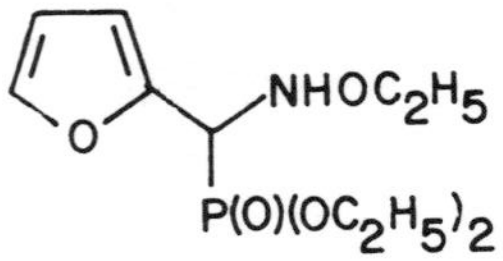 (2-furyl)CH(NHOC$_2$H$_5$)P(O)(OC$_2$H$_5$)$_2$	(C$_2$H$_5$O)$_3$P	E	39	62
[(ClCH$_2$CH$_2$O)$_2$P(O)CH(CH$_3$)NH]$_2$CO	(ClCH$_2$CH$_2$O)$_3$P	F	89	95
H$_2$O$_3$PCH(C$_6$H$_5$)NHCH$_2$C$_6$H$_5$	H$_3$PO$_3$	E	98	110

Compound	Reagent	Method	Yield	Ref.
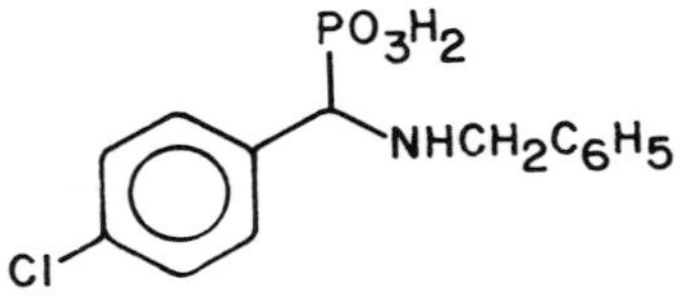 (structure: 4-Cl-C$_6$H$_4$CH(PO$_3$H$_2$)NHCH$_2$C$_6$H$_5$)	H$_3$PO$_3$	E	87	108
[H$_2$O$_3$PCH$_2$]$_2$NC$_{12}$H$_{25}$-n	H$_3$PO$_3$	D	91	106
H$_2$O$_2$PCH(CH$_2$CH$_2$OH)NHCH(C$_6$H$_5$)$_2$	H$_3$PO$_2$	F	42	121
[H$_2$O$_3$PCH$_2$]$_2$NC$_{15}$H$_{31}$-n	H$_3$PO$_3$	D	91	106
H$_2$O$_2$PCH(CH$_2$CH$_2$CO$_2$CH$_3$)NHCH(C$_6$H$_5$)$_2$	H$_3$PO$_2$	F	50	121
H$_2$O$_2$PCH(CH$_2$C$_6$H$_5$)NHCH(C$_6$H$_5$)$_2$	H$_3$PO$_2$	F	26	121
(C$_6$H$_5$O)$_2$P(O)CH$_2$NHCO$_2$CH$_2$C$_6$H$_5$	(C$_6$H$_5$O)$_3$P	E	54	105
(C$_6$H$_5$O)$_2$P(O)CH(CH$_3$)NHCSNHC$_6$H$_5$	(C$_6$H$_5$O)$_3$P	F	85	98
(C$_6$H$_5$O)$_2$P(O)CH(CH$_3$)NHCO$_2$CH$_2$C$_6$H$_5$	(C$_6$H$_5$O)$_3$P	F	46	104
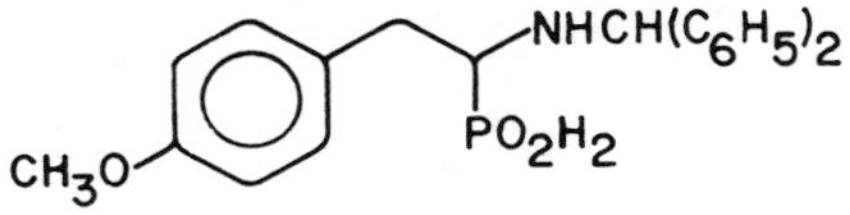 (structure: 4-CH$_3$O-C$_6$H$_4$CH$_2$CH(NHCH(C$_6$H$_5$)$_2$)PO$_2$H$_2$)	H$_3$PO$_2$	F	26	121
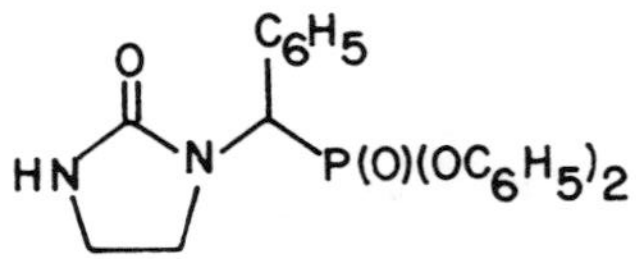 (structure: C$_6$H$_5$CH[P(O)(OC$_6$H$_5$)$_2$] on imidazolidinone)	(C$_6$H$_5$O)$_3$P	F	48	33
(C$_6$H$_5$O)$_2$P(O)CH(CH$_2$SCH$_3$)NHCSNHC$_6$H$_5$	(C$_6$H$_5$O)$_3$P	F	94	101
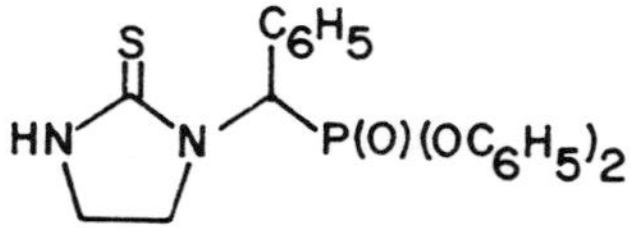 (structure: C$_6$H$_5$CH[P(O)(OC$_6$H$_5$)$_2$] on imidazolidine-thione)	(C$_6$H$_5$O)$_3$P	F	38	33

Table 2 (continued)
ABRAMOV REACTION OF IMINES

Product	Phosphorus reagent	Method	Yield (%)	Refs.
$(C_6H_5O)_2P(O)CH(C_3H_7\text{-}n)NHCSNHC_6H_5$	$(C_6H_5O)_3P$	F	89	98
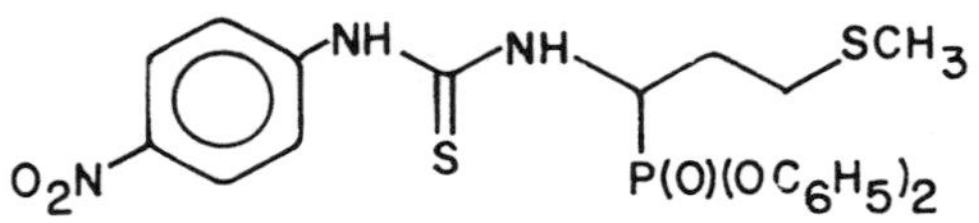	$(C_6H_5O)_3P$	F	56	100
$(C_6H_5O)_2P(O)CH(C_4H_9\text{-}n)NHCSNHC_6H_5$	$(C_6H_5O)_3P$	F	81	98
$(C_6H_5O)_2P(O)CH(CH_2CH_2SC_2H_5)NHCSNHC_6H_5$	$(C_6H_5O)_3P$	F	88	100
$(C_6H_5O)_2P(O)NH[CH_2SCH(CH_3)_2]NHCSNHC_6H_5$	$(C_6H_5O)_3P$	F	90	101
$(C_6H_5O)_2P(O)CH[CH_2CH(CH_3)_2]NHCO_2CH_2C_6H_5$	$(C_6H_5O)_3P$	F	54	104
$(C_6H_5O)_2P(O)CH(CH_2SC_4H_9\text{-}n)NHCSNHC_6H_5$	$(C_6H_5O)_3P$	F	95	101
$(C_6H_5O)_2P(O)CH(C_6H_5)NHCSNHC_6H_5$	$(C_6H_5O)_3P$	F	85	98
$(C_6H_5O)_2P(O)CH(C_6H_5)NHCO_2CH_2C_6H_5$	$(C_6H_5O)_3P$	F	51	104
$(C_6H_5O)_2P(O)CH(CH_2SCH_2C_6H_5)NHCSNHC_6H_5$	$(C_6H_5O)_3P$	F	97	101
$(C_6H_5O)_2P(O)CH(CH_2OCH_2C_6H_5)NHCO_2CH_2C_6H_5$	$(C_6H_5O)_3P$	F	51	103
$[(C_6H_5O)_2P(O)CH(CH_3)NH]_2CO$	$(C_6H_5O)_3P$	F	62	95
$[(C_6H_5O)_2P(O)CH(C_6H_5)NH]_2CO$	$(C_6H_5O)_3P$	F	71	95

Note: A, preformed imine used; ester linkages cleaved in work-up; B, imine formed *in situ*; ester linkages cleaved in work-up; C, nitrile used with phosphorus reagent; D, Mannich-type condensation performed *in situ*; E, preformed imine used; F, imine formed *in situ*.

III. THE PUDOVIK REACTION

A. Reactions and Mechanism

The addition of monobasic trivalent phosphorus acid species under basic conditions to carbonyl-type centers is a useful variation of the previously discussed Abramov reaction.

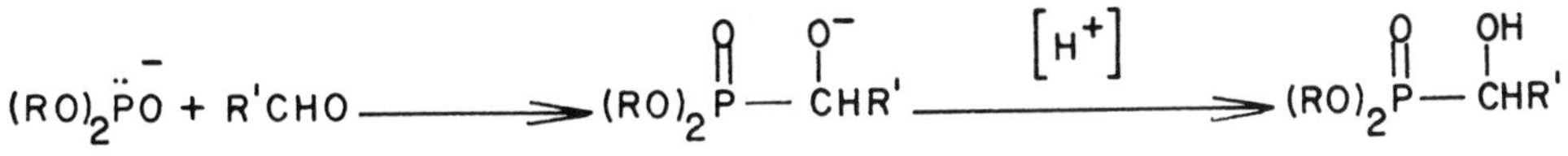

$$(RO)_2\ddot{P}O^- + R'CHO \longrightarrow (RO)_2\overset{O}{\overset{\|}{P}} - \overset{O^-}{\underset{|}{C}}HR' \xrightarrow{[H^+]} (RO)_2\overset{O}{\overset{\|}{P}} - \overset{OH}{\underset{|}{C}}HR'$$

With such an approach to carbon-phosphorus bond formation accompanied by expansion of phosphorus from a trivalent to a quinquevalent state, the problem of dealkylation is obviated. In the early work by Pudovik and Arbuzov using the sodium salt of dialkyl phosphites with an α,β-unsaturated ketone,[122] conjugate addition (hydrophosphinylation) was observed.

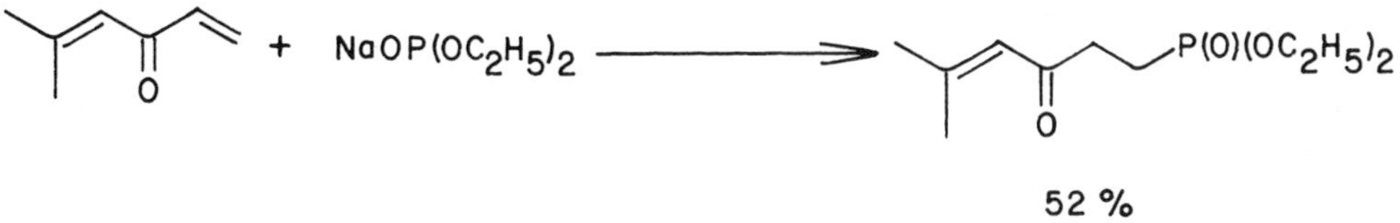

52 %

Later studies by Abramov using simple aldehydes and ketones demonstrated that stable products could be obtained by addition to the carbonyl carbon as well. Both of these investigations noted that only catalytic amounts of the phosphite salt in the presence of an equivalent amount of the free acid were necessary for reaction to go to completion.

The reaction appears to be of a simple "aldol" type with the anionic phosphorus center adding directly to the electron deficient carbonyl carbon. Investigations of the effect of solvent indicate that the rate of reaction increases directly with the ion solvating power of the medium.[123,124] If there were attack by oxygen of the ambient phosphorus anion, no product would be isolated to demonstrate such occurrence.

In addition to the observations concerning the requirement for only catalytic amounts of the phosphorus anion to be present, significant effort has been devoted to the investigation of the methods of generating the anion and the nature of the associated cationic species. Of course, simple alkali metal salts have been used commonly in numerous synthetic efforts.

Although the early efforts used salts of the dialkyl phosphites generated by direct treatment with sodium metal in an inert solvent, the more common procedure has involved the addition of a small amount of alcoholic alkoxide ion to the reaction mixture. This method has been used particularly in the syntheses of analogues of carbohydrate phosphates,[10-13] as well as for the preparation of antiviral agents,[28,125] and rigid bicyclic phosphonates of use in determining nuclear magnetic resonance coupling interactions.[126]

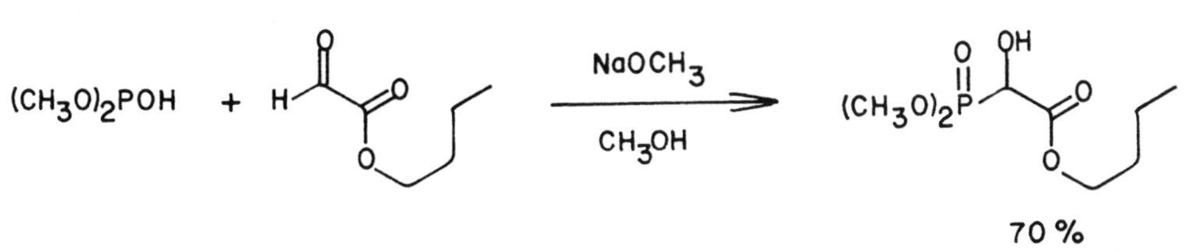

70 %

This method has also been found to give good yields of carbonyl adducts in α,β-unsaturated systems where competition by the conjugate addition route might be anticipated. [127-130] Carbonyl-addition represents the kinetically favored route, whereas conjugate addition is thermodynamically favored.

An interesting variation on the metal cation approach involves the generation of the Grignard[131] of diphenylphosphine oxide.[132] The anionic phosphorus reagent, when generated to the exclusion of other bases, gives carbonyl addition reaction in excellent yield.

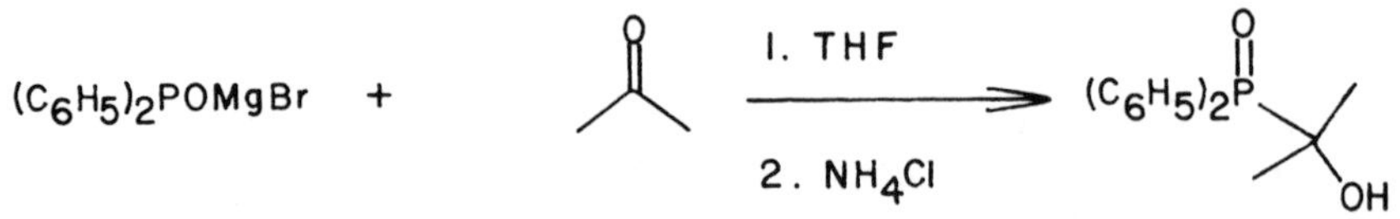

An alternative to the use of anionic bases for the generation of the phosphorus anions for the Pudovik reaction is the use of neutral amines. The most commonly used amine for such purposes is triethylamine, which acts efficiently when added in catalytic amounts. Benzene has been used as a solvent with this reaction system, but often no solvent other than the dialkyl phosphite is used. This approach has been used in the synthesis of α-hydroxyalkyl-phosphonic acids and esters which are analogues of natural phosphates,[9,133,134] herbicides,[24] and morphactin analogues.[135] The reaction using triethylamine is interesting in that it is found to proceed in good yield even in the presence of other hydroxyl groups,[136] a function which often presents difficulties for efficient reaction. In an investigation of the reaction of α-chloroacetophenone, triethylamine, out of the several amines studied, was found to provide the best yield, and reaction proceeded without formation of Perkow-type product.[137]

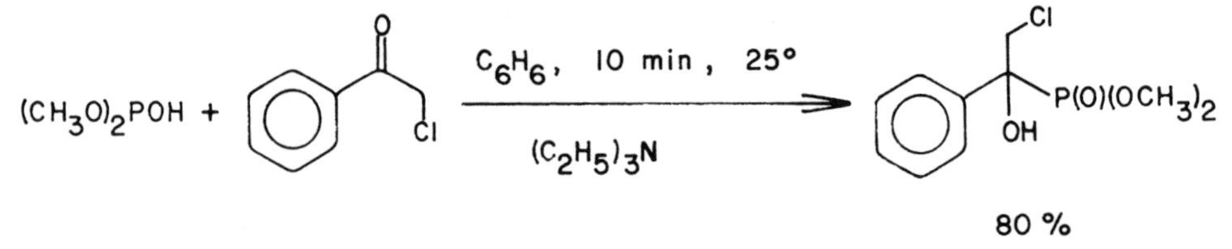

Triethylamine was found to be quite useful for the preparation of dibutyl hydroxymethyl-phosphonate. Reaction proceeds by heating paraformaldehyde with dibutyl phosphite and triethylamine in the absence of solvent.[138] In another investigation wherein reactions of ketones and aldehydes were compared using triethylamine as opposed to sodium methoxide, dimethyl phosphite added efficiently to the carbonyl carbon in the presence of triethylamine, but formed phosphate product in the methoxide catalyzed system.[133]

Attempts to perform the Abramov addition of trimethyl phosphite to a substituted 2-imino-1,3-propanedione species resulted in the formation of several products, none of which involved the generation of a carbon-phosphorus bond under the variety of conditions investigated.[139] The use, however, of triethylamine with dimethyl phosphite provided clean addition at a carbonyl carbon, leaving the imine site untouched.

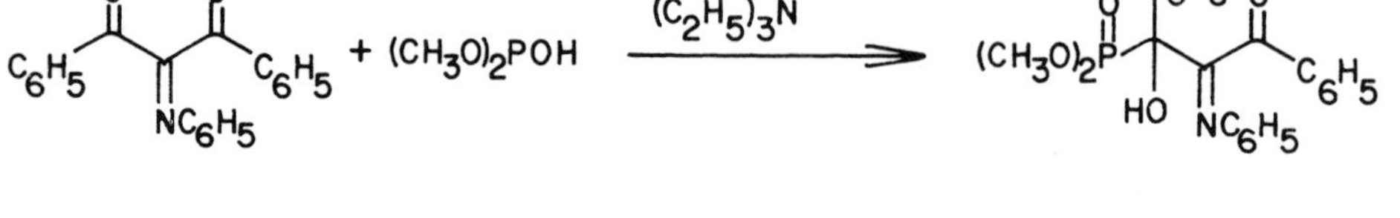

The preparation of a series of phosphine oxide analogues of morphactines has been accomplished under Pudovik conditions using triethylamine as the basic reagent. Secondary phosphine oxides have been added to fluorenone in the presence of an equivalent amount of triethylamine to generate the morphactine analogues, useful as herbicidal agents, in moderate to good yield.[140]

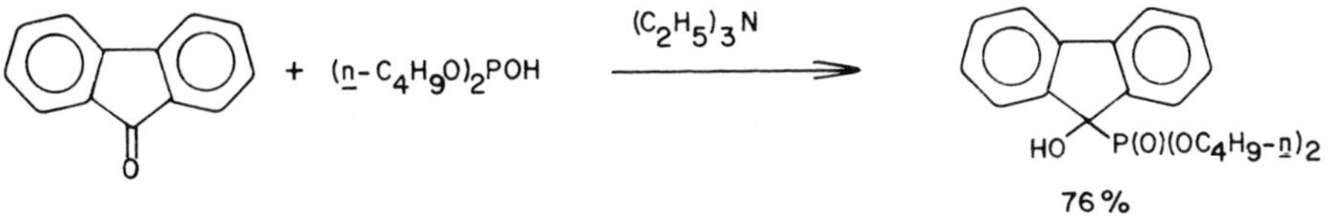

76%

Two particularly interesting reports have been made which are concerned with the accomplishment of asymmetric induction in the Pudovik reaction through the use of a chiral amine as the base. Dimethyl phosphite added to aromatic aldehydes in the presence of quinine to give the α-hydroxybenzylphosphonates in nearly quantitative yield with enantiomeric excess of the crude product being in the range of 10 to 28%.[141,142] Enantiomeric excesses of greater than 98% could readily be attained by further recrystallizations.

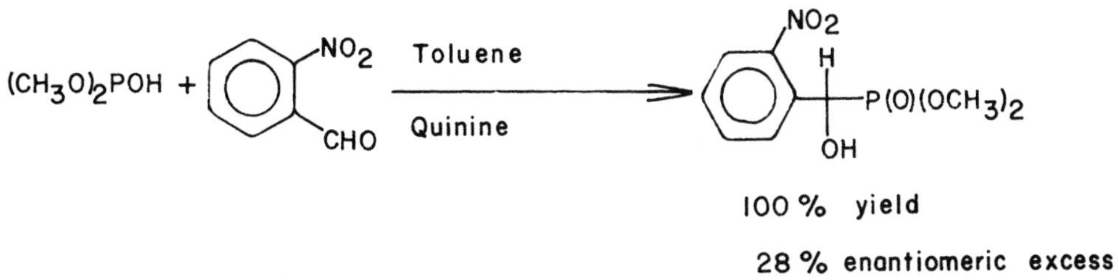

100% yield

28% enantiomeric excess

Other nitrogenous bases have also at times been used successfully for the Pudovik reaction. The addition of dimethyl phosphite to chloral in 98% yield has been reported using ammonia as the base.[30] With a variety of ketones, dialkyl phosphites have been reported to undergo addition in excellent yield with secondary alkyl amines as the catalytic agent.[143,144] Finally, basic nitrogenous sites present within the carbonyl reagent have been reported to facilitate efficient reaction with secondary phosphine oxides.[31]

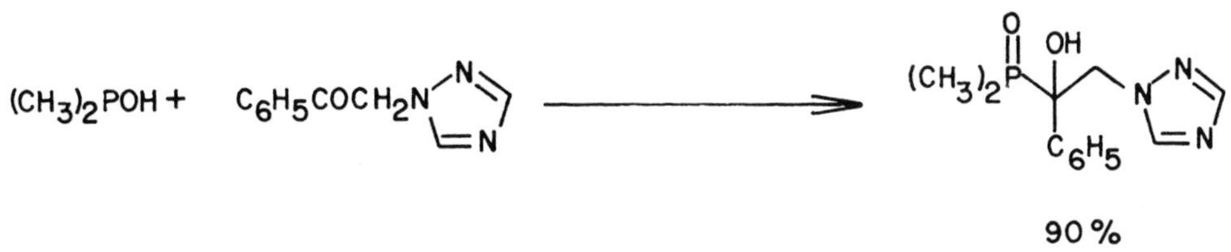

90%

In one instance the nonnucleophilic nitrogenous base 1,8-diazabicyclo[5.4.0]undec-7-ene has been used to facilitate the Pudovik reaction.[145] The reaction involved addition of a secondary phosphine oxide to an α-ketosulfonate. The reaction proceeds beyond the initial addition to the carbonyl function to generate the 1,2-epoxyphosphonate. An equivalent amount of the base was used, although it is not evident that this is absolutely necessary.

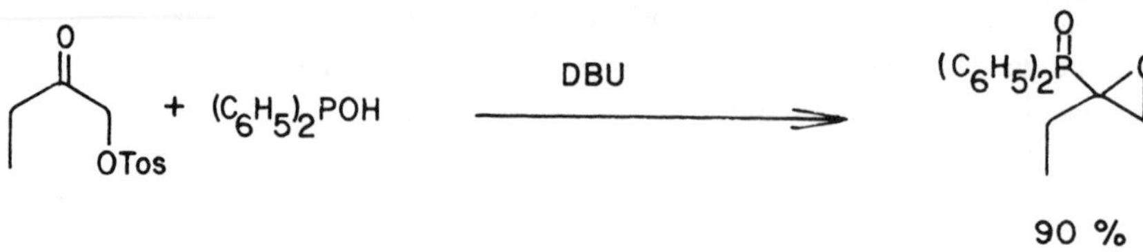

90 %

With particularly reactive carbonyl compounds, addition of dialkyl phosphites may be observed even in the absence of added base. This is found to occur with glyoxal,[146] glyoxalate esters,[29] and hexafluoroacetone,[147] the last also exhibiting the formation of varying amounts of phosphate product, resulting from addition at the carbonyl oxygen.

A particularly valuable development for the performance of Pudovik-type reactions has been the use of phase transfer agents. Carbonyl compounds bearing highly polar functionalities with resulting low solubility in standard organic media, but high solubility in aqueous solution, can be induced to undergo the Pudovik reaction with dialkyl phosphites by the addition to the reaction medium of quaternary ammonium salts with carbonate ion as the base. With this approach, good yields of adducts can be isolated from highly functionalized carbohydrate derivatives which otherwise present significant difficulties due to poor solubility characteristics.[148-151]

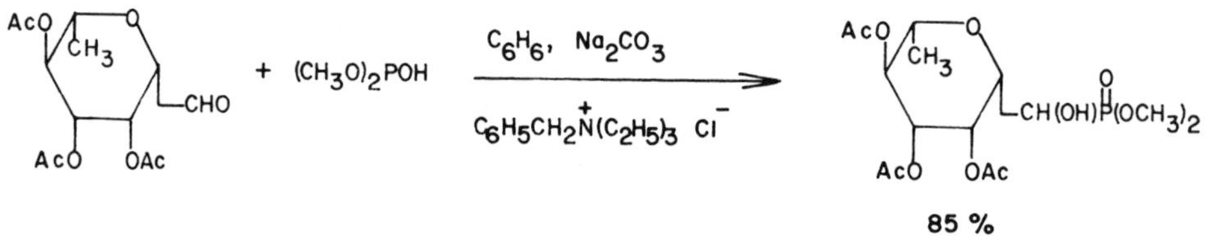

85 %

Another approach to the performance of the Pudovik addition which has been developed in recent years is the use of a solid phase material as a basic catalyst. The advantage of such an approach is the simplicity of product isolation. When used without solvent, the solid phase catalyst can be removed from the reaction mixture by simple filtration and the product isolated directly. Two reaction systems of this type have been reported recently, the first involving the addition to the carbonyl and dialkyl phosphite reagent of basic alumina.[152] The reaction is found to generate the target materials, the α-hydroxyalkylphosphonates, in moderate to excellent yield with stirring at room temperature. The reaction seems to proceed most favorably with aromatic aldehydes, although ketones react favorably as well. A requirement for the reaction, of course, is that the carbonyl compound and dialkyl phosphite form a liquid solution. Otherwise, sufficient contact with the catalyst for reaction is not attained.

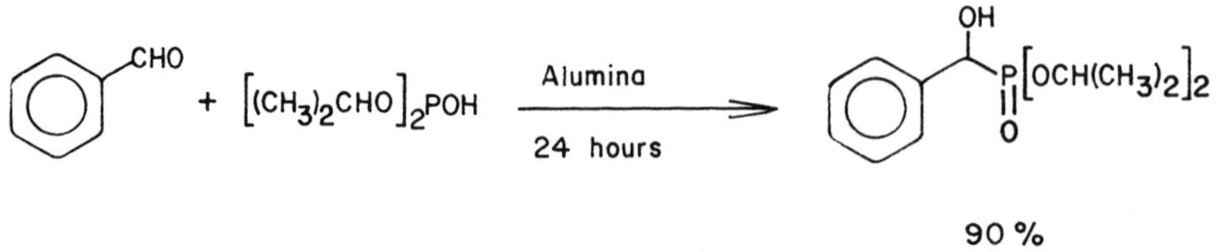

90 %

The second solid phase system for the facilitation of the Pudovik addition involves the addition of potassium fluoride or cesium fluoride to the mixture of carbonyl compound and dialkyl phosphite.[153,154] The isolated yields of adducts with this method are generally better than with the alumina-facilitated reaction system. However, the amount of salt required is large, usually from 2 to 5 M amounts compared to the reactants. Thus it is critical that the carbonyl compound and dialkyl phosphite form a fluid solution. This approach has been successful in the synthesis of a carbohydrate-related α-hydroxyphosphonate.[151]

80 %

As discussed earlier in this chapter, the addition of monobasic trivalent phosphorus acids to imine carbon has been of great interest in recent years for a variety of purposes, including the preparation of analogues of natural amino acids and herbicides, and other biologically active materials. The initial effort in this regard demonstrated that the reaction could be performed either in a "one pot" procedure with the imine being generated *in situ* with dialkyl phosphite present or with the imine being preformed prior to addition of phosphite.[155] Both approaches allow isolation of the adducts in excellent yield. Presumably, the imine itself is capable of acting as the facilitating base for the addition reaction.

$$(C_2H_5O)_2POH \quad + \quad H_2CO \quad + \quad (C_2H_5)_2NH \quad \longrightarrow \quad (C_2H_5O)_2\overset{\overset{O}{\|}}{P}CH_2N(C_2H_5)_2$$

94 %

The approach of using the preformed imine as the basic reagent for facilitating the phosphourus addition has been used in a large number of preparations of α-aminoalkylphosphonates for a variety of purposes. Included here are preparations of analogues of the naturally occurring amino acids,[94,156-162] herbicidal agents,[17,163-167] antibacterials,[168] reagents for *umpolung* processes,[169-171] and a variety of N-substituted α-aminoalkylphosphonates.[172-176]

In contrast to the reaction noted, previously performed in the presence of triethylamine, diethyl phosphite reacts at the imine carbon with an oxime bearing an additional carbonyl group when the reaction is performed without added base.[177] Oximes have also been found to undergo addition of hypophosphorous acid in the absence of added base.[113] The adducts may be considered as another class of analogues of the natural amino acids. Related monobasic analogues of the natural amino acids may also be prepared by the addition of monoalkyl phosphonites to substituted enamines which are tautomeric with imines.[178] Finally, there should be noted another preparation of phosphonic and phosphinic acid analogues of the natural amino acids via Pudovik-type addition to the imine linkage of thiazolines, followed by deprotection.[25] The product materials have found utility as pharmaceutical agents for bone disorders.

The symmetrical triazines bearing substitution at nitrogen have also been used as substrates for the Pudovik reaction of phosphite diesters, although there is formally no unsaturated linkage of carbon to nitrogen. The materials are trimers of formaldimines which are broken apart by the acidic phosphite diesters and serve as basic facilitators for the subsequent addition of phosphorus to carbon. This approach has been utilized in the preparation of materials for use as herbicides,[18,21,179] antibacterials,[180] and pharmaceuticals.[26]

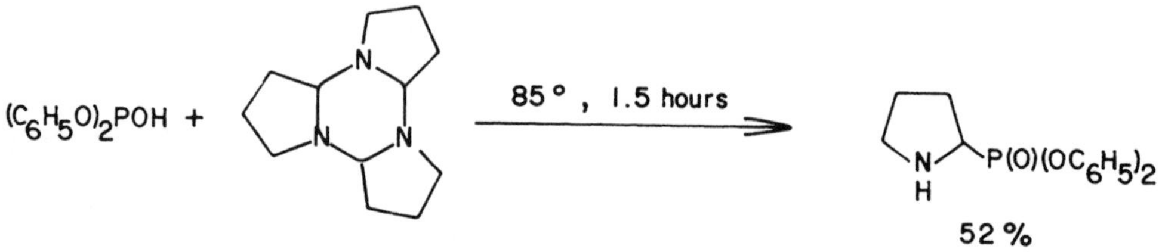

$(C_6H_5O)_2POH +$ $\xrightarrow{85°, \ 1.5 \ hours}$

52 %

In several instances additional basic reagents have been used as adjuncts for the Pudovik addition to imines. Triethylamine has been reported to provide excellent yields in addition of dialkyl phosphites and alkyl phosphonites with geminal diimines.[181,182] Similarly, the sodium salts of dialkyl phosphites, prepared via sodium hydride addition[16] or ethanolic sodium ethoxide,[183,184] have been used to undergo addition to imines and oximes.

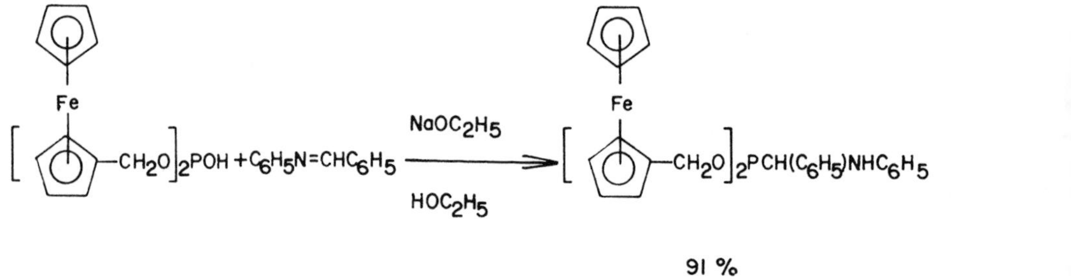

$$\left[\text{Fe}\cdots CH_2O \right]_2 POH + C_6H_5N=CHC_6H_5 \xrightarrow[HOC_2H_5]{NaOC_2H_5} \left[\text{Fe}\cdots CH_2O \right]_2 PCH(C_6H_5)NHC_6H_5$$

91 %

The *in situ* generation of imine followed by addition of a monobasic trivalent phosphorus acid, as originally developed by Fields,[155] has been applied to the synthesis of a large number of α-aminoalkylphosphonates. While some of these have been of interest as structural analogues of natural amino acids,[99,185-187] most of the target materials have been intended for use as herbicides.[19,20,22,23,135,188-191]

$$(CH_2O)_n + (C_2H_5O)_2POH + H_2NCH_2CO_2H \xrightarrow{\text{hydrolytic work-up}} H_2O_3PCH_2NHCH_2CO_2H$$

98 %

Two syntheses of optically active Pudovik-type adducts have been reported, starting with imines. A chiral imine was formed from optically active 1-phenylethylamine and benzaldehyde which underwent addition with diethyl phosphite in the absence of added base.[192] Asymmetric induction was also observed in the addition of a chiral dialkyl phosphite to a cyclic aldimine for the preparation of optically active phosphonic acid analogues of penicillamine.[193]

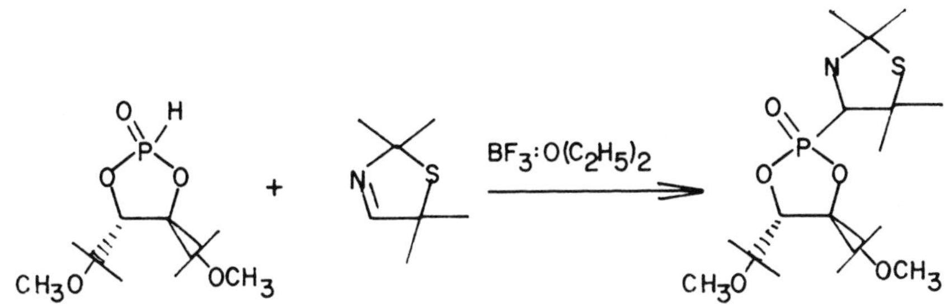

82 % yield

50 % enantiomeric excess

The use of acids as catalysts for the Pudovik reaction have been noted in several other instances. It might be anticipated thet addition of phosphorus at imine carbon could be facilitated by acid association at nitrogen as well as by generation of the phosphorus anion. Phosphonic acid analogues of the natural amino acids have been prepared in moderate to good yield by dialkyl phosphite addition to imine salts formed in the stannous chloride/hydrochloric acid reduction of nitriles.[194] In addition, phosphorous acid has been found to add to nitriles to generate geminal diphosphonic acids when heated in the presence of aluminum chloride.[115]

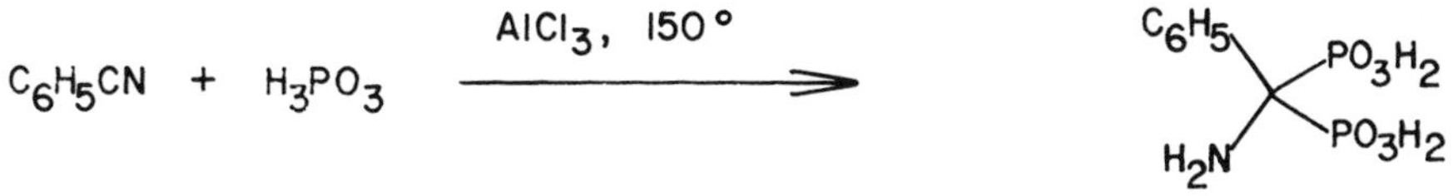

73 %

B. Comparison of Methods

Either the Abramov or Pudovik methods for the alkylation of trivalent phosphorus reagents may be used for the synthesis of phosphonates bearing α-heterosubstitution. The major advantage to be found in the use of the simply Pudovik approach and its variations is not in the yield of product, as comparable yields may be attained by either approach, but in the relative experimental simplicity of the procedure. In many instances, excellent yields may be attained with minimal preliminary syntheses or purifications.

In the preparation of materials which are to be used in carbonyl *umpolung* processes, variations of the Abramov approach wherein silylated linkages are involved would appear to be the most desirable methods for use. However, for the isolation of compounds bearing a free hydroxyl or free amino function at the position adjacent to the phosphorus, the use of partial esters of the phosphorus acid allows a minimum of experimental difficulties. Certainly for the preparation of analogues of the natural amino acids and related materials, the various methods related to the Pudovik approach are most feasible.

The use of the solid phase facilitating agents (alumina, alkali metal salts) exhibits particular experimental advantages as long as the other reagents form a fluid solution. If either the carbonyl compound or the phosphorus reagent remain as solids, however, the method is without value.

C. Experimental Procedures

Diethyl 1-Hydroxy-3-(pentylthio)propylphosphonate — Addition of a dialkyl phosphite to an aldehyde with basic catalysis.[24]

To *n*-amyl mercaptan (10.4 g, 0.10 mol) maintained under a nitrogen atmosphere was added a catalytic amount of triethylamine (0.5 g), and the mixture cooled to 20°C with an ice bath. To the cooled solution was added dropwise a solution of acrolein (5.6 g, 0.10 mol) in benzene (20 mℓ) maintaining the temperature at 15 to 20°C. After completion of the addition, the reaction mixture was stirred for an additional 20 min. To the reaction mixture was then added diethyl phosphite (14 g, 0.10 mol) along with a further catalytic amount of triethylamine (0.5 g). There was then added a solution of sodium ethoxide in ethanol (0.5 g, 1 *M*), whereupon a reaction exotherm resulted. Upon addition of a further catalytic amount of triethylamine (0.5 g), a second stronger reaction exotherm resulted. The mixture was stirred for 30 min after the exotherm had subsided. The solvent was removed by evaporation under reduced pressure and the residue was vacuum distilled in a molecular still (157 to 163°C/0.017 to 0.028 torr) to give pure diethyl 1-hydroxy-3-(pentyl-thio)propylphosphonic acid (23.9 g, 80%) which exhibited analytical data in accord with the proposed structure.

Methyl 2,3-*O*-Isopropylidene-5-deoxy-5-(dimethoxyphosphinoyl)hydroxymethyl-α-*D*-ribofuranoside — Reaction of an aldehyde with a dialkyl phosphite catalyzed by alkoxide.[13]

To a solution of methyl 5-deoxy-2,3-*O*-isopropylidene-α-D-ribo-hexodialdo-1,4-furano-side (2.8 g, 13 mmol) and dimethyl phosphite (1.65 g, 15 mmol) in benzene (10 mℓ) was added a 25% solution of sodium methoxide in methanol (5 drops), at which time an exotheric reaction began. The reaction was stirred for 3 hr at 40°C and an additional 12 hr at room temperature. Chloroform (100 mℓ) was added, and the solution washed with saturated, aqueous ammonium chloride until neutral and dried over sodium sulfate. The solvent was evaporated under reduced pressure to give a crude product that was purified by chromatography on a column (4 × 50 cm) of silica gel (200 g) which was eluted with 3:1 ethyl acetate-dichloromethane. The pure methyl 2,3-*O*-isopropylidene-5-deoxy-5-(dimethoxy-phosphinoyl)hydroxymethyl-α-D-ribofuranoside was isolated from the 1500 to 2250 mℓ fraction as an oil (3.0 g, 71%) which exhibited spectra and analytical data in accord with the proposed structure.

Diphenyl 1-(4-Nitrophenylamino)-3-phenyl-2-propenephosphonate — Reaction of a diaryl phosphite with an imine.[169]

A solution of cinnamaldehyde (13.2 g, 0.1 mol) and 4-nitroaniline (13.8 g, 0.1 mol) in benzene (300 mℓ) was gently refluxed. The water generated was removed from the reaction by means of a Dean-Stark trap. After the reaction system had cooled to room temperature, diphenyl phosphite (23.4 g, 0.1 mol) was added and the resulting mixture was refluxed for 1 hr. The solvent was removed by evaporation under reduced pressure, and the residual oil was dissolved in ether (300 mℓ) and cooled in the refrigerator until crystallization had occurred. The resulting precipitate was collected and recrystallized from ethyl acetate to give pure diphenyl 1-(4-nitrophenyl)-3-phenyl-2-propenephosphonate (20.2 g, 44%).

N-Phosphonomethylglycine — Condensation of formaldehyde, an amine, and diethyl phosphite with triethylamine catalysis.[23]

In a reaction vessel are mixed methanol (400 mℓ), paraformaldehyde (45 g, 1.5 mol), and a catalytic amount of triethylamine (5 mℓ). The mixture is heated at reflux until complete dissolution is attained. To the solution is then added glycine (75 g, 1 mol) followed by triethylamine (102 g, 1 mol). The temperature of the mixture is maintained at 60 to 70°C until complete dissolution is attained, and then diethyl phosphite (91 g, 0.66 mol) is added over a period of 10 min. The reaction mixture is stirred for 1.5 hr at 65 to 72° C. During cooling of the reaction mixture, sodium hydroxide (250 g) in the form of a 40% aqueous solution is added and the reaction mixture is then refluxed for another 1.5 hr. After this time the volatile organic materials are distilled and the remaining solution is acidified to a pH of 1.5 by the addition of concentrated hydrochloric acid. With cooling to 15°C the pure N-phosphonomethylglycine precipitates and is collected by filtration (71 g, 64%).

1-Aminomethylphosphonic Acid — Addition of diethyl phosphite to an imine salt formed by reduction of a nitrile.[194]

In a round-bottom flask equipped with a stirrer and protected against atmospheric moisture were placed dry diethyl ether (100 mℓ) and anhydrous tin(II) chloride (142 g, 0.75 mol). The mixture was chilled in an ice bath and saturated with dry hydrogen chloride. There was then added acetonitrile (2.05 g, 0.05 mol) and the mixture was stirred for 4 hr. After this time, the reaction mixture was allowed to stand for 4 days and the volatile materials were evaporated under reduced pressure. To the residue was then added diethyl phosphite (345 g, 2.5 mol) and the mixture was heated at 80 to 100°C for 3 hr. There was then added concentrated hydrochloric acid (100 mℓ) and the mixture stirred for 7 hr. The volatile materials were removed under reduced pressure and the residue was dissolved in dilute (1 N) hydrochloric acid. Gaseous hydrogen sulfide was passed through the solution at 60 to 70°C and the resultant precipitates removed by filtration. The solution was purified on an ion exchange column (Dowex® 50 W X-1) and the solvent removed from the eluent to give the pure 1-aminoethylphosphonic acid (2.95 g, 55%).

Table 3
PUDOVIK REACTION OF CARBONYL COMPOUNDS

Product	Phosphorus reagent	Method	Yield (%)	Refs.
[structure: polyhydroxy phosphonic acid, PO_3H_2 with OH, HO, OH, OH, OH]	$(C_2H_5O)_2POH$	A	63	10
$(CH_3)_2CHCH(OH)P(O)(OC_2H_5)_2$	$(C_2H_5O)_2POH$	A	34	10
$(CH_3O)_2P(O)CH(OH)CO_2C_4H_9\text{-}n$	$(C_2H_5O)_2POH$	B	66	153
$CH_3CH(CH_2Cl)CH(OH)P(O)(OC_2H_5)_2$	$(CH_3O)_2POH$	A	70	28
$(C_2H_5O)_2P(O)C(OH)(CH_3)CH(Cl)CH_3$	$(C_2H_5O)_2POH$	B	91	153
$(n\text{-}C_4H_9O)_2P(O)CH_2OH$	$(C_2H_5O)_2POH$	C	72	152
	$(n\text{-}C_4H_9O)_2POH$	D	96	138
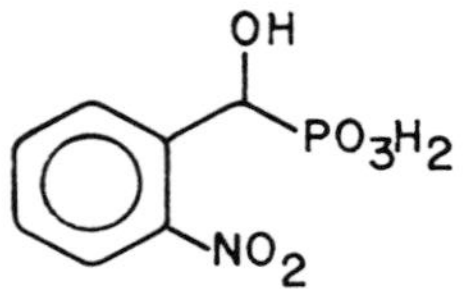 [structure: 2-nitrophenyl-CH(OH)PO_3H_2]	$(CH_3O)_2POH$	D	100	141
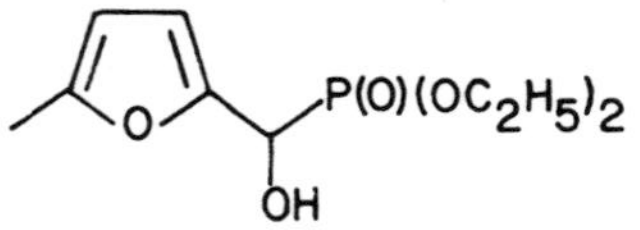 [structure: 5-methylfuran-2-yl-CH(OH)P(O)(OC$_2$H$_5$)$_2$]	$(C_2H_5O)_2POH$	A	100	125

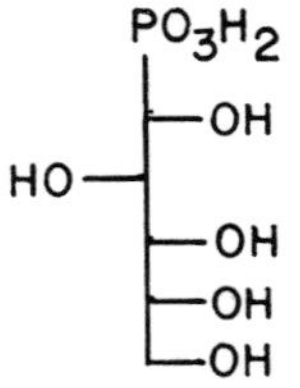

Structure	Reagent	Method	Yield (%)	
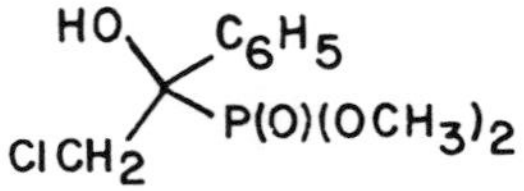 HO–C(C$_6$H$_5$)(ClCH$_2$)–P(O)(OCH$_3$)$_2$	(CH$_3$O)$_2$POH	D	80	137
C$_6$H$_5$CH(OH)P(O)(OC$_2$H$_5$)$_2$	(C$_2$H$_5$O)$_2$POH	B	98	153
C$_6$H$_5$CH=CHCH(OH)P(O)(OCH$_3$)$_2$	(CH$_3$O)$_2$POH	A	70	127
(CH$_3$O)$_2$P=O	(CH$_3$O)$_2$POH	D	78	134
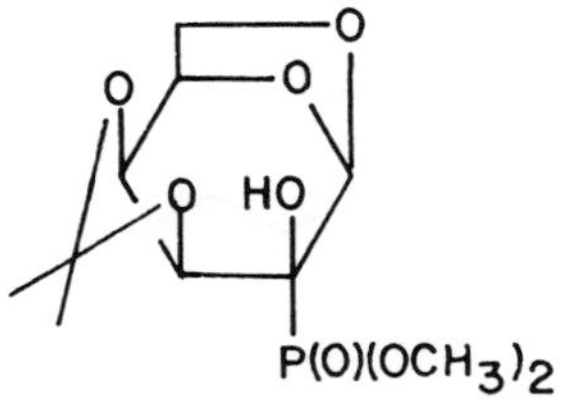 P(O)(OCH$_3$)$_2$	(CH$_3$O)$_2$POH	D	90	134
(CF$_3$)$_2$C(OH)P(O)(OC$_4$H$_9$-n)$_2$	(n-C$_4$H$_9$O)$_2$POH	E	95	147
	(C$_2$H$_5$O)$_2$POH	B	92	153

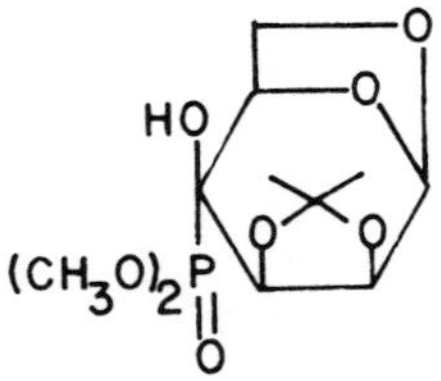

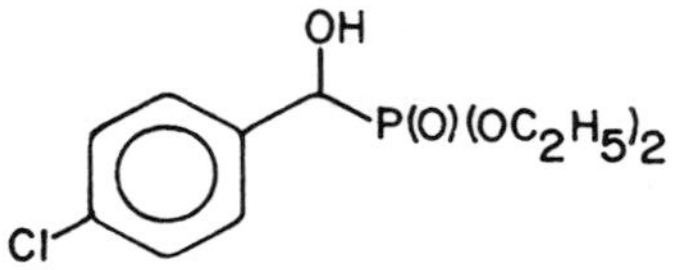

Table 3 (continued)
PUDOVIK REACTION OF CARBONYL COMPOUNDS

Product	Phosphorus reagent	Method	Yield (%)	Refs.
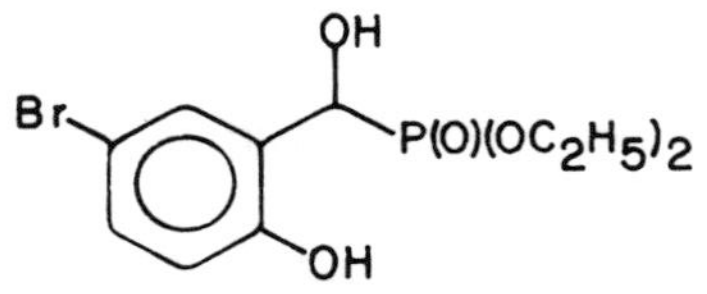	$(C_2H_5O)_2POH$	D	43	136
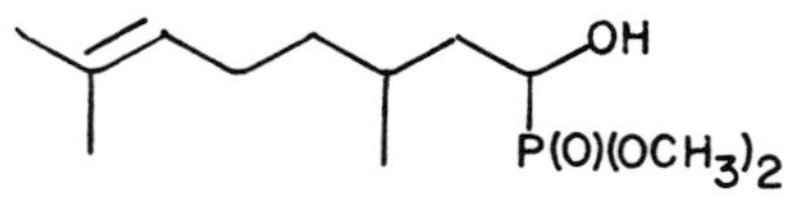	$(CH_3O)_2POH$	C	48	152
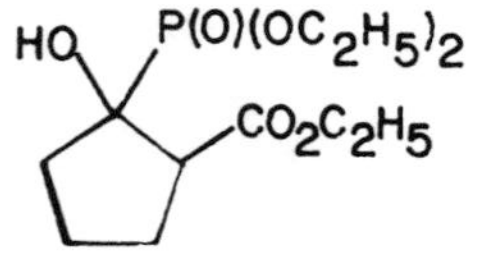	$(C_2H_5O)_2POH$	D	57	143
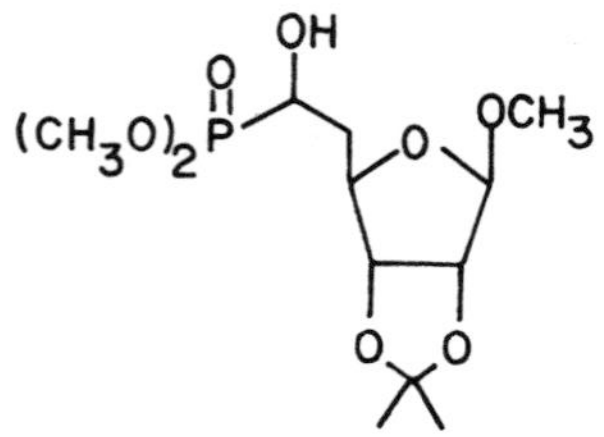	$(CH_3O)_2POH$	A	71	13

Structure	Reagent	Method	Yield	Ref
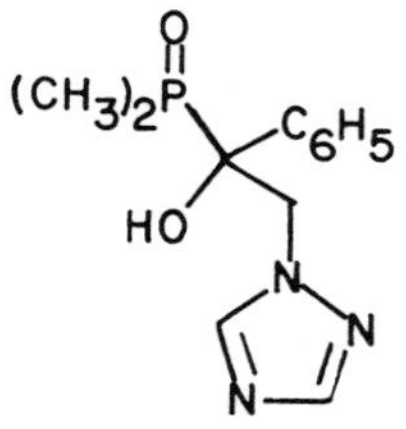 (CH₃)₂P(O)C(C₆H₅)(OH)CH₂-imidazole	$(CH_3)_2POH$	E	90	31
$n\text{-}C_5H_{11}SCH_2CH_2CH(OH)P(O)(OC_2H_5)_2$	$(C_2H_5O)_2POH$	A	80	24
$C_6H_5CH(OH)P(O)(OC_3H_7\text{-}i)_2$	$(i\text{-}C_3H_7O)_2POH$	C	90	152
$C_6H_5CH=CHCH(OH)P(O)(OC_2H_5)_2$	$(C_2H_5O)_2POH$	B	94	153
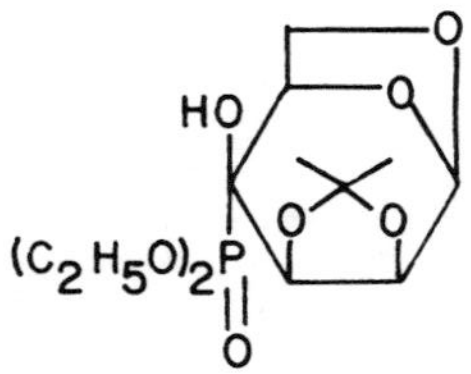 (bicyclic sugar structure with $(C_2H_5O)_2P(O)$, HO)	$(C_2H_5O)_2POH$	D	77	134
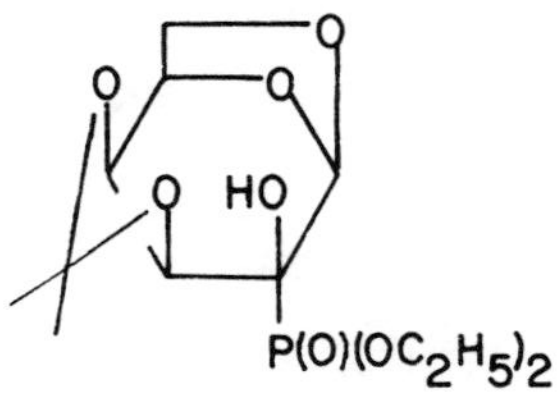 (bicyclic sugar structure with $P(O)(OC_2H_5)_2$, HO)	$(C_2H_5O)_2POH$	D	78	134

Table 3 (continued)
PUDOVIK REACTION OF CARBONYL COMPOUNDS

Product	Phosphorus reagent	Method	Yield (%)	Refs.
	$(C_2H_5O)_2POH$	B	80	151
	$(CH_3O)_2POH$	D	64	133
	$(CH_3O)_2POH$	A	24	12

Structure	Reagent			
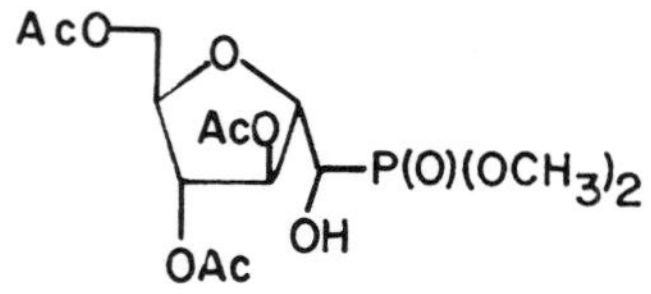 AcO, AcO, OAc, O, P(O)(OCH₃)₂, OH	(CH₃O)₂POH	F	60	150
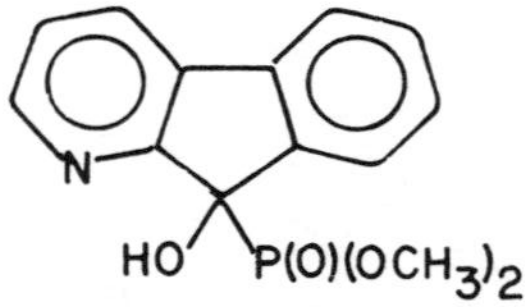 HO, P(O)(OCH₃)₂ (fluorenyl, N)	(CH₃O)₂POH	D	90	135
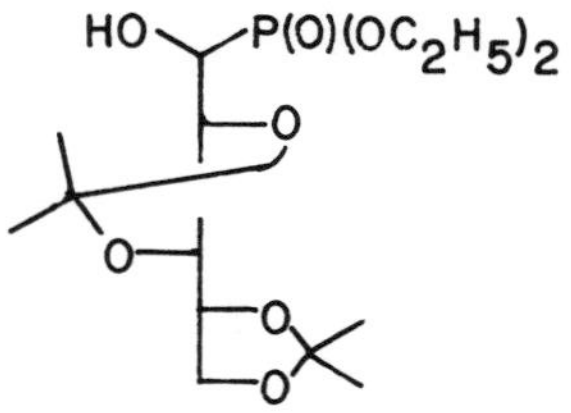 HO, P(O)(OC₂H₅)₂, O, O	(C₂H₅O)₂POH	A	70	11
(C₆H₅)₂P(O)C(CH₃)₂OH	(C₆H₅)₂POH	G	92	132
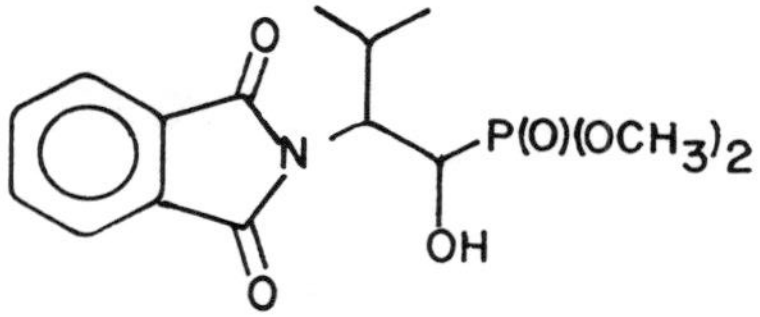 P(O)(OCH₃)₂, OH, N, O, O	(CH₃O)₂POH	D	69	133

Table 3 (continued)
PUDOVIK REACTION OF CARBONYL COMPOUNDS

Product	Phosphorus reagent	Method	Yield (%)	Refs.
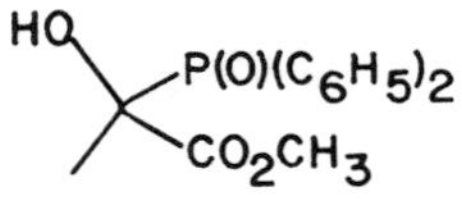	$(C_6H_5)_2POH$	E	51	195
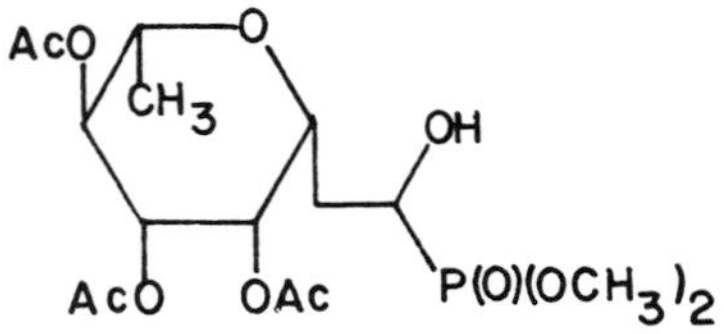	$(CH_3O)_2POH$	F	85	148
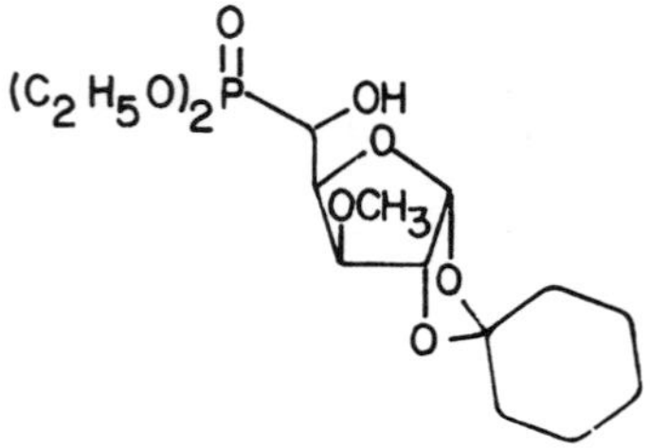	$(C_2H_5O)_2POH$	F	83	150

Structure	Reagent		Yield	m.p.
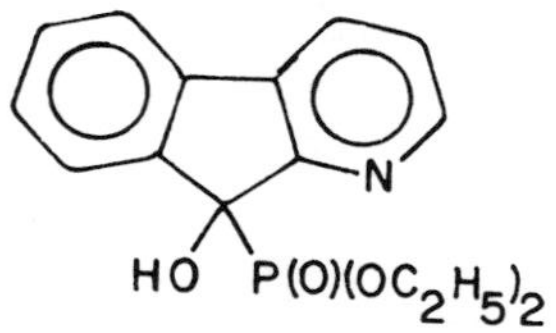 HO $P(O)(OC_2H_5)_2$ (fluorene/indole system)	$(C_2H_5O)_2POH$	D	40	135
$(C_6H_5CH_2)_2C(OH)P(O)(OCH_3)_2$	$(CH_3O)_2POH$	D	98	143
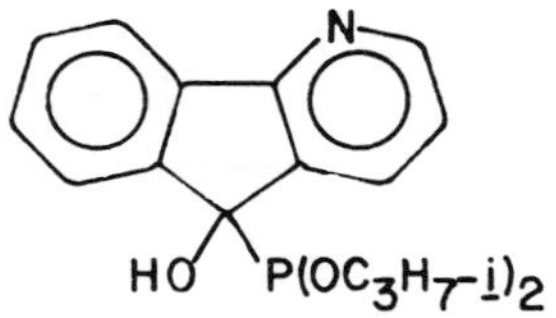 HO $P(OC_3H_7-i)_2$ (fluorene/indole system)	$(i\text{-}C_3H_7O)_2POH$	D	63	135
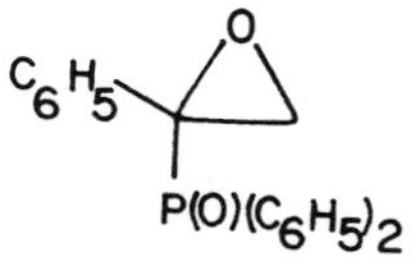 C_6H_5 epoxide $P(O)(C_6H_5)_2$	$(C_6H_5)_2POH$	D	83	145
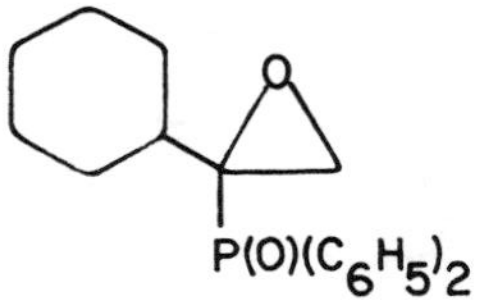 cyclohexyl epoxide $P(O)(C_6H_5)_2$	$(C_6H_5)_2POH$	D	80	145
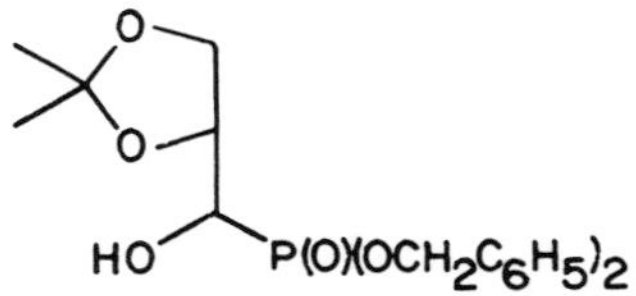 dioxolane $P(O)(OCH_2C_6H_5)_2$, HO	$(C_6H_5CH_2O)_2POH$	D	46	9

Table 3 (continued)
PUDOVIK REACTION OF CARBONYL COMPOUNDS

Product	Phosphorus reagent	Method	Yield (%)	Refs.
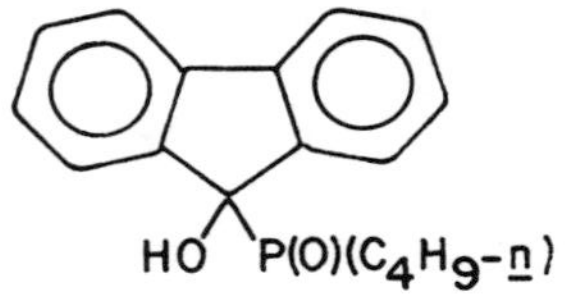 (fluorene) HO–C–P(O)[C$_4$H$_9$-n]$_2$	(n-C$_4$H$_9$)$_2$POH	D	70	140
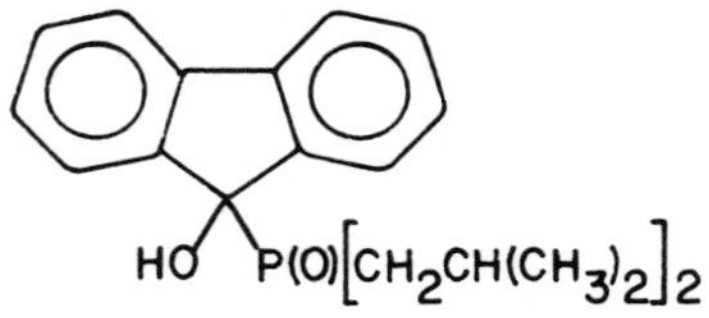 (fluorene) HO–C–P(O)[CH$_2$CH(CH$_3$)$_2$]$_2$	(i-C$_4$H$_9$)$_2$POH	D	60	140
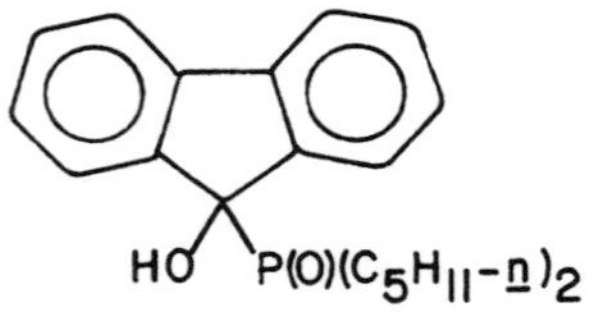 (fluorene) HO–C–P(O)(C$_5$H$_{11}$-n)$_2$	(n-C$_5$H$_{11}$)$_2$POH	D	76	140
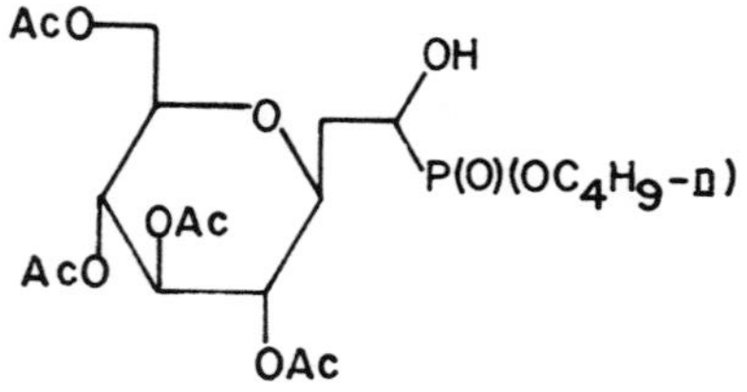 (acetylated sugar) with OH, P(O)(OC$_4$H$_9$-n), OAc, AcO, AcO, OAc	(n-C$_4$H$_9$O)$_2$POH	F	80	148

C$_2$H$_5$OP(H)OH

A   38   129

Note: A, alkoxide base used; B, KF or CsF used as facilitating reagent; C, alumina used as facilitating reagent; D, amine used as base; E, no base used; F, phase transfer catalysis used; G, Grignard form of phosphorus reagent used.

Table 4
PUDOVIK REACTION OF IMINES

Product	Phosphorus reagent	Method	Yield (%)	Refs.
$H_2O_3PCH_2NHCH_2CO_2H$	$(C_2H_5O)_2POH$	A	64	23
		A	96	166
$n\text{-}C_4H_9CH(NH_2)PO_3H_2$	$(C_2H_5O)_2POH$	A	60	159
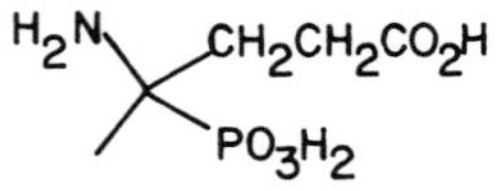 $H_2NC(CH_3)(PO_3H_2)CH_2CH_2CO_2H$	$(C_2H_5O)_2POH$	B	25	185
$C_2H_5CH(NH_2)PO_3H_2$	$(C_2H_5O)_2POH$	A	65	162
$C_6H_5CH_2CH(NH_2)PO_3H_2$	$(C_2H_5O)_2POH$	A	71	194
$(C_2H_5O)_2P(O)CH_2NHCH_2CO_2H$	$(C_2H_5O)_2POH$	C	88	191
$CH_3COCH(NHOH)P(O)(OC_2H_5)_2$	$(C_2H_5O)_2POH$	C	54	177
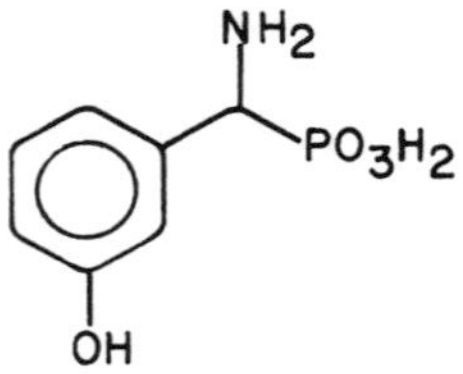 $(3\text{-}HOC_6H_4)CH(NH_2)PO_3H_2$	$(C_2H_5O)_2POH$	A	69	159
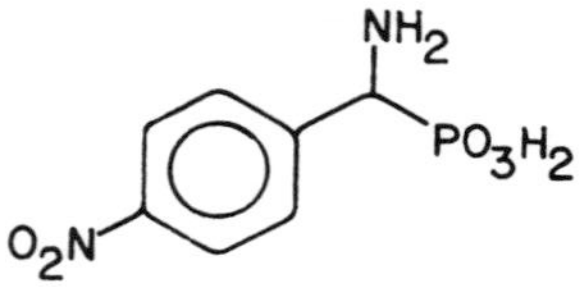 $(4\text{-}O_2NC_6H_4)CH(NH_2)PO_3H_2$	$(C_2H_5O)_2POH$	A	54	159

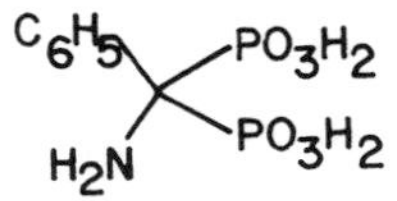

Structure	Reagent	Method		
C$_6$H$_5$C(PO$_3$H$_2$)(PO$_3$H$_2$)(NH$_2$)	H$_3$PO$_3$	D	73	115
C$_6$H$_5$CH(NH$_2$)P(O)(CH$_3$)OH	CH$_3$P(OC$_2$H$_5$)OH	A	73	161
(pyrrolidin-2-yl)P(O)(OC$_2$H$_5$)$_2$	(C$_2$H$_5$O)$_2$POH	E	52	26
(CH$_3$O)$_2$P(O)CH$_2$NHCH$_2$CH$_2$NHCH$_2$P(O)(OCH$_3$)$_2$	(CH$_3$O)$_2$POH	F	45	186
(C$_2$H$_5$O)$_2$P(O)CH(C$_2$H$_5$)NHC$_2$H$_5$	(C$_2$H$_5$O)$_2$POH	C	89	155
(CH$_3$)$_2$C(NHC$_2$H$_5$)P(O)(OC$_2$H$_5$)$_2$	(C$_2$H$_5$O)$_2$POH	C	97	155
1-(CH$_2$CO$_2$C$_2$H$_5$)-phospholane 2-(O)(OC$_2$H$_5$)	ClCH$_2$CH$_2$P(OC$_2$H$_5$)OH	E	27	179
C$_2$H$_5$CH(NHCH(C$_6$H$_5$)PO$_3$H$_2$	(C$_2$H$_5$O)$_2$POH	A	55	156
(piperidin-1-yl)CH$_2$P(O)(OC$_2$H$_5$)$_2$	(C$_2$H$_5$O)$_2$POH	F	87	155

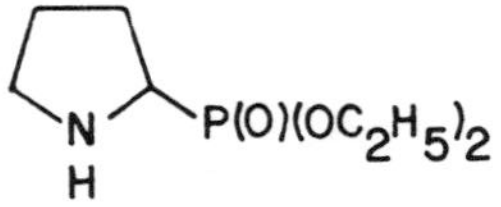

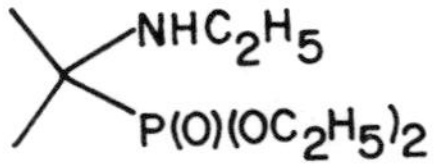

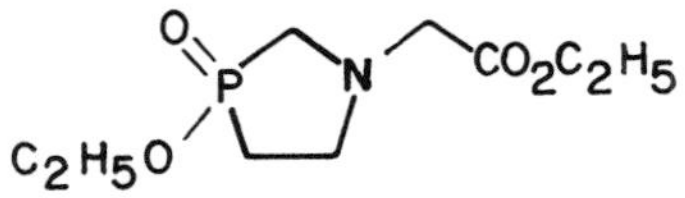

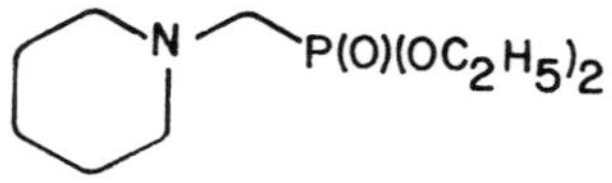

Table 4 (continued)
PUDOVIK REACTION OF IMINES

Product	Phosphorus reagent	Method	Yield (%)	Refs.
(PO$_3$H$_2$, NH–CO$_2$CH$_3$, NO$_2$)	(4-CH$_3$C$_6$H$_4$CH$_2$O)$_2$POH	A	67	158
(NH$_2$, PO$_3$H$_2$ naphthalene)	(C$_2$H$_5$O)$_2$POH	A	66	159
C$_6$H$_5$CH(NH$_2$)P(O)(OC$_2$H$_5$)$_2$	(C$_2$H$_5$O)$_2$POH	C	96	181
n-C$_4$H$_9$CH(NHCH$_2$C$_6$H$_5$)PO$_3$H$_2$	(C$_2$H$_5$O)$_2$POH	A	65	156
(C$_2$H$_5$O)$_2$P(O)C(CH$_3$)$_2$N(C$_2$H$_5$)$_2$	(C$_2$H$_5$O)$_2$POH	F	64	155
(NHCH$_3$, P(O)(OCH$_3$)$_2$)	(C$_2$H$_5$O)$_2$POH	C	74	176
[(CH$_3$)$_2$CHCH$_2$O]$_2$P(O)CH$_2$NHCH$_2$CO$_2$H	[(CH$_3$)$_2$CHCH$_2$O]$_2$POH	E	94	18

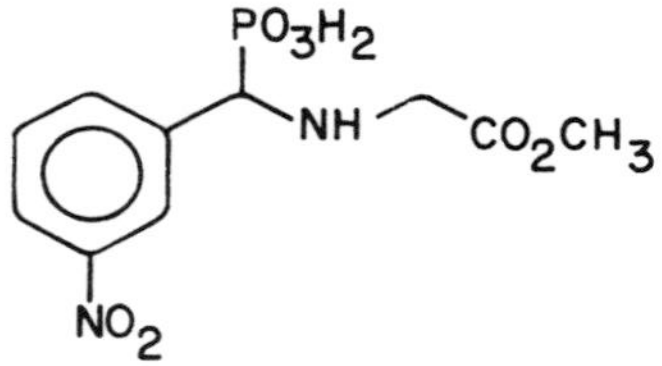

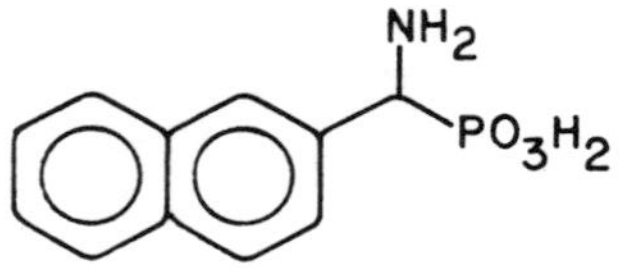

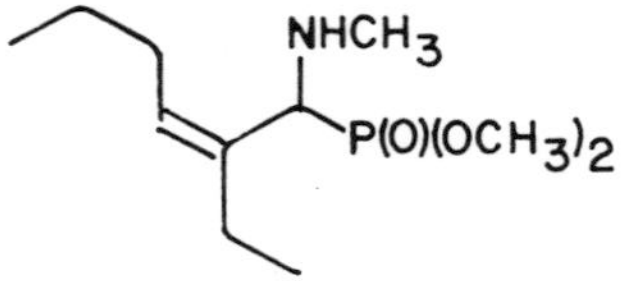

Product	Reagent			
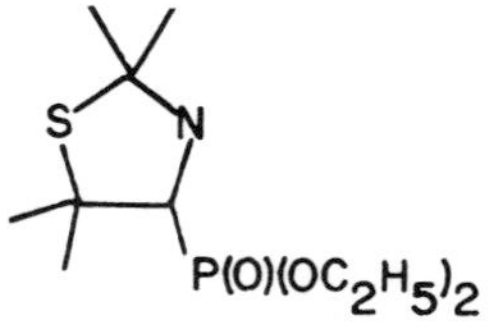 (thiazolidine)—P(O)(OC$_2$H$_5$)$_2$	(C$_2$H$_5$O)$_2$POH	C	79	25
C$_6$H$_5$CH$_2$N(CH$_2$CO$_2$H)CH$_2$P(O)(CCl$_3$)OH	CCl$_3$PO$_2$H$_2$	F	71	22
n-C$_4$H$_9$CH(NHC$_4$H$_9$-n)P(O)(OC$_2$H$_5$)$_2$	(C$_2$H$_5$O)$_2$POH	C	94	155
n-C$_3$H$_7$CH[N(C$_2$H$_5$)$_2$]P(O)(OC$_2$H$_5$)$_2$	(C$_2$H$_5$O)$_2$POH	F	79	155
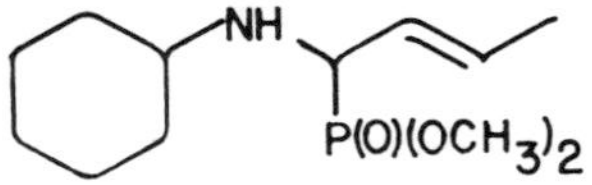 (cyclohexyl-NH)—P(O)(OCH$_3$)$_2$	(C$_2$H$_5$O)$_2$POH	C	92	176
C$_6$H$_5$CH$_2$NHCH$_2$P(O)(OC$_2$H$_5$)$_2$	(C$_2$H$_5$O)$_2$POH	E	68	180
C$_6$H$_5$CH$_2$N(CH$_2$CO$_2$H)CH$_2$P(O)(C$_2$H$_5$)OH	C$_2$H$_5$PO$_2$H$_2$	F	83	22
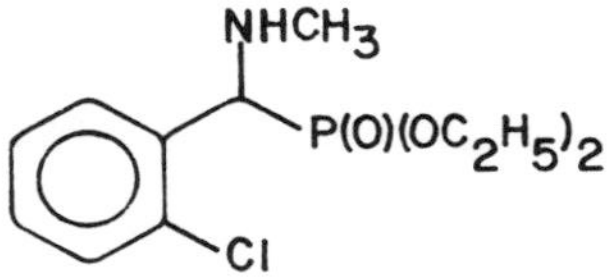 NHCH$_3$ / (2-Cl-phenyl)CH—P(O)(OC$_2$H$_5$)$_2$	(C$_2$H$_5$O)$_2$POH	C	88	155
n-C$_3$H$_7$CH(NHCH$_2$C$_6$H$_5$)P(O)(C$_2$H$_5$)OH	C$_2$H$_5$P(OC$_2$H$_5$)OH	A	41	156
C$_6$H$_5$CH(NH$_2$)P(O)(C$_6$H$_5$)OH	C$_6$H$_5$P(OC$_2$H$_5$)OH	A	81	161
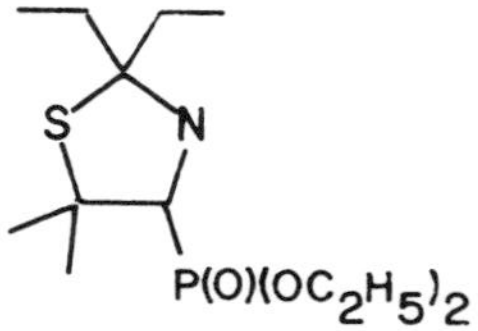 (2,2-diethyl-thiazolidine)—P(O)(OC$_2$H$_5$)$_2$	(C$_2$H$_5$O)$_2$POH	C	86	25

Table 4 (continued)
PUDOVIK REACTION OF IMINES

Product	Phosphorus reagent	Method	Yield (%)	Refs.
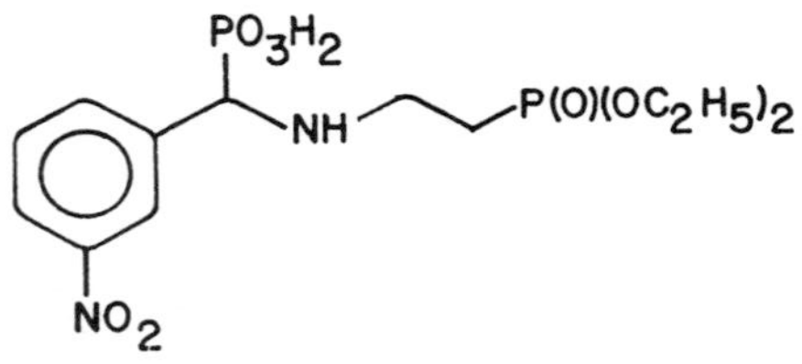	$(4\text{-}CH_3C_6H_4CH_2O)_2POH$	A	79	158
$C_6H_5CH(NHCH_2C_6H_5)PO_3H_2$	$(C_2H_5O)_2POH$	A	70	156
$C_6H_5CH_2N(CH_2CN)CH_2P(O)(OC_2H_5)_2$	$(C_2H_5O)_2POH$	C	54	20
$C_6H_5CH(NH_2)P(O)(OC_2H_5)C_6H_5$	$C_6H_5P(OC_2H_5)OH$	A	85	182
$(i\text{-}C_3H_7O)_2P(O)C(CH_3)_2NHP(O)(OC_3H_7\text{-}i)_2$	$(i\text{-}C_3H_7O)_2PONa$	C	14	184
$CH_2=CHCH_2NHCH_2P(O)(OC_6H_5)_2$	$(C_6H_5O)_2POH$	C	98	17
$C_6H_5CH_2N(CH_2CO_2H)CH_2P(O)(C_6H_5)OH$	$C_6H_5PO_2H_2$	F	53	22
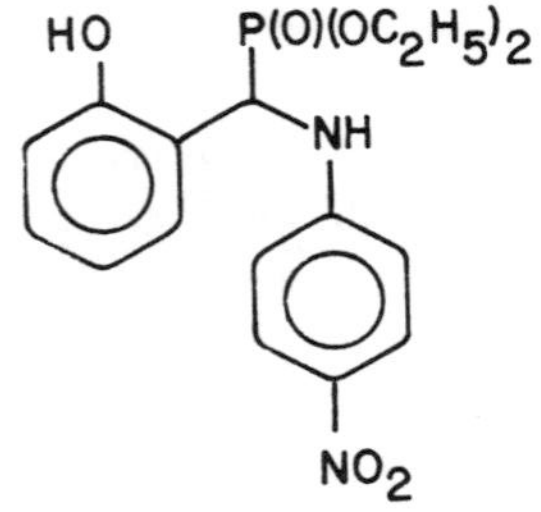	$(C_2H_5O)_2POH$	C	87	174
$n\text{-}C_4H_9CH(NHCH_2C_6H_5)P(O)(C_6H_5)OH$	$C_6H_5P(OC_2H_5)OH$	A	35	156

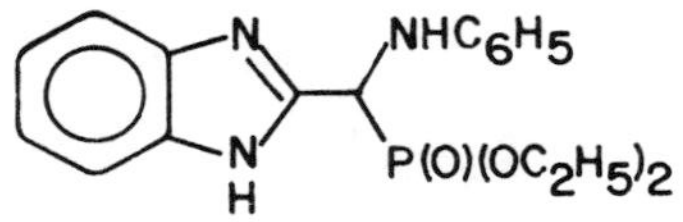

Structure	Reagent			
NHC$_6$H$_5$ / P(O)(OC$_2$H$_5$)$_2$ benzimidazole	(C$_2$H$_5$O)$_2$POH	F	68	187
fluorene, H$_2$N, P(O)(C$_3$H$_7$-i)	(i-C$_3$H$_7$)$_2$POH	C	63	140
fluorene, n-C$_4$H$_9$NH, P(O)(OC$_2$H$_5$)$_2$	(C$_2$H$_5$O)$_2$POH	C	10	135
fluorene, H$_2$N, P(O)(C$_4$H$_9$-n)$_2$	(n-C$_4$H$_9$)$_2$POH	C	60	140
fluorene, n-C$_4$H$_9$NH, P(O)(OC$_2$H$_5$)$_2$	(C$_2$H$_5$O)$_2$POH	C	59	165

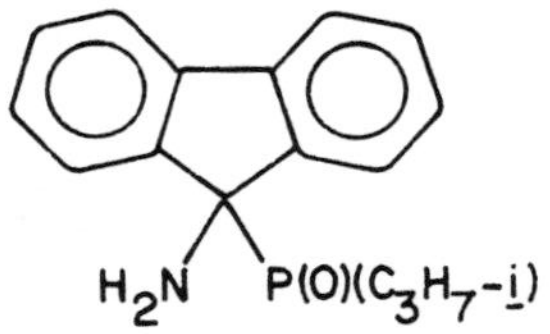

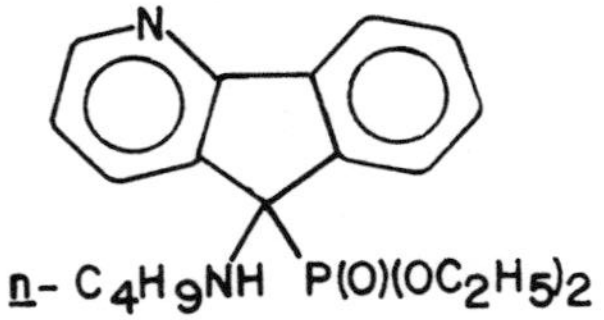

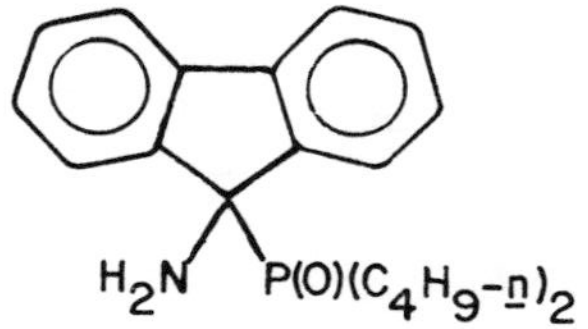

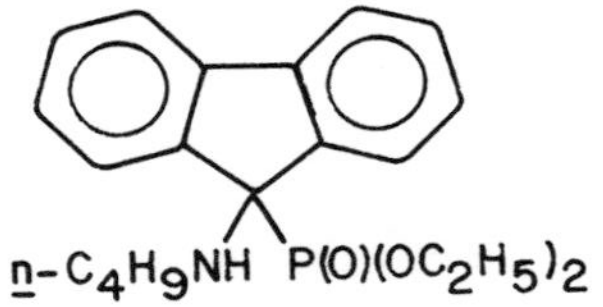

Table 4 (continued)
PUDOVIK REACTION OF IMINES

Product	Phosphorus reagent	Method	Yield (%)	Refs.
(2-nitrofluorenyl structure) $\underline{n}$-C$_4$H$_9$NH P(O)(OC$_2$H$_5$)$_2$	(C$_2$H$_5$O)$_2$POH	C	65	164
C$_6$H$_5$CH(NHCH$_2$C$_6$H$_5$)P(O)(C$_6$H$_5$)OH	C$_6$H$_5$P(OC$_2$H$_5$)OH	A	58	156
(cyclopentyl–C$_6$H$_5$, NH, P(O)(OC$_2$H$_5$)$_2$ structure)	(C$_2$H$_5$O)$_2$POH	C	83	160
(C$_6$H$_5$O)$_2$P(O)CH$_2$N(COCH$_2$CO$_2$C$_2$H$_5$)CH$_2$CO$_2$C$_2$H$_5$ (4-CH$_3$OC$_6$H$_4$O)$_2$P(O)CH$_2$N(CH$_2$CO$_2$C$_2$H$_5$)CON(CH$_3$)$_2$	(C$_6$H$_5$O)$_2$POH (4-CH$_3$OC$_6$H$_4$O)$_2$POH	E E	10 27	21 21
(structure with CO$_2$C$_2$H$_5$, POC$_6$H$_5$, OH, NH, HN, C$_6$H$_5$OP, CO$_2$C$_2$H$_5$)	(C$_6$H$_5$O)$_2$POH	F	62	19

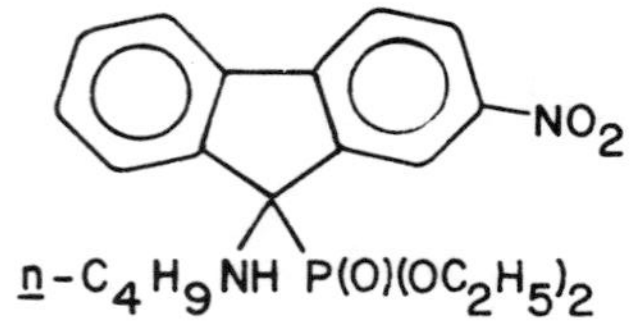

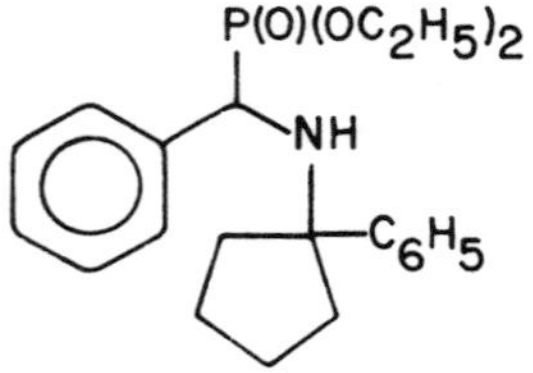

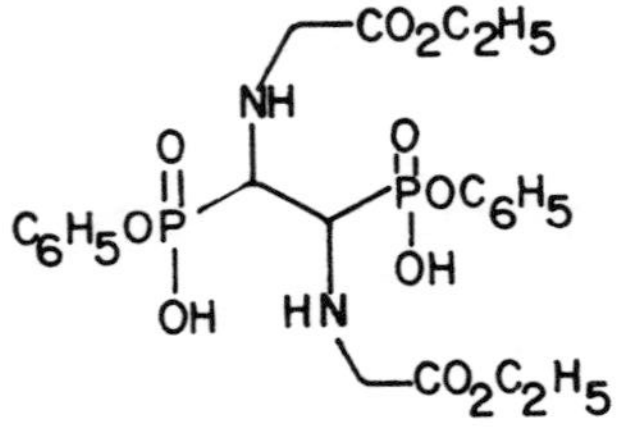

Structure	Reagent			
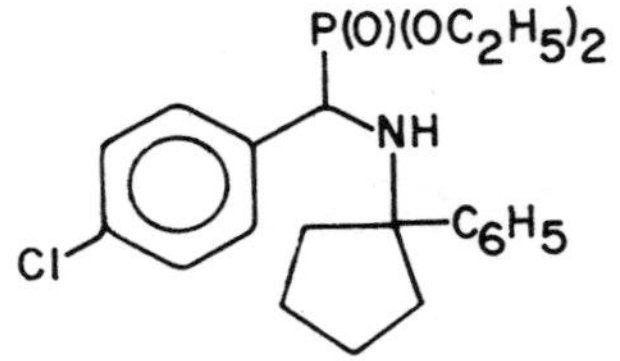 ![4-chlorophenyl derivative] P(O)(OC₂H₅)₂, NH, C₆H₅, Cl	(C₂H₅O)₂POH	C	90	160
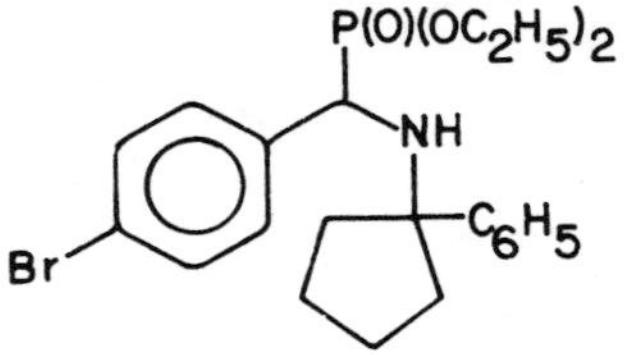 ![4-bromophenyl derivative] P(O)(OC₂H₅)₂, NH, C₆H₅, Br	(C₂H₅O)₂POH	C	64	160
[(CH₃)₂N]₂P(O)C(C₆H₅)₂NHC₆H₅	[(CH₃)₂N]₂POH	C	62	172
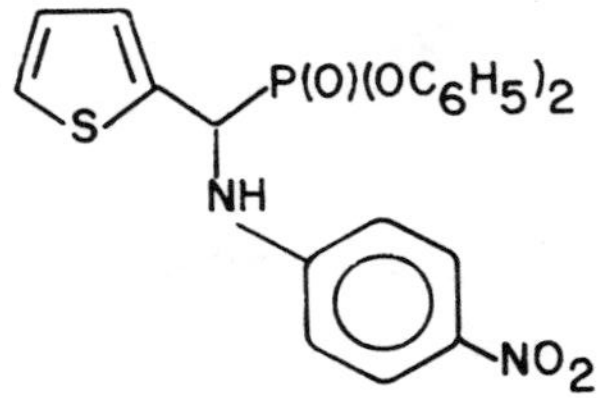 ![thiophene NO₂ derivative] P(O)(OC₆H₅)₂, NH, S, NO₂	(C₆H₅O)₂POH	F	75	171
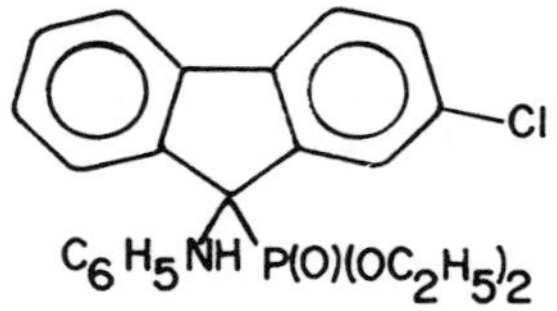 ![fluorene Cl derivative] C₆H₅NH P(O)(OC₂H₅)₂, Cl	(C₂H₅O)₂POH	C	60	164

Table 4 (continued)
PUDOVIK REACTION OF IMINES

Product	Phosphorus reagent	Method	Yield (%)	Refs.
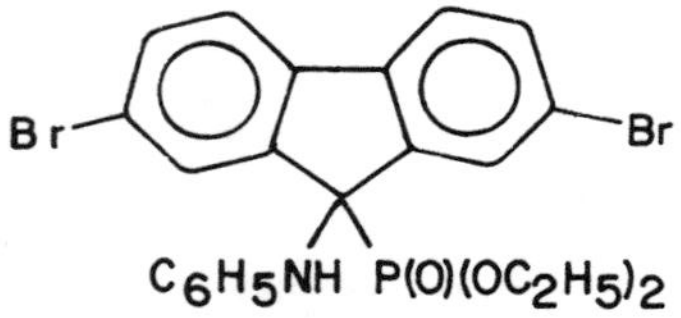	$(C_2H_5O)_2POH$	C	60	164
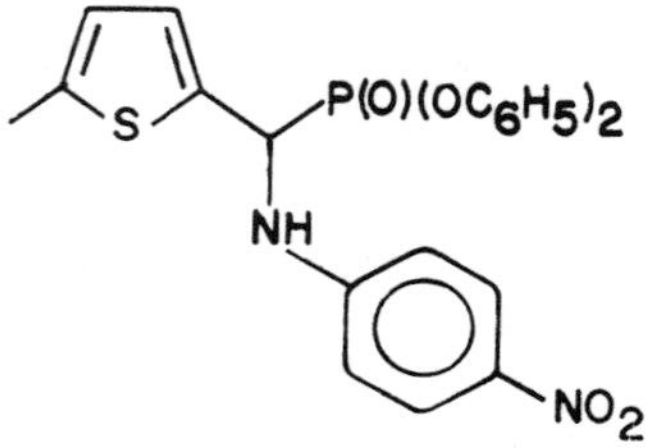	$(C_6H_5O)_2POH$	F	78	171
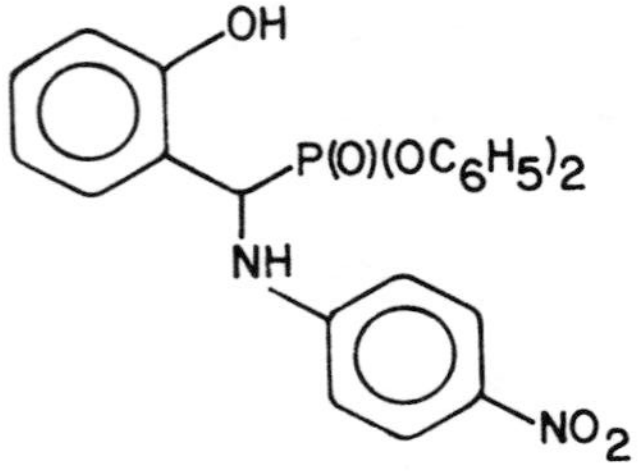	$(C_6H_5O)_2POH$	C	91	174

$(C_6H_5O)_2POH$	F	84	171
$[(C_2H_5)_2N]_2POH$	C	68	172

$[(C_2H_5)_2N]_2P(O)OC(C_6H_5)_2NHC_6H_5$

Note: A, preformed imine used; ester linkages cleaved in work-up; B, imine formed *in situ*; ester linkages cleaved in work-up; C, preformed imine used; D, nitrile used as substrate; E, triazine used as substrate; F, imine formed *in situ*.

REFERENCES

1. **Borowitz, I. J., Firstenberg, S., Borowitz, G. B., and Schuessler, D.,** On the kinetics and mechanism of the Perkow reaction. *J. Am. Chem. Soc.,* 94, 1623, 1972.
2. **Gaydou, E. M. and Bianchini, J.-P.,** Etude du mechanisme de la formation de phosphates d'enols a partir de composes carbonyles α-halogenes et de triacoylphosphites, *Can. J. Chem.,* 54, 3626, 1976.
3. **Chopard, P. A., Clark, V. M., Hudson, R. F., and Kirby, A. J.,** A short synthesis of several gambir alkaloids, *Tetrahedron,* 26, 1961, 1970.
4. **Sturtz, G.,** Action des phosphites sodes sur les cetones-halogenees, *Bull. Soc. Chim. Fr.,* p. 2333, 1964.
5. **Sarin, V., Tropp, B. E., and Engel, E.,** Isosteres of natural phosphates. 7. The preparation of 5-carboxy-4-hydroxy-4-methylpentyl-1-phosphonic acid, *Tetrahedron Lett.,* p. 351, 1977.
6. **Seebach, D.,** Methods of reactivity umpolung, *Angew. Chem. Int. Ed. Engl.,* 18, 239, 1979.
7. **Sturtz, G., Corbel, B., and Pangam, J.-P.,** Nouveaux synthons phosphores: bianions D'hydroxy-1-propen-2-ylphosphonamides, carbanions et α-d'acides carboxyliques potentiels, *Tetrahedron Lett.,* p. 47, 1976.
8. **Sekine, M., Mori, H., and Hata, T.,** Protection of phosphonate function by means of ethoxycarbonyl group. A new method for generation of reactive silyl phosphite intermediates, *Bull. Chem. Soc. Jpn.,* 55, 239, 1982.
9. **Adams, P. R, Harrison, R., and Inch, T. D.,** Dehydrogenation of a phosphonate substrate analogue by glycerol 3-phosphate dehydrogenase, *Biochem. J.,* 141, 729, 1974.
10. **Paulsen, H. and Kuhne, H.,** Synthesis of (1*S*)-dialkyl-D-arabitol-1-phosphonate and its derivatives, *Chem. Ber.,* 107, 2635, 1974.
11. **Paulsen, H. and Kuhne, H.,** Synthesis of α-hydroxy- and α-amino-phosphonates of acyclic monosaccharides, *Chem. Ber.,* 108, 1239, 1975.
12. **Evelyn, L., Hall, L. D., Lynn, L., Steiner, P. R., and Stokes, D. H.,** Some 3-*C*-(dimethoxy)phosphinyl derivatives of D-glucose, D-allose, and D-ribose, *Carbohydr. Res.,* 27, 21 1973.
13. **Mlotkowska, B., Tropp, B. E., and Engel, R.,** The preparation of methyl 5-deoxy-5-(dihydroxyphosphinoyl)hydroxymethyl-2,3-*O*-isopropylidene-β-D-ribofuranoside, a precursor to a hydroxymethylene analog of D-Ribose 5-Phosphate, *Carbohydr. Res.,* 117, 95, 1985.
14. **Glamkowskii, E. J., Gal, G., Purick, R., Davidson, A. J., and Sletzinger, M.,** A new synthesis of the antibiotic phosphonomycin, *J. Org. Chem.,* 35, 3510, 1970.
15. **Takaya, T. and Chiba, T.,** 7-Amino-3-phosphonocephalosporanic Acid Compounds, U.S. Patent 4, 399, 277, 1983.
16. **Takaya, T. and Chiba, T.,** 3-Phosphonocephalosporanic Acid Derivatives, and Pharmaceutical Composition Comprising the Same, U.S. Patent 4,291,031, 1981.
17. **Singh, R. K.,** *N*-Substituted *N*-(Phosphonomethyl)aminoethanal Derivatives as Herbicides, U.S. Patent 4,444,580, 1984.
18. **Issleib, K., Balszuweit, A., Wetzke, G., Moegelin, W., Kochmann, W., and Guenther, E.,** *N*-Phosphonomethylglycine and Its Diester, East German Patent 141,929, 1980.
19. Otsuka Chemical Co., Ltd., Phenyl 1,2-Di(alkoxycarbonylmethylamino)ethylene-1,2-diphosphonates, Japanese Patent 81 55,349, 1981.
20. **Felix, R. A.,** *O,O,*- Dialkyl-*N*-(benzyl or *t*-butyl)-*N*-cyanomethylaminomethylphosphonates, U.S. Patent 4,525,311, 1985.
21. **Dutra, G. A.,** Herbicidal *N*- Substituted Triesters of *N*-Phosphonomethylglycine, U.S. Patent 4,261,727, 1981.
22. **Maier, L.,** Herstellung und Eigenschaften von (*N*-Hydroxycarbonylmethyl-aminomethyl)alkyl- und aryl-phosphinsauren, $(HO_2CH_2NHCH_2)RP(O)OH$, und Derivaten, *Phosphorus and Sulfur,* 11, 139, 1981.
23. **Ehrat, R.,** Process for the Preparation of *N*- Phosphonomethyl Glycine, U.S. Patent 4,237,065, 1980.
24. **Gaertner, V. R.,** Herbicidal α-Hydroxy Phosphonates, U.S. Patent 4,475,943, 1984.
25. **Drauz, K., Koban, H. G., Martens, J., and Schwarze, W.,** Process for the Production of Substituted Phosphonic and Phosphinic Acids and Thiazolidinyl Phosphonic and Phosphinic Acid Ester Intermediates, U.S. Patent 4,524,211, 1985.
26. **Petrillo, E. W.,** Mercaptoacylpyrrolidine Phosphonic Acids and Related Compounds, U.S. Patent 4,186,268, 1980.
27. **Bentzen, C. L., Nguyen, L. M., and Niesor, E.,** Diphosphonate and phosphonophosphate Derivatives and Pharmaceutical Compositions Containing Them, United Kingdom Patent 2,079,285B, 1984.
28. **Lieb, F., Oediger, H., and Streible, G.,** Phosphono-hydroxy-acetic Acid and Its Salts, Their Production and Their Medicinal Use, U.S. Patent 4,340,599, 1982.
29. **Lieb, F., Oediger, H., and Streible, G.,** Phosphonohydroxyacetic Acids and Their Pharmaceutical Use, West German Patent application 2,941,384, 1981.
30. **Schulz, P., Vilceanu, R., Kuruneyi, L., and Suhateanu, T.,** Procedeu de sinteza a insecticidului O,O'-dimetil-(1-hidroxi-2,2,2-trichloretil)fosfonat, Romanian Patent 65876, 1973.

31. **Lindner, E., Schilling, B., and Varughese, K. A.,** Tertiary Ethylphosphine Oxides and Sulfides and Their Fungicidal Use, West German Patent 3,344,663, 1985.

32. **Sekine, M., Futatsugi, T., and Hata, T.,** A new method for the synthesis of L-ascorbic acid 2-O-phosphate by utilizing phosphoryl rearrangement, *J. Org. Chem.,* 47, 3453, 1982.

33. **Mikroyannidis, J. A.,** Synthesis of substituted N-[(phosphonyl)methyl]-2-imidazolidinones and N-[(phosphonyl)methyl]-2-pyrrolidinone, *Phosphorus and Sulfur,* 12, 249, 1982.

34. **Kuno, J., Ishizuka, Y., Nakagawa, S., and Yoshida, S.,** Aminomethylphosphonates, Japanese Patent 60,209,593, 1985.

35. **Szczepaniak, W. and Siepak, J.,** Chelating ion exchanger of the organo-phosphorus complexone type, *Chem.-Anal. (Warsaw),* 18, 1019, 1973.

36. **Wilson, D. A. and Crump, D. K.,** Alkylene Phosphonic Acid Scale Inhibitor Compositions, U.S. Patent 4,540,508, 1985.

37. **Abramov, V. S.,** Reaction of aldehydes with phosphites, *Dokl. Akad. Nauk S.S.S.R.,* 95, 991, 1954.

38. **Ramirez, F., Patwardhan, A. V., and Heller, S. R.,** The Reaction of trialkyl phosphites with aliphatic aldehydes. P^{31} and H^1 nuclear magnetic resonance spectra of tetraoxy phosphoranes, *J. Am. Chem. Soc.,* 86, 514, 1964.

39. **Ramirez, F., Bhatia, S. B., and Smith, C. P.,** Reaction of trialkyl phosphites with aromatic aldehydes, *Tetrahedron,* 23, 2067, 1967.

40. **Ramirez, F., Gulati, A. S., and Smith, C. P.,** Reactions of five-membered cyclic triaminophosphines with hexafluoroacetone, trifluoroacetophenone, and fluorenone. Attack by phosphorus on carbonyl oxygen and isolation of crystalline 2,2,2-triamino-1,3,2-dioxaphospholanes, *J. Am. Chem. Soc.,* 89, 6283, 1967.

41. **Gazizov, T. K., Pyndyk, A. M., Sudarev, Y. I., Podobedov, V. B., Kovalenko, V. I., and Pudovik, A. N.,** Mechanism of the reactions of trialkyl phosphites with α-halocarbonyl compounds (transl.), *Bull. Acad. Sci. U.S.S.R., Chem. Sci.,* 27, 2319, 1978.

42. **Foss, V. L., Lukashev, N. V., Tsvetkov, Y. E., and Lutsenko, I. F.,** Reactions of bis(tetraethyl-phosphorodiamidous) anhydride with carbonyl compounds, *J. Gen. Chem. U.S.S.R.* (Engl.) 52, 1942, 1982.

43. **Pudovik, A. N., Gazizov, T. K., and Sudarev, Y. I.,** Reaction of diethyl trimethylsilyl phosphite with chloral (transl.), *J. Gen. Chem. U.S.S.R.,* 43, 2072, 1973.

44. **Gaziaov, T. K., Sudarev, Y. I., Kibardin, A. M., Belyalov, R. U., and Pudovik, A. N.,** Mechanism of Perkow Reaction (transl.), *J. Gen. Chem. U.S.S.R.,* 51, 5, 1981.

45. **Schulz, P., Vilceanu, R., and Kurunczi, L.,** Procedeu de sinteza a insecticiduliu $O,O\,'$-dimetil-(1-hidroxi-2,2,2-tricloretil)-fosfonat, Romanian Patent 63864, 1974.

46. **Laczko, I., Bass, E., Cselovszki, J., Fekete, L., Virag, I., Altay, L., and Farkas, I.,** Di-(O-alkyl)-1-hydroxy-2,2,2-trichloroethylphosphonates, Hungarian Patent 16,875, 1979.

47. **Oleksyszyn, J., Mastalerz, P., and Tyka, R.,** α-Aminophosphinic Acid, Polish Patent 105,728, 1980.

48. **Strepikheev, Y.-A., Kovalenko, L. V., Batalina, A. V., and Livshits, A. I.,** Mechanism of the Abramov reaction catalyzed by secondary amines, (transl.), *J. Gen. Chem. U.S.S.R.,* 46, 2364, 1976.

49. **Odinets, I. L., Novikova, Z. S., and Lutsenko, I. F.,** Reactions of tetraalkyl methylenebisphosphonites with aromatic aldehydes, (transl.), *J. Gen. Chem. U.S.S.R.,* 85, 488, 1985.

50. **Mikroyannidis, J. A.,** New synthesies of 1,2-dihydroxy-1,2-bisphosphonylethanes, *Phosphorus and Sulfur,* 20, 323, 1984.

51. **Winkler, T. and Bencze, W. L.,** Perkow reaction induced C,C-bond formation, *Helv. Chim. Acta,* 63, 402, 1980.

52. **Okamoto, Y. and Azuhata, T.,** The reaction of diethyl acyl phosphites with α,β-unsaturated carbonyl compounds, *Bull. Chem. Soc. Jpn.,* 57, 2693, 1984.

53. **Hertzog, K., Neumann, P., and Schuelke, U.,** Combined Production of Substituted Alkyl-1,1-bisphosphonic Acids and Alkyl Phosphites, East German Patent 222,599, 1985.

54. **Nesterov, L. V., Krepysheva, N. E., Sabirova, R. A., and Romanova, G. N.,** Phosphorus derivatives. VII. Reactions of dialkyl trialkylsilyl phosphites with aldehydes and ketones, (transl.), *J. Gen. Chem. U.S.S.R.,* 41, 2474, 1971.

55. **Kibardin, A. M., Gazizov, T. K., and Pudovik, A. N.,** Reactions of diethyl trimethylsilyl phosphite and of acetyl diethyl phosphite with fluorinated ketones (transl.), *J. Gen. Chem. U.S.S.R.,* 45, 1947, 1975.

56. **Birum, G. H. and Richardson, G. A.,** Silicon-phosphorus Compounds, U.S. Patent 3,113,139, 1963.

57. **Evans. D. A., Hurst, K. M., Truesdale, L. K., and Takacs, J. M.,** The carbonyl insertion reactions of mixed tervalent phosphorus-organosilicon reagents, *Tetrahedron Lett.,* p. 2495, 1977.

58. **Evans, D. A., Hurst, K. M., and Takacs, J. M.,** New silicon-phosphorus reagents in organic synthesis. Carbonyl and conjugate addition reactions of silicon phosphite esters and related systems, *J. Am. Chem. Soc.,* 100, 3467, 1978.

59. **Koenigkramer, R. E. and Zimmer, H.,** α-Heterosubstituted phosphonate carbanions -IX. Diethyl 1-phenyl-1-trimethylsiloxymethane phosphonate as an acyl anion equivalent; a novel method for the preparation of α-hydroxyketones, *Tetrahedron Lett.,* p. 1017, 1980.

60. **Koenigkramer, R. E. and Zimmer, H.,** Benzoins *via* an acyl anion equivalent. Novel one-pot preparation of benzo[b]furans *via* benzoins using hydriodic acid, *J. Org. Chem.,* 45, 3994, 1980.

61. **Gazizov, M. B., Zakharov, V. M., Khairullin, R. A., and Moskva, V. V.,** Reactions of phosphorus trichloride with aldehydes in presence of trialkyl phosphites (transl.), *J. Gen. Chem. U.S.S.R.,* 84, 2493, 1984.

62. **Osipova, M. P., Kuzmina, L. V., and Kukthin, V. A.,** Reaction of triethyl phosphite with 2-furaldehyde oxime (transl.), *J. Gen. Chem. U.S.S.R.,* 52, 392, 1982.

63. **Sommer, L. H.,** *Stereochemistry, Mechanism and Silicon,* McGraw-Hill, New York, 1965, 176.

64. **Novikova, Z. S., Moshoshina, S. N., Sapozhnikova, T. A., and Lutsenko, I. F.,** Reaction of diethyl trimethylsilyl phosphite with carbonyl compounds (transl.), *J. Gen. Chem. U.S.S.R.,* 41, 2655. 1971.

65. **Evans, D. A., Takacs, J. M., and Hurst, K. M.,** Phosphonamide stabilized allylic carbanions. New homoenolate anion equivalents, *J. Am. Chem. Soc.,* 101, 371, 1979.

66. **Pudovik, A. N., Gazizov, T. K., and Kibardin, A. M.,** Reaction of diethyl trimethylsilyl phosphite with hexafluoroacetone (transl.), *J. Gen. Chem. U.S.S.R.,* 44, 1170, 1974.

67. **Pudovik, A. N., Betyeva, E. S., and Alfonsov, V. A.,** Reactions of trimethylsilyl tetraethylphosphoro-diamidite with benzaldehyde (transl.), *J. Gen. Chem. U.S.S.R.,* 45, 921, 1975.

68. **Hata, T., Hashizume, A., Nakajima, M., and Sekine, M.,** A convenient method of ketone synthesis utilizing the reaction of diethyl trimethylsilyl phosphite with carbonyl compounds, *Tetrahedron Lett.,* p. 363, 1978.

69. **Sekine, M., Nakajima, M., and Hata, M.,** Silyl phosphites. 16. Mechanism of the Perkow reaction and the Kukhtin-Ramirez reaction. Elucidation by means of a new type of phosphoryl rearrangements utilizing silyl phosphites, *J. Org. Chem.,* 46, 4030, 1981.

70. **Sekine, M., Nakajima, M., Kume, A., Hashizume, A., and Hata, T.,** Versatile utility of α-(trime-thylsiloxy)alkylphosphonates as key intermediates for transformation of aldehydes into several carbonyl derivatives, *Bull. Chem. Soc. Jpn.,* 55, 224, 1982.

71. **Pudovik, A. N., Kibardin, A. M., Pashinkin, A. P., Sudarev, Y. I., and Gazizov, T. K.,** Reactions of diethyl trimethylsilyl phosphite with carbonyl compounds (transl.), *J. Gen. Chem. U.S.S.R.,* 44, 500, 1974.

72. **Novikova, Z. S. and Lutsenko, I. F.,** Reaction of trialkylsilyl diethyl phosphites with unsaturated com-pounds, (transl.), *J. Gen. Chem. U.S.S.R.,* 40, 2110, 1971.

73. **Pudovik, A. N., Batyeva, E. S., and Zameletdinova, G. U.,** Reaction of diethyl trimethylsilyl phosphite with diethyl acetylphosphonate (transl.) *J. Gen. Chem. U.S.S.R.,* 43, 676, 1973.

74. **Konovalova, I. V., Burnaeva, L. A., Saifullina, N. S., and Pudovik, A. N.,** Reactions of diethyl trimethylsilyl phosphite with α-keto Carboxylic esters and carbonitriles, (transl.), *J. Gen. Chem. U.S.S.R.,* 46, 17, 1976.

75. **Ofitserova, E. K., Ivanova, O. E., Ofitserova, E. N., Konovalova, I. V., and Pudovik, A. N.,** Reactions of dicarbonyl compounds with dialkyl silyl phosphites and trialkyl phosphites. Steric structure of products (transl.), *J. Gen. Chem. U.S.S.R.,* 51, 390, 1981.

76. **Liotta, D., Sunay, U., and Ginsberg, S.,** Phosphonosilylations of cyclic enones, *J. Org. Chem.,* 47, 2227, 1982.

77. **Hata, T., Nakajima, M., and Sekine, M.,** Facile synthesis of β-alkyl-substituted esters from α,β-unsaturated aldehydes, *Tetrahedron Lett.* p. 2047, 1979.

78. **Sekine, M., Nakajima, M., and Hata, T.,** A new type of P-C and C-C bond cleavage reactions in α-trimethylsiloxy-β-oxo phosphonates and α-ethoxy-β-oxo phosphonates, *Bull. Chem. Soc. Jpn.,* 55, 218, 1982.

79. **Hata, T., Hashizume, A., Nakajima, M., and Sekine, M.,** α-Hydroxybenzyl anion equivalent: a con-venient method for the synthesis of α-hydroxy ketones utilizing α-trimethylsiloxybenzylphosphonate, *Chem. Lett.,* p. 519, 1979.

80. **Belokrinitskii, M. A. and Orlov, N. F.,** Reaction of alkyl halides with triorganosilyl derivatives of phosphorous acid, *Kremniiorg. Mater.,* p. 145, 1971.

81. **Orlov, N. F., Kaufman, B. L., Sukhi, L., Selsarand, L. N., and Sudakova, E. V.,** Synthesis of triorganosilyl derivatives of phosphorous acid, *Khim. Prim. Soedin.,* p. 111, 1966.

82. **Sekine, M., Yamamoto, I., Hashizume, A., and Hata, T.,** The reaction of tris(trimethylsilyl) phosphite with carbonyl compounds, *Chem. Lett.,* p. 485, 1977.

83. **Sekine, M., Okimoto, K., and Hata, T.,** Silyl phosphites. 6. Reactions of tris(trimethylsilyl) phosphite with halocarbonyl compounds, *J. Am. Chem. Soc.,* 100, 1001, 1978.

84. **Sekine, M., Okimoto, K., Yamada, K., and Hata, T.,** Reactions of silyl phosphites with α-halo carbonyl compounds. Elucidation of the mechanism of the Perkow reaction and related reactions with confirmed experiments *J. Org. Chem.,* 46, 2097, 1981.

85. **Sekine, M., Tetsuaki, T., Yamada, K., and Hata, T.,** A facile synthesis of phosphoenolpyruvate and its analogue utilizing *in situ* generated trimethylsilyl bromide, *J. Chem. Soc. Perkin 1,* p. 2509, 1982.

86. **Sekine, M., Yamagata, H., and Hata, T.,** A general and convenient method for the synthesis of unesterified carbamoyl- and thiocarbamoyl-phosphonic acids, *Tetrahedron Lett.,* p. 3013, 1979.

87. **Alfonsov, V. A., Zameletdinova, G. U., Batyeva, E. S., and Pudovik, A. N.,** New path in reaction of silyl phosphites with acetyl chloride (transl.), *J. Gen. Chem. U.S.S.R.,* 54, 414, 1984.

88. **Thottathil, J. K., Ryono, D. E., Przybyla, C. A., Moniot, J. L., and Neubeck, R.,** Preparation of phosphinic acids: Michael additions of phosphonous acids/esters to conjugated systems, *Tetrahedron Lett.,* p. 4741, 1984.

89. **Hata, T. and Sekine, M.,** Silyl- and stannyl-esters of phosphorous oxyacids — intermediates for the synthesis of phosphate derivatives of biological interest, in *Phosphorus Chemistry Directed Towards Biology,* Stec, W. J., Ed., Pergamon, New York, 1980, 197.

90. **Zon, J.,** Asymmetric addition of tris(trimethylsilyl) phosphite to chiral aldimines, *Pol. J. Chem.,* 55, 643, 1981.

91. **Kowalik, J., Sawka-Dobrowolska, W., and Glowiak, T.,** Synthesis, molecular structure, and absolute configuration of an optically active (1-amino-2-phenylethyl)phosphonic acid monohydrate, *J. Chem. Soc. Chem. Commun.,* p. 446, 1984.

92. **Kowalik, J., Kupczyk-Subotkowska, L., and Mastalerz, P.,** Preparation of dialkyl 1-aminoalkylphosphonates by reduction of dialkyl 1-hydroxyiminoalkanephosphonates with zinc in formic acid, *Synthesis,* p. 57, 1981.

93. **Huber, J. W. and Gilmore, W. F.,** Optically active α-amino-phosphonic acids from ureidophosphonates, *Tetrahedron Lett.,* p. 3049, 1979.

94. **Glowiak, T. and Sawka-Dobrowolska, W.,** Absolute configuration of optically active aminophosphonic acids, *Tetrahedron Lett.,* p. 3965, 1977.

95. **Birum, G. H.,** Urylenediphosphonates. A general method for the synthesis of α-ureidophosphonates and related structures, *J. Org. Chem.,* 39, 209, 1974.

96. **Oleksyszyn, J., Tyka, R., and Mastalerz, P.,** Guanidinophosphonic acids, *Synthesis,* p. 571, 1977.

97. **Oleksyszyn, J. and Tyka, R.,** 1-(S-Alkylisothioureido)-benzylphosphonic acids as special guanidylation agents. A general method for the synthesis of 1-guanidionbenzylphosphonic acids substituted in the N'-position, *Pol. J. Chem.,* 52, 1949, 1978.

98. **Kudzin, Z. H. and Stec, W. J.,** Synthesis of 1-aminoalkanephosphonates *via* thioureidoalkanephosphonates, *Synthesis,* p. 469, 1978.

99. **Yuan, C., Qi, Y., and Xiang, C.,** Organophosphorus compounds. XII. Synthesis of α-aminoalkylphosphonic acids, *Huaxue Xuebao,* 43, 243, 1985.

100. **Tam, C. C., Mattocks, K. L., and Tischler, M.,** Synthesis of phosphomethionine and related compounds, *Synthesis,* p. 188, 1982.

101. **Kudzin, Z. H.,** Phosphocysteine derivatives; thioureidoalkanephosphonates *via* acetals, *Synthesis,* p. 643, 1981.

102. **Tyka, R. and Oleksyszyn, J.,** α-Aminophosphonic Acids, Polish 105,825, 1980.

103. **Lejczak, B., Kafarski, P., Soroka, M., and Mastalerz, P.,** Synthesis of the phosphonic acid analog of serine, *Synthesis,* p. 577, 1984.

104. **Oleksyszyn, J., Subotkowska, L., and Mastalerz, P.,** Diphenyl 1-aminoalkanephosphonates, *Synthesis,* p. 985, 1979.

105. **Oleksyszyn, J. and Subotkowska, L.,** Aminomethanephosphonic acid and its diphenyl ester, *Synthesis,* p. 906, 1980.

106. **Moedritzer, K. and Irani, R. R.,** The direct synthesis of α-aminomethylphosphonic acids. Mannich-type reactions with orthophosphorous acid, *J. Org. Chem.,* 31, 1603, 1966.

107. **Szczepaniak, W. and Siepak, J.,** Phosphoroorganic complexones. I. Synthesis and dissociation constants of α-(N-benzylamino)alkylphosphonic acids, *Rocz. Chem.,* 47, 929, 1973.

108. **Redmond, D.,** Chemistry of phosphorous acid: new routes to phosphonic acids and phosphate esters, *J. Org. Chem.,* 43, 992, 1978.

109. **Redmore, D.,** N-Benzyl-α-amino phosphonic acids, *J. Org. Chem.,* 43, 996, 1978.

110. **Redmore, D.,** α-Aminophosphonic Acids, U.S. Patent 4,235,809, 1980.

111. **Oleksyszyn, J. and Gruszecka, E.,** Amidoalkylation of phosphorous acid, *Tetrahedron Lett.,* p. 3537, 1981.

112. **Schuelke, U. and Kayser, R.,** 1-Aminoalkyl-1,1-bisphosphonic Acids, East German Patent 222,598, 1985.

113. **Biryukov, A. I., Osipova, T. I., Khomutov, R. M., and Khurs, I. N.,** Preparation of Aminophosphonic Acid for Inhibition of Enzymes, U.S.S.R. Patent 717,062, 1980.

114. **Chai, B. J. and Muggee, F. D.,** Method of Preparing Phosphonates from Nitriles, U.S. Patent 4,239,695, 1980.

115. **Sommer, K.,** 1-Aminoalkane-1,1-diphosphonic Acids, West German Patent application 2,754,821, 1979.

116. **Schuelke, U., Kayser, R., Loeper, M., and Ludewig, D.,** Hydroxyalkylphosphonic Acids, East German Patent 222,597, 1985.

117. **Yoneda, S., Kawase, T., and Yoshida, Z.-i.,** Syntheses of [1-(alkylthio)]- and (1-Mer-capto)cycloalkanephosphonic esters by the reactions of cycloalkanethiones with trialkyl phosphites, *J. Org. Chem.,* 43, 1980, 1978.

118. **Kawase, T., Yoneda, S., and Yoshida, Z.-i.,** Syntheses of alkylphosphonic esters by the reactions of aliphatic thiones with trialkyl phosphites, *Bull. Chem. Soc. Jpn.,* 52, 3342, 1979.

119. **Zimin, M. G., Burilov, A. R., and Pudovik, A. N.,** Reactions of silyl phosphites with thioketones (transl.), *J. Gen. Chem. U.S.S.R.,* 51, 1841, 1981.

120. **Morton, D. W. and Neilson, R. H.,** Reactions of (silylamino)phosphines with ketones and aldehydes, *Organometallics,* 1, 289, 1982.

121. **Baylis, E. K., Campbell, C. D., and Dingwall, J. G.,** 1-Aminoalkylphosphonous acids. Part 1. Isosteres of the protein amino acids, *J. Chem. Soc. Perkin 1,* p. 2845, 1984.

122. **Pudovik, A. N. and Arbuzov, B. A.,** Addition of dialkyl phosphites to unsaturated compounds. I. Addition of dialkyl phosphites to 2,2-dimethylvinyl vinyl ketone, *Zh. Obshch. Khim. S.S.S.R.,* 21, 382, 1951.

123. **Vorontsova, N. A., Vlasov, O. N., and Melnikov, N. N.,** Influence of solvents on the kinetics of the condensation of diethyl hydrogen phosphite with chloral (transl.), *J. Gen. Chem. U.S.S.R.,* 48, 2415, 1978.

124. **Gartman, G. A., Pak, V. D., and Simonova, E. V.,** Influence of the solvent on the kinetics and mechanism of the reactions of anils with dimethyl hydrogen phosphite, (transl.), *J. Gen. Chem. U.S.S.R.,* 49, 2295, 1979.

125. **Castagnino, E., Corsano, S., and Strappavecchia, G. P.,** The preparation of a novel 3-oxo-cyclopenten-2-phosphonate derivative, useful intermediate for 2-alkyl-substituted cyclopentenones synthesis, *Tetrahedron Lett.,* 93, 1985.

126. **Adiwidaja, G., Meyer, B., and Thiem, J.,** Synthesis and crystal structure of endo-2–dimethylphosphono-exo-2-hydroxy-(−)-camphene for the determination of ^3J(CCCP) vicinal coupling constants, *Z. Naturforsch. Teil B,* 34B, 1547, 1979.

127. **Arbuzov, B. A., Zoroastrova, V. M., Tudril, G. A., and Fuzhenkova, A. V.,** Reaction of benzalacetone with dimethyl phosphonate (transl.), *Izv. Akad. Nauk S.S.S.R.,* p. 2541, 1974.

128. **Vysotskii, V. I., Pavel, G. V., Chuprakova, K. G., Shchukin, V. A. and Tilichenko, M. N.,** Reactions of 1,5-diketones. XXXV. Reactions of 2-[α-(α-methylenephenacyl)benzyl]-cyclohexanone with dialkyl hydrogen phosphites (transl.), *J. Gen. Chem. U.S.S.R.,* 49, 1717, 1979.

129. **Vysotskii, V. I., Skobun, A. S., and Tilichenko, M. N.,** Reactions of 1,5-diketones. XXXVI. Reactions of 1,5-diketones with alkyl hydrogen hypophosphites, (transl.), *J. Gen. Chem. U.S.S.R.,* 49, 1721, 1979.

130. **Arbuzov, B. A., Fuzhenkova, A. V., Tudril, G. A., and Zoroastrova, V. M.,** Reaction of dimethyl phosphite with α,β-unsaturated ketones (transl.), *Izv. Akad. Nauk S.S.S.R.,* 1285, 1975.

131. **Warren, S. and Williams, M. R.,** Electrophilic substitution at phosphorus: dealkylation and decarboxylation of phosphinylformate esters, *Chem. Commun.,* p. 180, 1969.

132. **Cann, P. F., Warren, S., and Williams, M. R.,** Electrophilic substitutions at phosphorus: reactions of diphenylphosphinyl systems with carbonyl compounds, *J. Chem. Soc. Perkin 1,* p. 2377, 1972.

133. **Tone, T., Okamoto, Y., and Sakurai, H.,** Preparation of 1-hydroxy-2-aminoethylphosphonic acid and its alkyl-substituted derivatives, *Chem. Lett.,* p. 1349, 1978.

134. **Paulsen, H. and Greve, W.,** Synthese von α-Hydroxyphosphonaten der 1,6-Anhydrohexosen Untersuchungen uber die PCOH-Kopplung, *Chem. Ber.,* 106, 2124, 1973.

135. **Wieczorek, J. S., Boduszek, B., and Gancarz, R.,** Phosphonic derivatives of azafluorenes, *J. Prakt, Chem.,* 326, 349, 1984.

136. **Tewari, R. S. and Shukla, R.,** Organophosphorus compounds. Addition reaction of *O,O*-dialkyl hydrogen phosphites with substituted aromatic and long chain aliphatic aldehydes, *Labdev,* Part A, 9, 112, 1971.

137. **Ruveda, M. A. and deLicastro, S. A.,** Synthesis of dimethyl α-hydroxy phosphonates from dimethyl phosphite and phenacyl chloride and cyanide, *Tetrahedron,* 28, 6018, 1972.

138. **Kluge, A. F. and Cloudsdale, I. S.,** Phosphonate reagents for the synthesis of enol ethers and one-carbon homologation to aldehydes, *J. Org. Chem.,* 44, 4847, 1974.

139. **Polezhaeva, N. A., Ilyasov, A. V., Nafikova, A. A., Ismaev, I. E., Elshiva, E. V., and Arbuzov, B. A.,** Reaction of 1,3-diphenyl-2-(phenylimino)-1,3-propanedione with trimethyl phosphite (transl.), *J. Gen. Chem. U.S.S.R.,* 55, 1545, 1985.

140. **Gancarz, R., Wielkopolski, W., Jaskulska, E., Kafarski, P., Lejczak, B., Mastalerz, P., and Wieczorek, J. S.,** Phosphonic analogues of morphactins. IV. 9-Aminofluoren-9-ylphosphine oxides, *Pestic. Sci.,* 16, 234, 1985.

141. **Wynberg, H. and Smaardijk, A. A.,** Asymmetric catalysis in carbon-phosphorus bond formation, *Tetrahedron Lett.,* p. 5899, 1983.

142. **Smaardijk, A. A., Noorda, S., van Bolhuis, F., and Wynberg, H.,** The Absolute configuration of α-hydroxyphosphonates, *Tetrahedron Lett.,* p. 493, 1985.

143. **Pudovik, A. N., Zimin, M. G., and Sobanov, A. A.**, Reactions of dialkyl hydrogen phosphites with ketones activated by electronegative groups (transl.), *J. Gen. Chem. U.S.S.R.*, 42, 2170, 1972.

144. **Bentzen, C. L., Mong, N. L., and Niesor, E.**, Diphosphonate Derivatives and Pharmaceutical Compositions Containing Them, United Kingdom Patent 2,079,285, 1982.

145. **Yamashita, M., Tsunekawa, K., Sugiura, M., and Oshikawa, T.**, Novel preparation of diphenylphosphine oxides *via* direct deoxygenation of 1,2-epoxyethyldiphenylphosphine oxides, *Synthesis,* p. 65, 1985.

146. **Mikroyannidis, J. A., Tsolis, A. K., and Gourghiotis, D. J.**, Synthesis and chemical properties of substituted 2-hydroxy-2-phosphonylethanals and 1,2-dihydroxy-1,2-bisphosphonylethanes, *Phosphorus and Sulfur,* 13, 279, 1982.

147. **Janzen, A. F. and Pollitt, R.**, Reaction of dialkyl phosphonates with hexafluoroacetone, *Can. J. Chem.,* 48, 1987, 1970.

148. **Zhdanov, Y. A., Glebova, Z. I., and Uzlova, L. A.**, Interfacially catalyzed Abramov reaction in carbohydrate chemistry (transl.), *J. Gen. Chem. U.S.S.R.*, 51, 1428, 1981.

149. **Zhdanov, Y. A., Kistyan, G. K., Uzlova, L. A., Polenov, V. A., and Yakshina, N. V.**, Synthesis of phosphorus analogs of uronic acids with *D*-gluco and *L*-ido configurations (transl.), *J. Gen. Chem. U.S.S.R.*, 53, 1271, 1983.

150. **Zhdanov, Y. A., Glebova, Z. I., Kistyan, G. K., Polenov, V. A., and Uzlova, L. A.**, Synthesis of polyhydroxy phosphonic esters (transl.), *J. Gen. Chem. U.S.S.R.*, 54, 1478, 1984.

151. **Zhdanov, Y. A., Glebova, Z. I., Kistyan, G. K., and Uzlova, L. A.**, Application of potassium fluoride in the synthesis of carbohydrate α-hydroxyphosphonates (transl.), *J. Gen. Chem. U.S.S.R.*, 54, 1079, 1984.

152. **Texier-Boullet, F. and Foucaud, A.**, Synthesis of 1-hydroxyalkanephosphonic esters on alumina, *Synthesis,* p. 916, 1982.

153. **Texier-Boullet, F. and Foucaud, A.**, A convenient synthesis of dialkyl 1-hydroxyalkanephosphonates using potassium or caesium fluoride without solvent, *Synthesis,* p. 165, 1982.

154. **Texier-Boullet, F.**, Synthesis of α-hydroxyphosphonates in a heterogeneous solid-liquid medium and at the surface of inorganic solids. Study of factors of their formation in some two-phase systems. *Bull. Soc. Sci. Bretagne,* 56, 57, 1984.

155. **Fields, E. K.**, The synthesis of esters of substituted amino phosphonic acids, *J. Am. Chem. Soc.,* 74, 1528, 1952.

156. **Tyka, R.**, Novel synthesis of α-aminophosphonic acids, *Tetrahedron Lett.,* p. 677, 1970.

157. **Rachon, J. and Wasielewski, C.**, Preparation of diethyl (α-aminobenzyl)phosphonates, *Z. Chem.,* 13, 254, 1973.

158. **Lukszo, J., Kowalik, J., and Mastalerz, P.**, Advantages of using di(*p*-methylbenzyl) hydrogen phosphite in the synthesis of aminophosphonates from aldimines, *Chem. Lett.,* p. 1103, 1978.

159. **Lukszo, J. and Tyka, R.**, New protective groups in the synthesis of 1-aminoalkanephosphonic acids and esters, *Synthesis,* p. 239, 1977.

160. **Lukszo, J. and Tyka, R.**, New synthesis of diethyl *N*-acyl-α-aminophosphonates, *Pol. J. Chem.,* 52, 321, 1978.

161. **Wasielewski, C., Antczak, K., and Rachon, J.**, Phosphoranaloge von Aminosauren und Peptiden; eine einfache Methode zur Darstellung der α-Aminobenzylphosphinsaure, *Z. Chem.,* 19, 253, 1979.

162. **Tyja, R. and Lukszo, J.**, α-Aminophosphonic Acids, Polish Patent 99,012, 30 October 1978.

163. **Wieczorek, J. S. and Gancarz, R.**, Aminophosphonic acid derivatives of fluorene, *Rocz. Chem.,* 50, 2171, 1976.

164. **Gancarz, R. and Wieczorek, J. S.**, Synthesis of phosphonic analogues of morphactines, *J. Prakt Chem.,* 322, 213, 1980.

165. **Wieczorek, J. S. and Gancarz, R.**, α-Aminophosphonic Acid Esters, Polish Patent 105,252, 1980.

166. **Brendel-Hajoczki, M., Gulyas, I., Gyoker, I., Zsupan, K., Csorvassy, I., Salamon, Z., Somogyi, G., Szent-Kiralyi, I., and Timar, T.**, *N*-Phosphonomethylglycine, French Patent 2,460,959, 1981.

167. **Gancarz, R., Kafarski, P., Lejczak, B., Mastalerz, P., Wieczorek, J. S., Przybylka, E., and Czerwinski, W.**, Phosphonic analogues of morphactines, *Phosphorus and Sulfur,* 18, 373, 1983.

168. Grelan Pharmaceutical Co., Ltd., Piperazinylalkylphosphonic Acids, Japanese Patent 81 61,395, 1981.

169. **Zimmer, H. and Neue, D. M.**, Syntheses with α-heterosubstituted phosphonate carbanions. VI. Desoxybenzoins and indoles, *Chimia,* 31, 330, 1977.

170. **Seemuth, P. D. and Zimmer, H.**, α-Hetero-substituted carbanions. 7. Synthesis of deoxybenzoins and benzo[b]furans, *J. Org. Chem.,* 43, 3063, 1978.

171. **Zimmer, H., Koenigkramer, R. E., Cepulis, R. L., and Neue, D. M.**, Syntheses with α-heterosubstituted phosphonate carbanions. 10. Autoxidation of the anion, *J. Org. Chem.,* 45, 2018, 1980.

172. **Nifantev, E. E. and Shilov, I. V.**, Study of tetraalkylphosphorodiamidous acids. Aminoalkylation (transl.), *J. Gen. Chem. U.S.S.R.*, 42, 502, 1972.

173. **Issleib, K., Dopfer, K.-P., and Balszuweit, A.**, Contribution to the synthesis of α-aminoalkane phosphonic acid dialkyl esters, *Z. Anorg. Allg. Chem.,* 444, 249, 1978.

174. **Zelenova, T. P., Patlina, S. I., Vasyaniva, L. K., and Nifantev, E. E.,** Synthesis and NMR spectra of (α-anilino-*o*-hydroxybenzyl)phosphonic esters (transl.), *J. Gen. Chem. U.S.S.R.,* 49, 484, 1979.

175. **Issleib, K., Doepfer, K. P., and Balszuweit, A.,** Synthesis of 1-aminoalkane phosphonic acids *via* benhydrylic Schiff bases, *Z. Naturforsch. B,* 36B, 1392, 1981.

176. **Sobanov, A. A., Bakhtiyarov, I. V., Badeeva, E. K., Zimin, M. G., and Pudovik, A. N.,** Interaction of incomplete esters of phosphorus acids with α,β-unsaturated imines (transl.), *J. Gen. Chem. U.S.S.R.,* 55, 22, 1985.

177. **Osipova, M. P., Lukin, P. M., and Kukhtin, V. A.,** α-(Hydroxyamino) phosphonates (transl.), *J. Gen. Chem. U.S.S.R.,* 52, 393, 1982.

178. **Soroka, M. and Mastalerz, P.,** Phosphonic and phosphinic analogs of aspartic acid and asparagine, *Rocz. Chem.,* 48, 1119, 1974.

179. **Maier, L.,** Organische Phosphorverbindungen. 74. Zur Kenntnis der Umsetzung von Cyanomethyldichlorphosphin und 2-Chlorathyldichlorphosphin mit Benzylglycin und Formaldehyde in saurer Losung, *Phosphorus and Sulfur,* 11, 149, 1981.

180. **Ratcliffe, R. W. and Christensen, B. G.,** Total synthesis of β-lactam antibiotics. I. α-Thioformamidodiethylphosphonoacetates, *Tetrahedron Lett.,* p. 4645, 1973.

181. **Rogozhin, S. V., Davankov, V. A., and Belov, Y. P.,** Optically active diethyl α-aminobenzylphosphonate (transl.), *Izv. Akad. Nauk S.S.S.R.,* 926, 1973.

182. **Belov, Y. P., Rakhnovich, G. B., Davankov, V. A., Godovikov, N. N., Aleksandrov, G. G., and Struchkov, Y. T.,** Synthesis and determination of the absolute configuration of *R*-α-aminobenzylphenylphosphinic acid, (transl.), *Izv. Akad. Nauk S.S.S.R.,* 832, 1980.

183. **Boev, V. I. and Dombrovskii, A. V.,** Synthesis and properties of bis(ferrocenylmethyl) hydrogen phosphite (transl.), *J. Gen. Chem. U.S.S.R.,* 49, 1093, 1979.

184. **Zimin, M. G., Burilov, A. R., and Pudovik, A. N.,** Reactions of dialkyl phosphites with oximes (transl.), *J. Gen. Chem. U.S.S.R.,* 50, 595, 1980.

185. **Gruszecka, E., Soroka, M., and Mastalerz, P.,** Phosphonic analogs of α-methylaspartic and α-methylglutamic acids, *Pol. J. Chem.,* 53, 2327, 1979.

186. **Benckiser, J. A.,** Verfahren zur Herstellun von Aminoalkylenphosphonsauren, West German Patent 2,741,504, C3, 1981.

187. **Mastevosyan, G. L., Matyushicheva, R. M., Vodovatova, S. N., and Zavlin, P. M.,** Phosphorylated benzimidazoles. VI. Phosphorylated imidazoles and benzimidazoles (transl.), *J. Gen. Chem. U.S.S.R.,* 51, 636, 1981.

188. Monsanto, Co., Oxazolidinonephosphonates, JP 79,157,559, 1979.

189. **Ehrat, R.,** *N*-Phosphonomethylglycine, West German Patent Application 2,942,898, 1980.

190. **Gaertner, V. R.,** Oxazolidinone Phosphonates and Their Use as Herbicides, European Patent 0 007 684, 1983.

191. **Nagubandi, S.,** Process for Preparing Phosphonomethylated Amino Acids, U.S. Patent, 4,439,373, 1984.

192. **Arbuzov, B. A., Fuzhenkova, A. V., and Galyautdinov, I. V.,** Reaction of dimethyl phosphite with tetracyclone, (transl.), *Izv. Akad. Nauk S.S.S.R.,* 381, 1978.

193. **Hoppe, I., Schollkopf, U., Nieger, M., and Egert, E.,** Asymmetric addition of a chiral cyclic phosphite to a cyclic imine — synthesis of phosphonic acid analogues of *D*- and *L*- penicillamine, *Angew. Chem. Int. Ed. Eng.,* 24, 1067, 1985.

194. **Gancarz, R. and Wieczorek, J. S.,** A useful method for the preparation of 1-aminoalkanephosphonic acids, *Synthesis,* p. 625, 1977.

195. **Pudovik, A. N., Guryanov, I. V., Romanov, G. V., and Lapin, A. A.,** Reaction of diphenylphosphine with the methyl ester of pyroracemic acid (transl.), *J. Gen. Chem. U.S.S.R.,* 41, 710, 1971.

Chapter 4

CONJUGATE ADDITION REACTIONS OF TRIVALENT PHOSPHORUS REAGENTS

I. INTRODUCTION

The previous chapter was concerned with the nucleophilic addition of trivalent phosphorus reagents (in either an anionic or neutral state) to carbonyl-type centers. A logical extension of this type of reaction involves the application of the same phosphorus reagents to reactions with α,β-unsaturated carbonyl compounds, and the variety of related Michael-type substrates. As facile addition occurs at ordinary carbonyl centers, conjugate addition might similarly be anticipated with α,β-unsaturated compounds.

In fact, addition of both categories of nucleophilic trivalent phosphorus reagents is observed to occur in a conjugate manner with a wide range of Michael-type substrates. Addition occurs under quite mild conditions and generates the organophosphorus products bearing functionality at the 2-position relative to phosphorus.

II. REACTIONS AND MECHANISM

The reaction of trialkyl phosphites with α,β-unsaturated ketones in the presence of a proton source was first investigated by Harvey and Jensen using ethanol as a solvent.[1] The initial addition of the nucleophilic trivalent phosphorus center in a conjugate manner to the unsaturated ketone would be anticipated to proceed quite readily. The troublesome point in the completion of such a conjugate addition reaction would be expected to be at the same stage as difficulties occur in the simple Abramov reaction involving trialkyl phosphites, that is, with the dealkylation of the zwitterionic adduct.

The use of a protic solvent such as an alcohol provides a solution to this difficulty in that it provides a means for dealkylation to occur and the reaction to proceed to completion. Not only is the alcohol a source of a proton for the anionic site of the zwitterion, but alcohol (alkoxide ion) provides a nucleophile for the required dealkylation step. A relatively innocuous ether by-product is produced in the process.

Further efforts[2] demonstrated that superior yields of the target phosphonates usually could be obtained using phenol as the proton source-dealkylating agent rather than simple alcohols. Also, using simple alcohols often resulted in the isolation of mixtures of products, including acetals (ketals) and enol ethers as well as the simple carbonyl compounds. The latter complication may be minimized by the use of t-butyl alcohol.

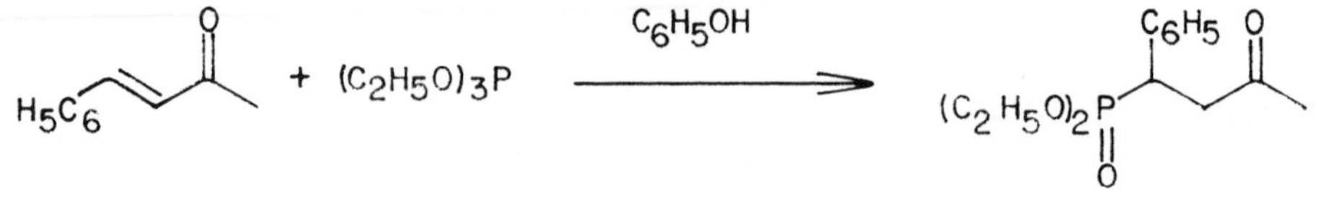

While this fundamental reaction of α,β-unsaturated ketones has been performed success-fully in other instances using alcohol or phenol as reaction adjuncts,[3-5] a variety of associated reagents have also been used. These include carboxylic acids[6] as part of the reaction medium or materials added in the work-up of the initial adduct, such as water[7] or hydrogen chloride.[8] Hydrogen bromide has been used to facilitate the addition of triphenylphosphine to α,β-unsaturated aldehydes and ketones, but of course no dealkylation is involved in this reaction.[9] In one instance a dialkyl phosphorochloridate was added as a trap of the intermediate zwitterion, resulting in the formation of an enol-phosphate derivative of the phosphonate.[10]

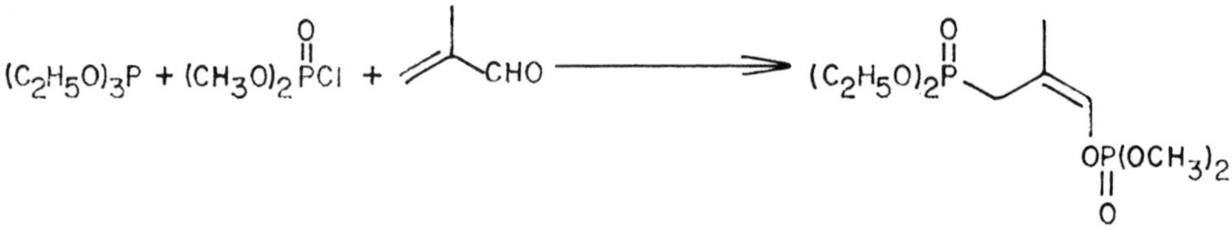

The second step (dealkylation) is accomplished more easily in the conjugate addition reaction than with the carbonyl-carbon addition (Abramov) reaction. Intramolecular dealky-lation of the quasi-phosphonium ion intermediate has fewer steric restrictions in this situation, and intramolecular processes can be observed to occur. An interesting reaction which pro-ceeds readily in the absence of an adjunct reactant involves the use of dialkyl acyl phosphites. Intramolecular deacylation occurs to generate the enol-ester product, although with some of the substrates investigated, reaction is found to occur primarily at the carbonyl-carbon.[11]

Fuchsone undergoes 1,6-addition of trialkyl phosphites in good yield.[12] With this Michael substrate, dealkylation of the intermediate quasi-phosphonium ion center necessarily occurs via an intermolecular process.

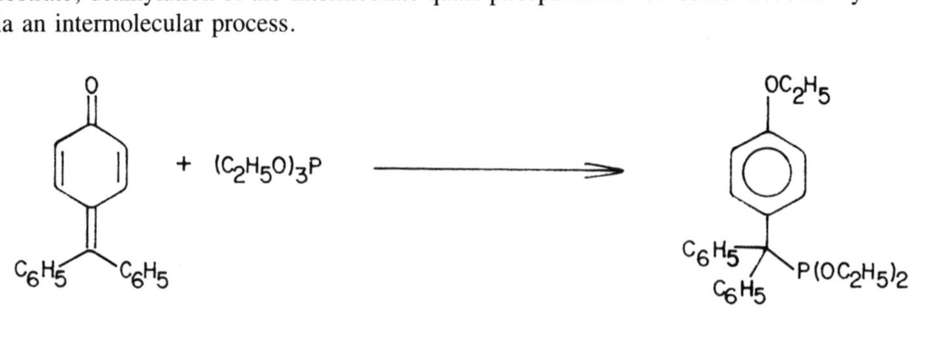

Quinones generally react with trialkyl phosphites to generate aromatic phosphate esters.[13] However, with an additional acyl function substituted on the ring, hydrophosphinylation reaction occurs to give a mixture of products.

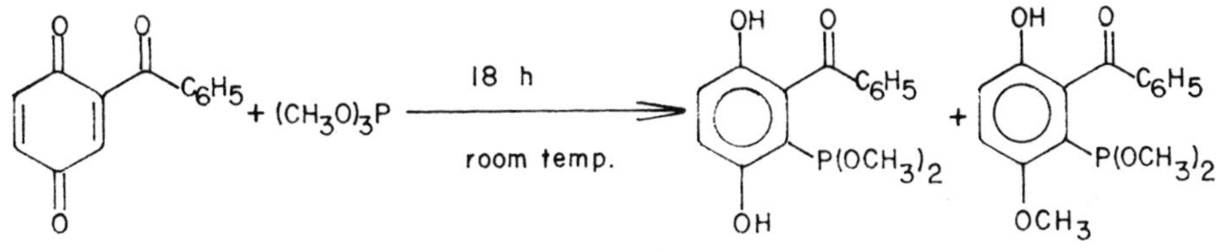

Unsaturated esters are found to react with trialkyl phosphites quite well under the same conditions as do aldehydes and ketones.[2] In one instance the substrate for the hydrophosphinylation reaction was generated *in situ* by an initial reaction of trialkyl phosphite with an α-halo-β-alkoxy ester.[14] Alcohol, involved in completing the second step of the reaction (dealkylation), was generated in the initial addition-elimination process. Another reaction of an unsaturated ester, which might at first appear to be a (rare) Michaelis-Arbuzov reaction directly on a vinylic halide, more likely occurs via an addition-elimination route.[15] The β-bromoacrylate esters are found to undergo reaction with trialkyl phosphites to give the vinylic phosphonates. Addition of phosphorus to carbon occurs initially forming a resonance-stabilized zwitterion which subsequently loses halide ion. The halide ion then completes the final dealkylation process. It is unlikely that initial carbon-phosphorus bond formation would occur under these conditions were not the ester linkage present in conjugation with the olefinic bond of the vinylic halide.

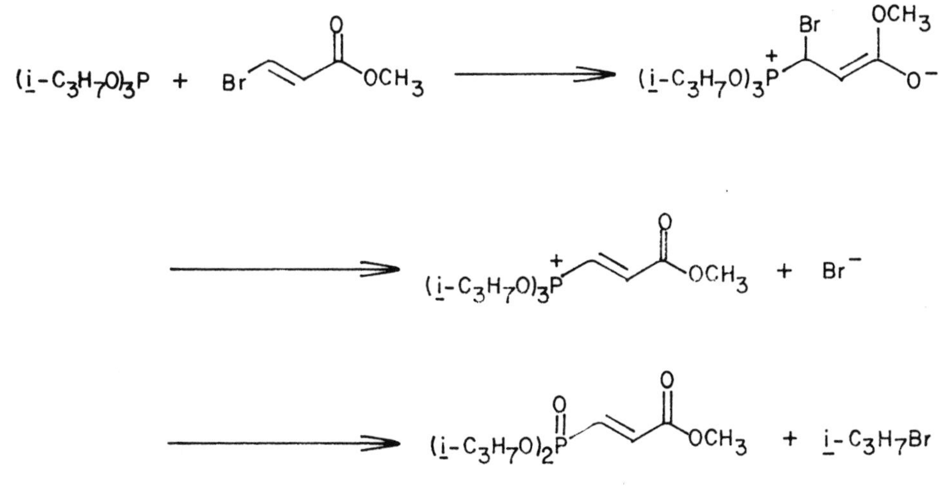

Hydrophosphinylation of an α,β-unsaturated ester formed *in situ* by a condensation reaction has also been observed with a phosphoramidite used in place of a trialkyl phosphite.[16] It is the nitrogen functionality which is lost preferentially from the phosphorus center rather than an ester.

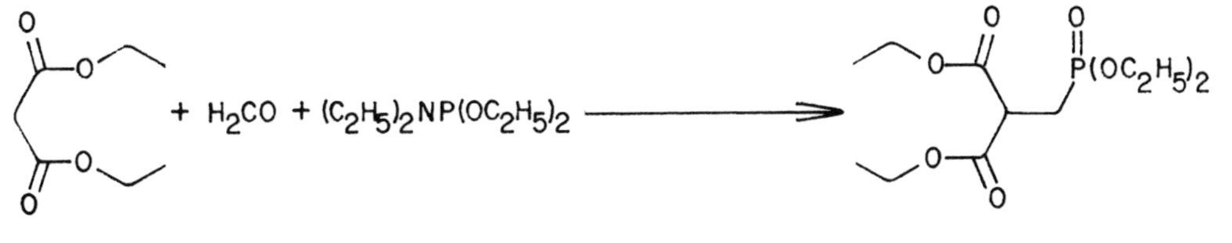

The use of phosphoramidites with α,β-unsaturated amides has also been investigated. In reaction with *N*-substituted succinimide, exclusive loss of the nitrogen substituent from phosphorus was noted,[17] although with acrylamide the nitrogen substituent was retained.[18] With hexamethyl phosphorous triamide, a zwitterionic adduct was found to be generated which did not undergo further reaction.[19] When trialkyl phosphites were used with the α,β-unsaturated amides, either in the presence of a hydroxylic solvent[2] or under conditions of intramolecular dealkylation,[19] the materials underwent the simple hydrophosphinylation reaction.

40%

The α,β-unsaturated nitriles[2,5,20] and carboxylic acids[21-26] are also observed to undergo the hydrophosphinylation reaction readily. With carboxylic acids there appears to be involved an initial esterification of the carboxylic acid site followed by addition of the phosphorus reagent to the olefinic site. Either excess trialkyl phosphite or the by-product, dialkyl phosphite, from the esterification, can be used to accomplish the hydrophosphinylation step.

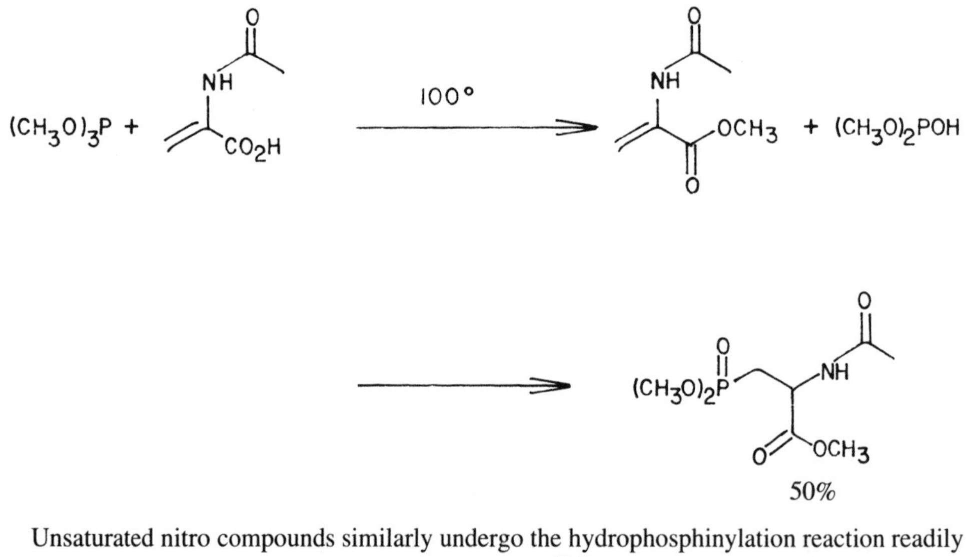

50%

Unsaturated nitro compounds similarly undergo the hydrophosphinylation reaction readily in alcohol solvent, often in near quantitative yield.[27-30] Amine hydrochlorides rather than alcohols have also been used in acetonitrile solution to facilitate the addition reactions.[31,32]

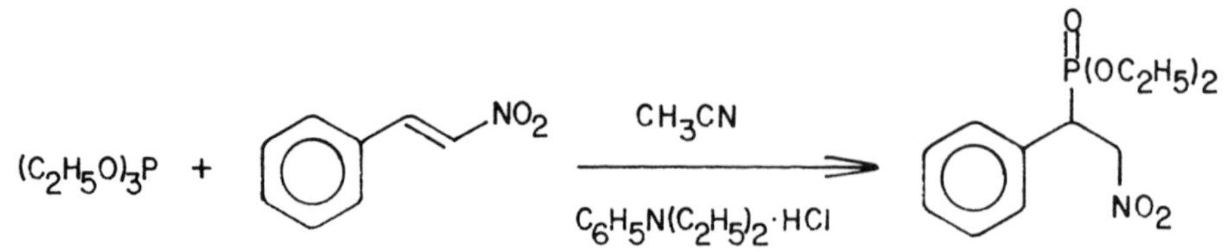

80%

In two instances where the hydrophosphinylation reaction was performed in the absence of a proton source, elimination, either of an adjacent halide or a nitro function, occurred subsequent to the initial addition reaction.[33,34] The functionality eliminated was presumably involved further in the final dealkylation of the quasi-phosphonium intermediate.

Fully esterified forms of trivalent phosphorus acids other than trialkyl phosphites have also been used successfully in these reactions. Alkyl esters of phosphonous[5,31,33] and phosphinous[2] acids have been used, as well as phenyl esters of phosphonous acids.[31] The isolated yields of adducts in these instances were comparable with those from reactions using trialkyl phosphites with the variety of substrate types.

As with addition at carbonyl carbon, the use of monobasic forms of trivalent phosphorus acids obviates certain of the difficulties inherent in the accomplishment of the dealkylation stage of the reaction. The addition of dialkyl phosphites to α, -unsaturated ketones in the presence of alkoxide base was observed quite early to yield readily the 3-oxophosphonates.[35,36] As might be anticipated when alternative sites for Michael addition were presented to the phosphorus reagent, that bearing fewer substituents at the β-position was observed to be the favored site for addition. Following this initial work, a variety of unsaturated ketones and aldehydes were observed to undergo the anionic variation of the hydrophosphinylation reaction with dialkyl phosphites in the presence of alkoxide ion.[37-39]

Conflicting reports have appeared regarding the addition of dialkyl phosphites to methyl styryl ketone. The differences are likely the result of the differing temperatures at which the reactions were performed. At room temperature the reaction proceeds to form the simple 1,2-type adduct via reaction at the carbonyl carbon center.[40] At higher temperatures (refluxing benzene) the conjugate addition product is isolated.[41] Addition at the carbonyl-carbon is the kinetically favored process, whereas conjugate addition occurs under conditions of thermodynamic control.

As noted earlier with the reaction of 3-bromoacrylates, conjugated enol-ethers also undergo reaction which appears to be a direct displacement, but is more likely an addition-elimination process. Dialkyl phosphites in the presence of alkoxide bases undergo reaction with conjugated enol-ethers to generate the conjugated vinylic phosphonates in good yield.[42] Addition of the phosphorus center to the vinylic carbon site is facilitated by the formation of the resonance stabilized anion, which subsequently loses alkoxide ion to complete the reaction.

In one instance, reaction of a 1,2-benzoquinone with dialkyl phosphites proceeded in the absence of added base.[8] A concerted formation of the activated complex involving a pair of reacting quinone species has been postulated to account for product formation, although initial addition to a carbonyl center followed by hydrogen transfer and phosphorus migration need also be considered.

As with the additions of fully esterified trivalent phosphorus reagents to conjugated sub-strates, the Michael-type addition reactions of the anionic forms of dialkyl phosphites with unsaturated esters,[22,43,44] amides,[45-47] and nitriles[43,48] proceed readily in good to excellent yield.

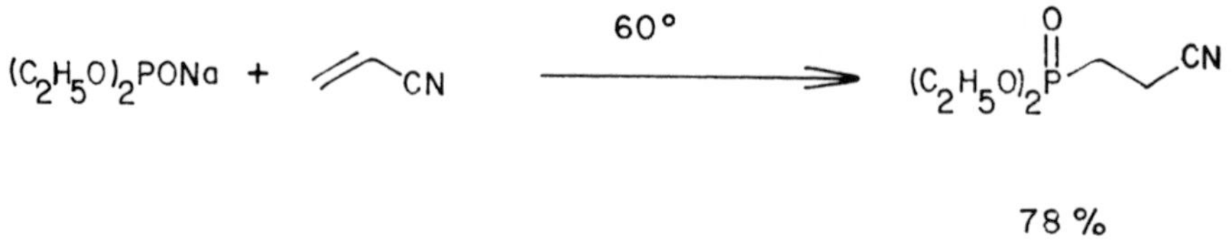

$(C_2H_5O)_2PONa$ + [structure: CH$_2$=CH-CN] $\xrightarrow{60°}$ $(C_2H_5O)_2P(=O)$—CH$_2$CH$_2$—CN

78 %

One report has been made of acryloyl chloride being used as a Michael substrate for reaction with dialkyl phosphites.[49] The primary adducts were not isolated directly, but treated with trimethylsilyl azide and heated to accomplish a Curtius rearrangement and to generate the 2-phosphonoisocyanates in moderate yield.

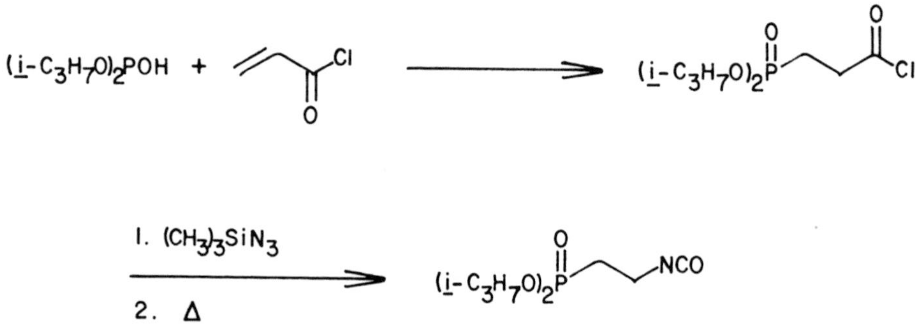

$(i\text{-}C_3H_7O)_2POH$ + [structure: CH$_2$=C(CH$_3$)-C(=O)Cl] $\longrightarrow$ $(i\text{-}C_3H_7O)_2P(=O)$—CH$_2$—C(=O)Cl

$\xrightarrow[\text{2. } \Delta]{\text{1. } (CH_3)_3SiN_3}$ $(i\text{-}C_3H_7O)_2P(=O)$—CH$_2$—NCO

Addition-elimination reactions of enol-ethers of unsaturated esters and nitriles, as noted previously with the ketones, also generate the vinylic phosphonates in good yield.[42] However, additions to conjugated enamides proceeds without elimination. Simple hydrophosphinyla-tion is observed to occur with enamides in the syntheses of carbon-phosphorus analogues of aspartate and asparagine.[26]

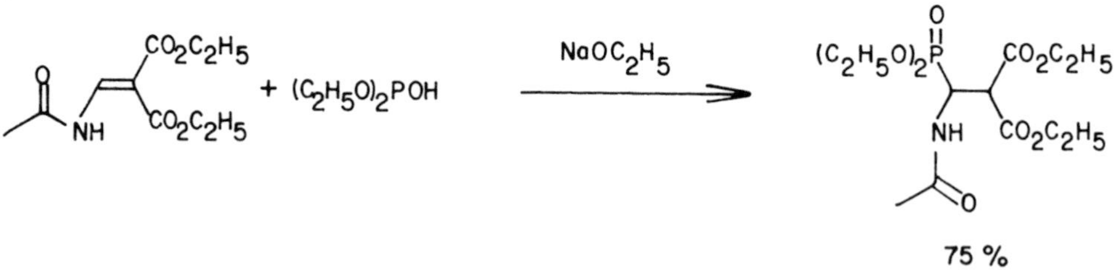

[structure: CH$_3$C(=O)NH-C(CO$_2$C$_2$H$_5$)=CH-CO$_2$C$_2$H$_5$] + $(C_2H_5O)_2POH$ $\xrightarrow{NaOC_2H_5}$ [structure: $(C_2H_5O)_2P(=O)$-CH(CO$_2$C$_2$H$_5$)-CH(NH-C(=O)CH$_3$)-CO$_2$C$_2$H$_5$]

75 %

Several reports have been made concerning the addition of dialkyl phosphites to unsaturated nitro compounds. With the use of triethylamine as the base, dialkyl phosphites have been found to add stereospecifically to unsaturated nitro derivatives of a carbohydrate series.[50]

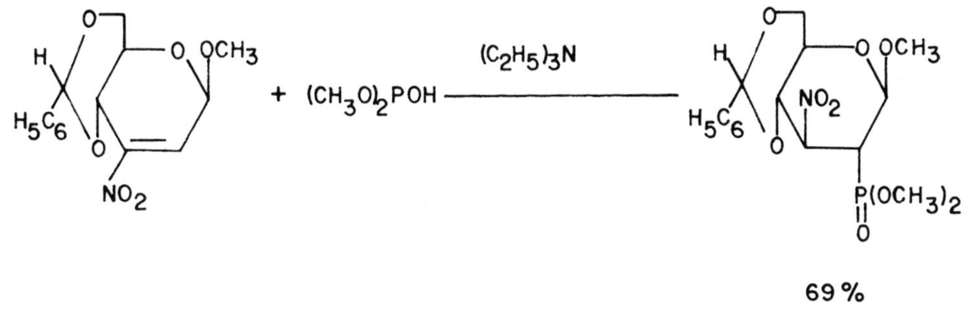

69 %

Several other unsaturated nitro systems have been found to undergo addition of dialkyl phosphites in the absence of added base.[51,52]

In addition to dialkyl phosphites, other monobasic trivalent phosphorus reagents have been used in hydrophosphinylation reactions. For example, monoalkyl phosphonites have been added to acrylonitrile[53] and to enamides for the synthesis of analogues of amino acids.[26,54] The anion from diphenylphosphine oxide has also been found to undergo addition at the carbonyl carbon of aldehydes, independent of its method of generation.[55] However, treatment of either unsaturated ketones or esters with the anion from either diphenyl- or dibenzylphosphine oxide results in Michael addition reaction.[55-57]

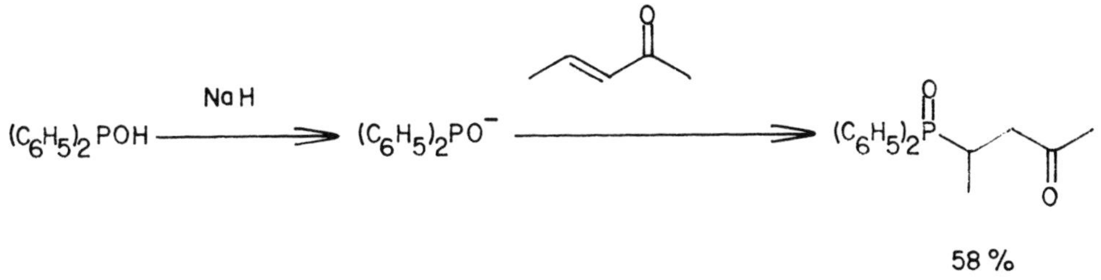

58 %

Two reports have been made of dibasic trivalent phosphorus acid species undergoing conjugate addition reactions. Methyl hypophosphite has been found to undergo additions to two molecules of substrate (acrylate esters, acrylamide, acrylonitrile) exothermically in the presence of sodium methoxide.[58] The ammonium salt of mono(2-ethylhexyl) phosphite has been found to undergo addition in good yield to a variety of unsaturated substrates in the presence of alkoxide ion.[59]

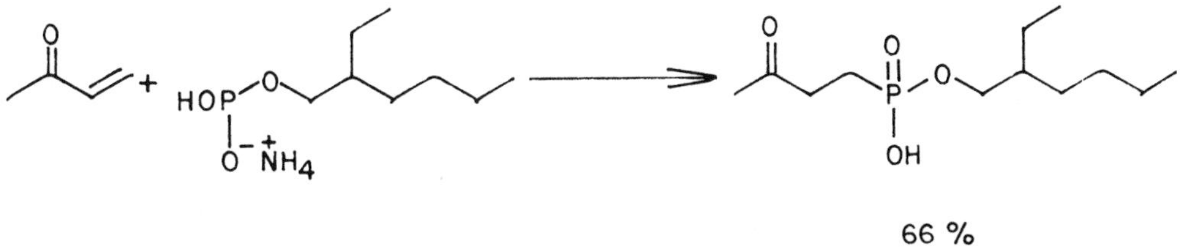

66 %

Two reports have been made regarding addition of unesterified hypophosphorous acid to unsaturated substrates. Acrylamides undergo conjugate addition with hypophosphorous acid quite readily to generate directly the free phosphonous acids in good yield.[60] In another report, aqueous hypophosphorous acid has been noted to undergo addition to vinyl acetate to generate 1-hydroxyethylphosphonic acid in excellent yield.[61] No mechanistic details were reported regarding this latter preparation.

As with other nucleophilic reactions of trivalent phosphorus reagents previously discussed, some of the most striking developments with conjugate addition reactions in recent years have been accomplished using phosphorus reagents bearing silyl ester linkages. Products of the reactions of the silyl esters are of particular value for the performance of *umpolung* reactions. Their application to syntheses of phosphorus molecules of biological significance has recently been reviewed.[62] Although the phosphorus reagents to be discussed in this category all may also be considered as "fully esterified" trivalent phosphorus reagents, the presence of one or more silyl esters lends special characteristics to the reaction system.

Trimethylsilyl diethyl phosphite has been observed to participate in conjugate addition reaction with a variety of Michael substrates. With acrylate esters, addition occurs readily at the β-position with transfer of the silyl ester linkage to the carboxyl oxygen.[63-65] The ketene silyl acetal derivatives thus generated require subsequent treatment with a proton donor in order that isolation of the simple ester product may be accomplished.

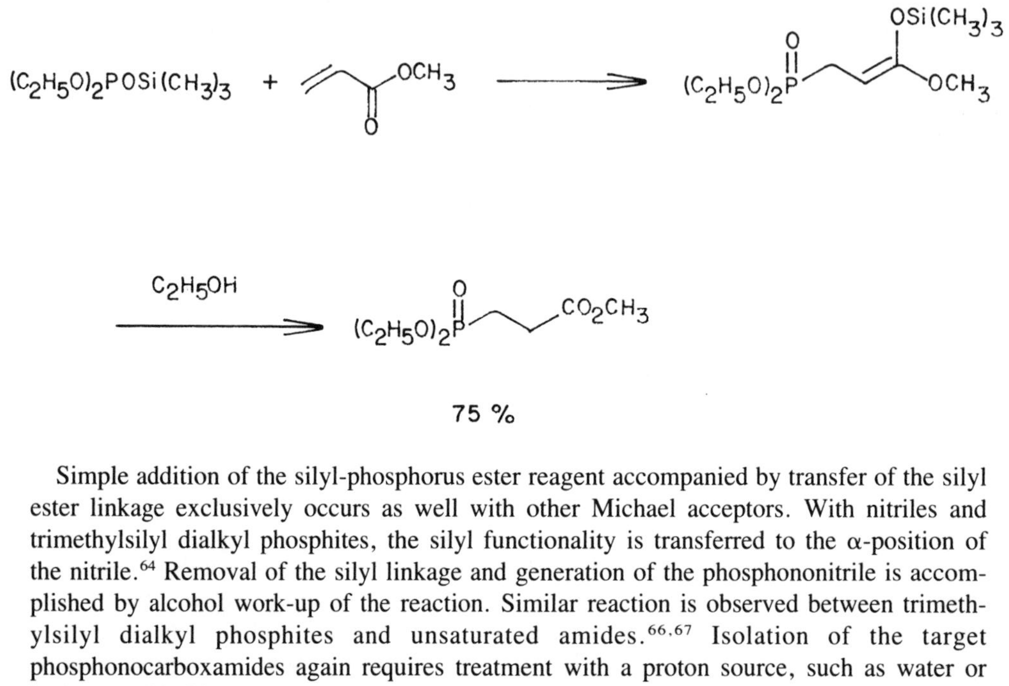

75 %

Simple addition of the silyl-phosphorus ester reagent accompanied by transfer of the silyl ester linkage exclusively occurs as well with other Michael acceptors. With nitriles and trimethylsilyl dialkyl phosphites, the silyl functionality is transferred to the α-position of the nitrile.[64] Removal of the silyl linkage and generation of the phosphononitrile is accomplished by alcohol work-up of the reaction. Similar reaction is observed between trimethylsilyl dialkyl phosphites and unsaturated amides.[66,67] Isolation of the target phosphonocarboxamides again requires treatment with a proton source, such as water or hydrogen chloride.

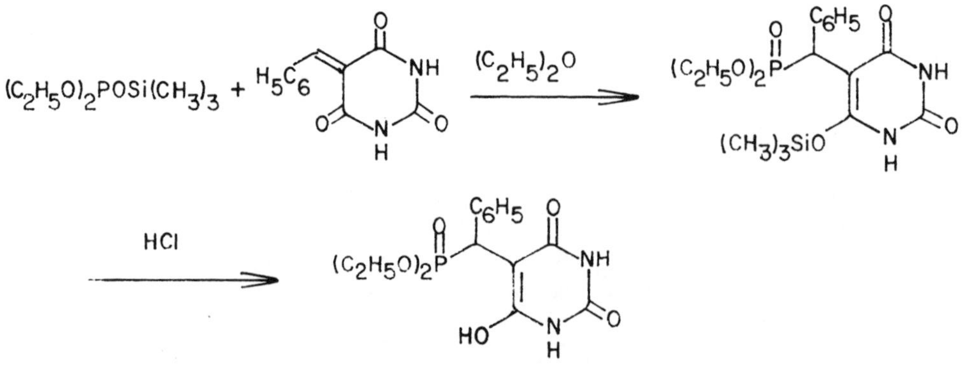

The reaction of α,β-unsaturated ketones and aldehydes with trialkylsilyl dialkyl phosphites proceeds quite readily, but with some complications. Although in isolated instances low yields of conjugate addition products are isolated in the reactions of aldehydes, the major (or sole) material formed is usually the (Abramov) product of addition at the carbonyl carbon.[68-72] With ketones, conjugate addition is usually found to be the dominant process.

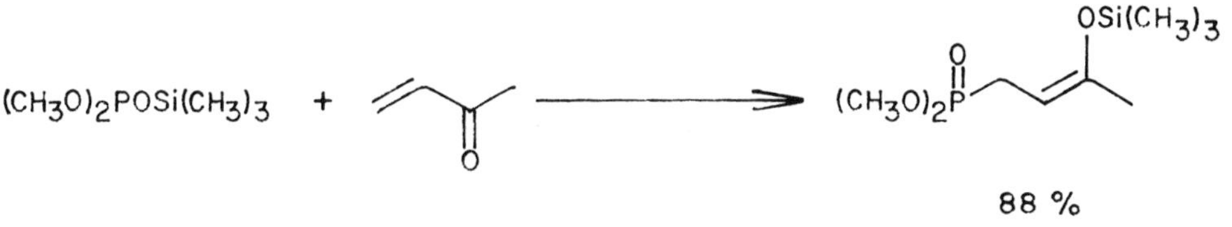

88 %

In all instances, transfer of the silyl ester function occurs exclusively. In most instances this can be envisioned as occurring via an intramolecular process. However, with a series of cyclic enones which have been investigated,[69] exclusive transfer of the silyl ester linkage again occurs, but via an intermolecular route. With the cyclic enones the ratio of 1,2- to 1,4-addition products could be varied greatly through changing of the reaction conditions. With only one substrate were conditions not found which would result in predominantly conjugate addition.

The related phosphorus reagent, triethylsilyl N,N,N',N'- tetramethyl phosphorodiamidite, exhibits reactivity which is quite similar to that of the phosphite. With aldehydes, addition at the carbonyl carbon center predominates, whereas with ketones conjugate addition is found to be most important.[70,71,73]

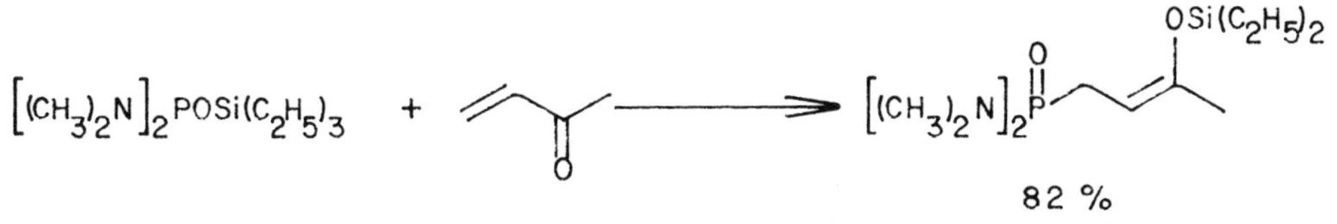

82 %

In certain instances, conjugate addition products can be isolated from reaction with unsaturated aldehydes through the use of a different reagent system. Treatment of the unsaturated aldehyde with a trialkyl phosphite (or other fully esterified trivalent phosphorus acid) in the presence of a silyl halide is found often to result in formation of conjugate addition product.[70,71] The silyl reagent serves to trap the anionic site of the intermediate zwitterion with the halide performing the subsequent dealkylation.

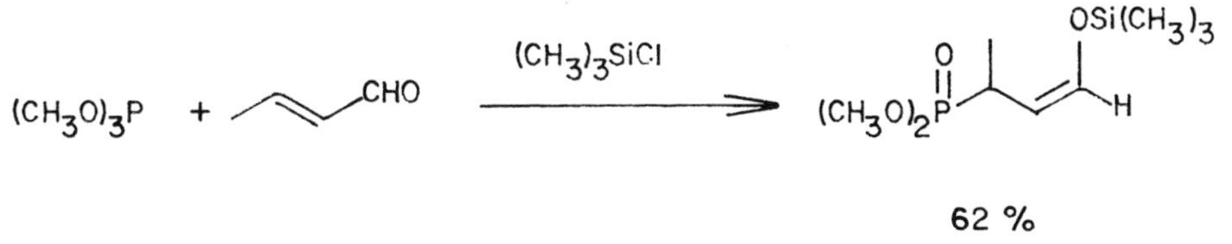

62 %

Trialkyl phosphites in the presence of silyl halides have also been used for conjugate addition to unsaturated esters,[74] nitriles,[75] and carboxylic acids.[74] It should be noted that these reactions proceed via addition of the trialkyl phosphite to the Michael substrate followed by silyl reagent trapping of the intermediate. Under the reaction conditions employed, the trialkyl phosphites are not attacked by the silyl reagents, and trialkylsilyl dialkyl phosphites are not intermediates.

Tris(trimethylsilyl) phosphite has also been used in conjugate addition reactions for the synthesis of phosphonic acids.[76] As with the other simple phosphite reagents, addition to unsaturated aldehydes occurs completely in a 1,2-manner (Abramov reaction) whereas ketones and esters exhibit conjugate addition products in good yield.

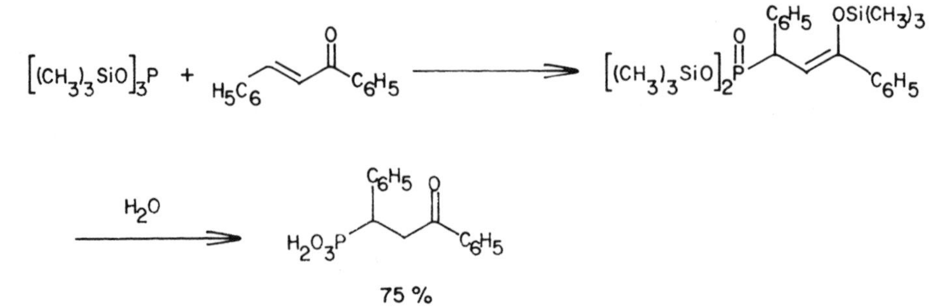

75 %

A single report has appeared concerning the addition of silyl esters of trivalent phosphorus acids to unsaturated nitro compounds.[33] Both trimethylsilyl dimethyl phosphite and trimethylsilyl methyl phenylphosphonite are found to undergo facile addition to 3,3,3-trichloro-1-nitropropene.

In addition to the reagents previously mentioned, other categories of reactants have been used in certain instances for conjugate addition reactions. Hexamethylphosphorous triamide has been used as a nucleophilic reagent in one Michael addition reaction. Conjugate addition to a benzylidene side chain was observed to occur upon its reaction with a substituted isoxazoline-5-one derivative.[77] No yields of adduct were reported.

Numerous other Michael substrates have been investigated in the hydrophosphinylation reaction other than the simple carbonyl compounds, esters, acids, amides, nitriles, and nitro compounds noted above. One category of compound which has received particular attention is the series of substituted cyclopentadienones. A wide range of reaction conditions has been used in the studies of cyclopentadienone addition, including the use of trialkyl phosphites directly,[78] dialkyl phosphites with amines[79,80] or alkoxide,[79] fully[81] or partially esterfied phosphonites,[82] and thermal reactions of dialkyl phosphites.[79,82] In no instance were products of addition at the carbonyl carbon observed. In most instances products were observed as the result of phosphorus addition at the carbon adjacent to the carbonyl center, that is via a 1,6-addition type process.

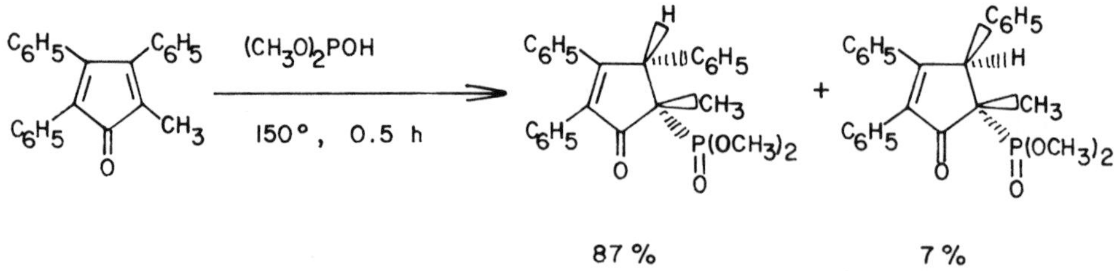

87 %   7 %

In one instance a mixture of products could be isolated in which 1,4-addition was observed.[80] Tetraphenylcyclopentadienone was found to react with dimethyl phosphite by heating in the absence of base to give both 1,4- and 1,6-addition products.

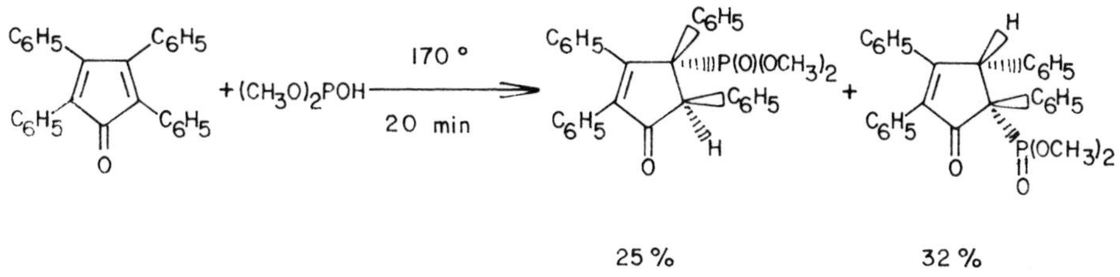

C_6H_5 ... $+ (CH_3O)_2POH \xrightarrow[\text{20 min}]{170°}$ ... $P(O)(OCH_3)_2$ ... $+$ ...

25 %   32 %

Several reports have been made of the use of imine compounds as Michael substrates for the hydrophosphinylation reaction. Dialkyl phosphites and dialkyl phosphine oxides undergo conjugate addition to N-alkyl- and N-sulfonylimines in moderate to good yield in the absence of base.[83,84] With oximes, conjugate addition is reported to occur at low temperature to give the N-ethoxyenamine, whereas at higher temperatures the nitrile is isolated.[85]

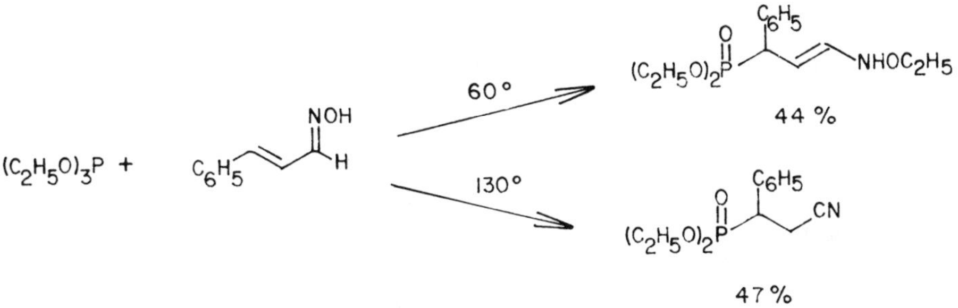

$(C_2H_5O)_3P +$ ... $\xrightarrow{60°}$ ... $(C_2H_5O)_2P$ ... $NHOC_2H_5$ 44 %

$\xrightarrow{130°}$ ... $(C_2H_5O)_2P$ ... CN 47 %

Vinyl substituted aromatics have also been found to be capable of serving as Michael substrates in hydrophosphinylation reactions. In addition to the other substrates previously noted, the ammonium salt of a monoalkyl phosphite is observed to add to the side-chain of vinylpyridines in moderate yield.

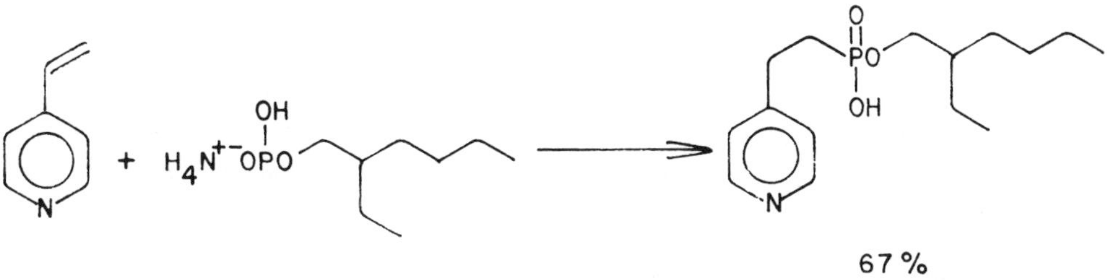

... $+ H_4N^+{}^-OPO$ ... $\longrightarrow$ ... 67 %

In one unusual instance, phosphorus has been found to add to the aromatic ring of substituted 5-nitropyrimidines.[86] Overall, dialkyl phosphites appear to be performing a displacement of the nitro group from the ring to give the substituted 5-phosphonopyrimidine. However, the reaction is best envisioned as occurring via "conjugate" addition to the ring at the site adjacent to the nitro group, followed by migration with nitrite ion loss to regenerate the aromatic ring.

60 %

III. COMPARISON OF METHODS

A variety of methods have been considered for the addition of trivalent phosphorus reagents in a Michael manner to α,β-unsaturated substrates. The best yields of Michael adducts are obtained by the addition of a monobasic trivalent phosphorus reagent to the Michael substrate in the presence of a silylating agent (trialkylsilyl halide, or N,N-bis-trimethylsilylacetamide) and a base, usually a tertiary amine.[70,71] Other approaches have been studied but the results with them are generally less favorable. The adducts formed in these reactions may be used directly for a variety of purposes, or the silyl and ester functions may be removed to isolate the simple phosphorus acid.

In some instances it may be desirable to avoid the use of silyl reagents. Of the several approaches investigated in this category, that which provides the most consistently high yields of adduct involves the addition of a fully esterified trivalent phosphorus reagent to the Michael substrate in the presence of a hydroxylic solvent.[2] While the early efforts would indicate that the more acidic hydroxylic adjunct reagent, phenol, provides the best results, excellent yields are at times obtained through the use of *t*-butyl alcohol. The use of "Pudovik conditions" (monobasic trivalent phosphorus reagent in the presence of an alkoxide base) has been made extensively, but generally gives poorer yields than the method using a fully esterified trivalent phosphorus reagent.

IV. EXPERIMENTAL PROCEDURES

Ethyl Methyl(2-carbomethoxy-3-phenylpropyl)phosphinate — Addition of a monobasic phosphinous ester to an unsaturated ester in the presence of a silylating agent.[87]

To a solution of ethyl methylphosphonite (0.66 g, 0.006 mol) in methylene chloride (7.5 mℓ) was added methyl 2-benzylacrylate (0.9 g, 0.0047 mol) and bis-trimethylsilylacetamide (1.03 g, 0.00474 mol). The reaction mixture was stirred at room temperature overnight. The reaction mixture was then washed with water followed by extraction with ether, and the volatile materials were removed under reduced pressure. The residue was subjected to chromatographic purification using neutral alumina (activity 3) from which was isolated the pure ethyl methyl(2-carbomethoxy-3-phenylpropyl)phosphinate (1.14 g, 80.7%) which exhibited spectral and analytical data in accord with the proposed structure.

2-Dimethoxyphosphinyl-2-methoxy-2-phenylacetaldehyde Oxime — Reaction of an unsaturated nitro compound with a trialkyl phosphite in the presence of an alcohol.[27]

A solution of β-nitrostyrene (30 g, 0.2 mol) in *t*-butyl alcohol (300 mℓ) was placed in a three-neck flask equipped with a condenser, a pressure-equalizing dropping funnel, and a thermometer; trimethyl phosphite (62 g, 0.5 mol) was added. An initial cooling of about 4°C was observed followed by an increase in temperature to a maximum of 65 to 75°C within 20 min. After 3 hr the solvent was removed under reduced pressure and the residue was cooled. The red oil was induced to crystallize by the addition of a seed crystal and the scratching of the vessel walls. After standing overnight, the crystals were filtered, washed with toluene (2 × 20 mℓ), and recrystallized from ethylene glycol dimethyl ether to give, in two crops, the pure 2-dimethoxyphosphinyl-2-methoxy-2-phenylacetaldehyde oxime (18.55 g, 34%) which exhibited spectral and analytical data in accord with the proposed structure.

3-Diethoxyphosphinyl-2-methylpropionamide — Addition of a monobasic trivalent phosphorus acid to an unsaturated amide in the presence of an alkoxide base.[45]

A mixture of diethyl phosphite (15.18 g, 0.11 mol) and 2-methylacrylamide (8.5 g, 0.1 mol) was heated at 60 to 70°C and an exothermic reaction was initiated by the further dropwise addition of a 3 M ethanolic sodium ethoxide solution (5 mℓ). After the completion of the exotherm, the reaction was further heated for 1 hr at 110°C. The reaction mixture was then cooled, diluted with ethanol, and neutralized with concentrated hydrochloric acid. The solution was then filtered and the solvent evaporated under reduced pressure. The residue solidified upon standing and was then recrystallized from benzene to yield the pure 3-diethoxyphosphinyl-2-methylpropionamide (17.84 g, 80%) of mp 75 to 77° which exhibited spectra and analytical data in accord with the proposed structure.

di-*n*-Butyl di-*n*-Butoxyphosphinylsuccinate — Addition of a monobasic trivalent phosphorus reagent to a unsaturated ester in the presence of base.[43]

To an agitated mixture of di-*n*-butyl phosphite (194 g, 1.0 mol) and sodamide (5 g, 0.13 mol) in a flask provided with a reflux condenser were added di-*n*-butyl maleate (228 g, 1.0 mol) dropwise over a period of 30 min while maintaining the reaction temperature at 50°C by cooling with a water bath. Further stirring for 1.25 hrs without cooling was performed to allow the reaction to proceed to completion. The reaction mixture was then neutralized with glacial acetic acid and filtered. The neutralized mixture was fractionally distilled under vacuum using a Claisen-type still to give the pure di-*n*-butyl di-*n*-butoxyphosphinylsuccinate (358.7 g, 85%) as an oily liquid of bp 190°C/1.2 torr.

Table 1
HYDROPHOSPHINYLATION REACTIONS

Product	Phosphorus reagent	Substrate type	Method	Yield (%)	Refs.
$H_2O_2PCH_2CH_2CONH_2$	H_3PO_2	Amide	A	73	60
$H_2O_3PCH(NH_2)CH_2CO_2H$	$(C_2H_5O)_3P$	Ester	B	75	26
$H_2O_3PCH_2CH(NH_2)CO_2H$	$(C_2H_5O)_3P$	Acid	B	75	26
$(CH_3O)_2P(O)CH_2CH_2NO_2$	$(CH_3O)_3P$	Nitro	C	88, 100	28, 29
$CH_3P(O)(OH)CH(NH_2)CH_2CO_2H$	$CH_3P(OC_2H_5)_2$	Ester	B	65	26
$CH_3P(O)(OH)CH_2CH(NH_2)CO_2H$	$CH_3P(OC_2H_5)_2$	Acid	B	65	26
$(CH_3O)_2P(O)CH(CCl_3)CH_2NO_2$	$(CH_3O)_2POSi(CH_3)_3$	Nitro	A	86	33
$CH_3P(O)(OC_2H_5)CH_2CH_2CN$	$CH_3P(OH)(OC_2H_5)$	Nitrile	D	63	87
$(C_2H_5O)_2P(O)CH_2CH_2CO_2H$	$(C_2H_5O)_3P$	Acid	B	82	24
$(C_2H_5O)_2P(O)CH_2CH_2CONH_2$	$(C_2H_5O)_3P$	Amide	C	42	2
$(C_2H_5O)_2P(O)CH_2CH_2CN$	$(C_2H_5O)_2PONa$	Nitrile	E	78	48
$(CH_3O)_2P(O)CH_2CH(CH_3)CO_2CH_3$	$(CH_3O)_2POH$	Ester	E	78	35
$CH_3OP(O)[CH_2CH_2CONH_2]_2$	CH_3OPOH_2	Amide	E	52	58
$(C_2H_5O)_2P(O)CH(CH_3)CH_2CHO$	$(C_2H_5O)_3P$	Aldehyde	C	82	2
$(C_2H_5O)_2P(O)CH_2CH(CH_3)CONH_2$	$(C_2H_5O)_2POH$	Amide	E	77	45
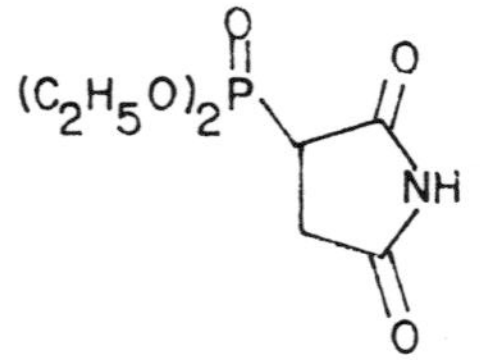 *(succinimide structure below)*	$(C_2H_5O)_2POSi(CH_3)_3$	Amide	A	48	67
$(CH_3O)_2P(O)CH_2CH(CO_2CH_3)NHCOCH_3$	$(CH_3O)_3P;\ (CH_3O)_2POH$	Acid	A	50	22
$(C_2H_5O)_2P(O)CH(CN)CH_2CN$	$(C_2H_5O)_3P$	Nitrile	C	55	2
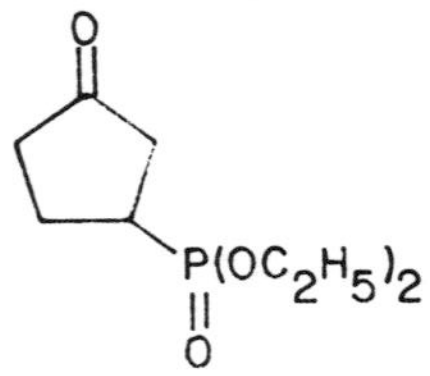 *(cyclopentanone structure below)*	$(C_2H_5O)_2POSi(CH_3)_3$	Ketone	A	100	69

$(C_2H_5O)_2P$ — succinimide structure (NH ring with two C=O)

cyclopentanone structure with $P(OC_2H_5)_2$, $\|$ O

Substrate	Reagent	Type	Method	Yield	Yield
$(C_2H_5O)_2P(O)CH_2CH_2CO_2C_2H_5$	$(C_2H_5O)_3P$	Ketone	C	72	2
$(CH_3O)_2P(O)CH(OCH_3)CH_2CO_2CH_3$	$(C_2H_5O)_2PONa$	Ester	E	84	48
$(i\text{-}C_3H_7O)_2P(O)CH_2CH_2CN$	$(C_2H_5O)_3P$	Haloester	A	61	14
$(CH_3O)_2P(O)CH_2CH=C(CH_3)OSi(CH_3)_3$	$(i\text{-}C_3H_7O)_2POH$	Nitrile	E	46	35
$(n\text{-}C_3H_7O)_2P(O)CH_2CH_2COCl$	$(CH_3H)_3P$	Ketone	D	79	70
$(CH_3O)_2P(O)CH_2CH=CHOP(O)(OC_2H_5)_2$	$(n\text{-}C_3H_7O)_2POH$	Acid Chloride	A	39	49
$(i\text{-}C_3H_7O)_2P(O)CH=CHCO_2CH_3$	$(CH_3O)_3P$	Aldehyde	F	73	10
	$(i\text{-}C_3H_7O)_3P$	Haloester	A	38	15
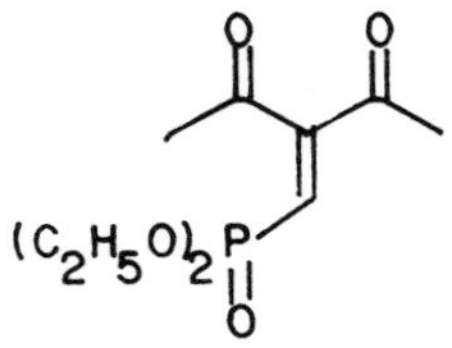	$(C_2H_5O)_2POH$	Ketone	E	80	42

Structure: $(C_2H_5O)_2P$ group

Reagent	Type	Method	Yield	Yield
$(C_2H_5O)_3P$	Ester	C	70	4
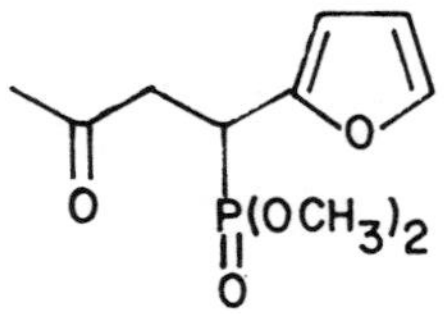 $(CH_3O)_2POH$	Ketone	E	63	38

$(C_2H_5O)_2P(O)CH_2CH_2COCH_2OCOCH_3$

Furan structure with $P(OCH_3)_2$ group

Substrate	Reagent	Type	Method	Yield	Yield
$(C_2H_5O)_2P(O)CH(CO_2CH_3)CH_2CO_2CH_3$	$(C_2H_5O)_2POSi(CH_3)_3$	Ester	A	52	65
$(C_2H_5O)_2P(O)CH(CH_3)CH_2CO_2C_2H_5$	$(C_2H_5O)_2POSi(CH_3)_3$	Ester	A	62	65
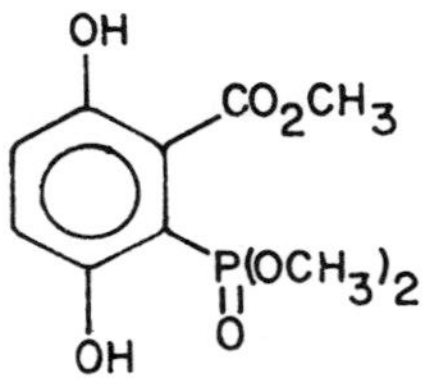	$(CH_3O)_3P$	Ketoester	D	48	13

Phenol structure with CO_2CH_3 and $P(OCH_3)_2$ groups

Table 1 (continued)
HYDROPHOSPHINYLATION REACTIONS

Product	Phosphorus reagent	Substrate type	Method	Yield (%)	Refs.
$CH_3O)_2P(O)CH(C_6H_5)CH_2NO_2$	$(CH_3O)_3P$	Nitro	A	26	31
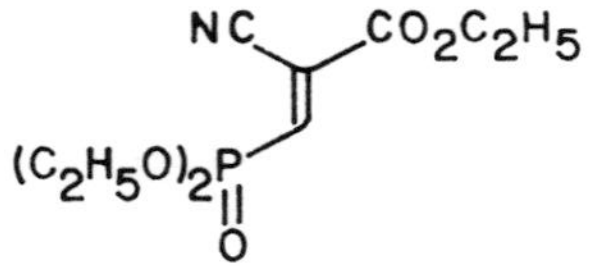 (structure: $NC-C(CO_2C_2H_5)=$ with $(C_2H_5O)_2P(=O)$)	$(C_2H_5O)_2POH$	Ketonitrile	E	79	42
$(C_2H_5O)_2P(O)CH_2CH(CN)Si(CH_3)_3$	$(C_2H_5O)_2POSi(CH_3)_3$	Nitrile	A	62	75
$(C_2H_5O)_2P(O)CH_2CH(CN)Si(CH_3)_3$	$(C_2H_5O)_2POSi(CH_3)_3$	Nitrile	A	70	35
$(C_2H_5O)_2P(O)CH_2CH_2CO_2Si(CH_3)_3$	$(C_2H_5O)_3P$	Acid	D	96	74
$(CH_3O)_2P(O)CH(CH_3)CH=C(CH_3)Si(CH_3)_3$	$(CH_3O)_2POSi(CH_3)_3$	Ketone	A	64	70
$C_6H_5P(O)(OCH_3)CH(CCl_3)CH_2NO_2$	$C_6H_5P(O)(OCH_3)OSi(CH_3)_3$	Nitro	A	50	33
$(C_2H_5O)_2P(O)CH_2CH_2COCH=C(CH_3)_2$	$(C_2H_5O)_2POH$	Ketone	E	52	35
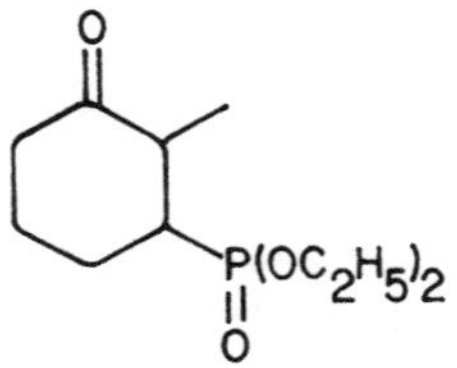	$(C_2H_5O)_2POSi(CH_3)_3$	Ketone	A	94	69
(structure: 2-methyl-cyclohexanone with $P(OC_2H_5)_2$, $=O$)					
$(C_2H_5O)_2P(O)CH_2CH_2CH_2CO_2C_4H_9\text{-}n$	$(C_2H_5O)_2POSi(CH_3)_3$	Ester	A	82	65
$(n\text{-}C_4H_9O)_2P(O)CH_2CH_2CN$	$(n\text{-}C_4H_9O)_2POH$	Nitrile	E	74	43
	$(CH_3O)_3P$	Nitro	A	70	31

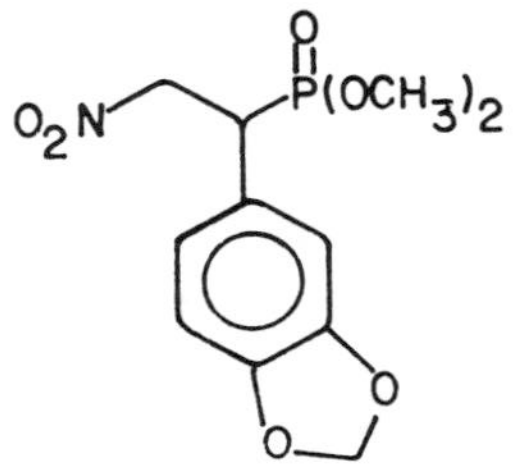

(C₂H₅O)₂P(O)CH₂CH=C(CH₃)OSi(CH₃)₃	(C₂H₅O)₂POSi(CH₃)₃	Ketone	A	91	68
(C₂H₅O)₂P(O)CH(CO₂C₂H₅)CH₂CO₂C₂H₅	(C₂H₅O)₂POH	Ester	E	82	44
(C₂H₅O)₂P(O)CH₂CH(CO₂C₂H₅)₂	(C₂H₅O)₂PN(C₂H₅)₂	Ester	G	81	16
CH₃COCH₂CH₂P(O)(OH)OCH₂CH(C₂H₅)C₄H₉-n	H₄NOP(OH)OCH₂CH(C₂H₅)C₄H₉-n	Ketone	E	66	59
(n-C₄H₉O)₂P(O)CH₂CH(CN)CH₃	(n-C₄H₉O)₂POH	Nitrile	E	87	43
(C₂H₅O)₂P(O)CH(C₆H₅)CH₂NO₂	(C₂H₅O)₂POH	Nitro	A	10	51
[structure: (CH₃O)₂P(O)CH(3,4-methylenedioxyphenyl)CH₂NO₂]	(C₂H₅O)₃P	Nitro	A	80	32
(CH₃O)₂P(O)CH₂CH=C(CH₃)OSi(C₂H₅)₃	(CH₃O)₃P	Ketone	D	76	70
(C₂H₅O)₂P(O)CH₂CH=C(OC₂H₅)OSi(CH₃)₃	(C₂H₅O)₂POSi(CH₃)₃	Ester	A	65	63
(n-C₃H₇O)₂P(O)CH₂CH(CN)Si(CH₃)₃	(C₂H₅O)₃P	Nitrile	D	63	75
(i-C₃H₇O)₂P(O)CH₂CH(CN)Si(CH₃)₃	(C₂H₅O)₂POSi(CH₃)₃	Nitrile	A	55	75
(C₂H₅O)₂P(O)CH₂C(CH₃)=CHOP(O)(OC₂H₅)₂	(C₂H₅O)₃P	Aldehyde	F	68	10
(C₂H₅O)₂P(O)CH(CH₃)CH=CHOP(O)(OC₂H₅)₂	(C₂H₅O)₃P	Aldehyde	F	52	10
C₆H₅(CH₂)₄P(O)(OH)CH₂CH₂CN	C₆H₅(CH₂)₄P(OH)₂	Nitrile	A	99	53
(C₂H₅O)₂P(O)CH(C₆H₅)CH₂CN	(C₂H₅O)₃P	Oxime	A	47	85
(C₂H₅O)₂P(O)CH(C₆H₅)CH₂CONH₂	(C₂H₅O)₂POH	Amide	E	81	45
[structure: (C₂H₅O)₂P(O)CH(4-CH₃O-C₆H₄)CH₂NO₂]	(C₂H₅O)₃P	Nitro	A	51	31

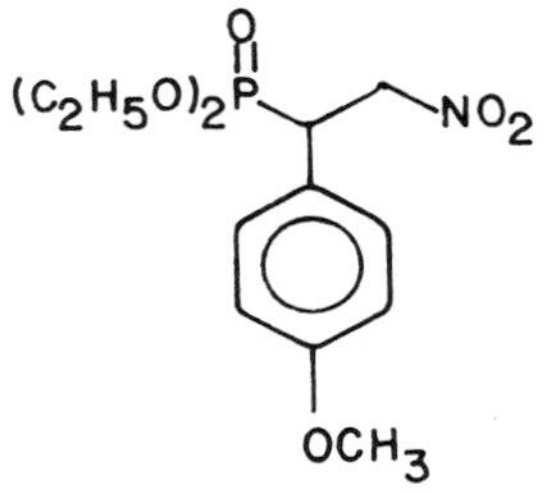

Table 1 (continued)
HYDROPHOSPHINYLATION REACTIONS

Product	Phosphorus reagent	Substrate type	Method	Yield (%)	Refs.
$CH_2P(O)(OC_2H_5)CH(NHCOCH_3)CH(CO_2C_2H_5)_2$	$CH_3P(OC_2H_5)OH$	Ester	E	31	54
$(C_2H_5O)_2P(O)CH_2CH=C(OC_4H_9\text{-}n)OSi(CH_3)_3$	$(C_2H_5O)_3P$	Ester	D	79	74
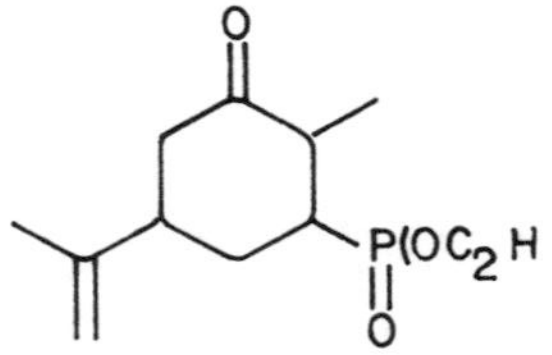	$(C_2H_5O)_2POSi(CH_3)_3$	Ketone	A	82	69
$(C_2H_5O)_2P(O)CH(C_6H_5)CH_2COCH_3$	$(C_2H_5O)_3P$	Ketone	C	89	2
$(C_6H_5O)_2P(O)CH_2CH_2NO_2$	$(C_6H_5O)_3P$	Nitro	C	65	28
$[(CH_3)_2N]_2P(O)CH_2CH=C(CH_3)OSi(C_2H_5)_3$	$[(CH_3)_2N]_2POSi(C_2H_5)_3$	Ketone	A	82	70
$(n\text{-}C_4H_9O)_2P(O)CH_2CH(CN)Si(CH_3)_3$	$(C_2H_5O)_3P$	Nitrile	D	67	75
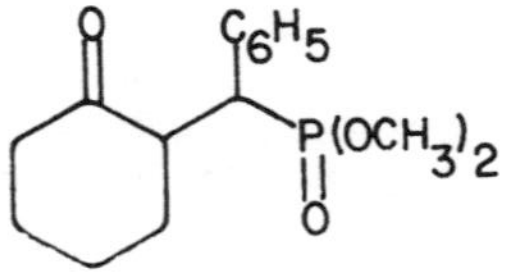	$(C_3O)_2POH$	Ketone	E	79	38
$H_2O_3PCH(C_6H_5)CH_2COC_6H_5$	$[(CH_3)_3SiO]_3P$	Ketone	A	75	76
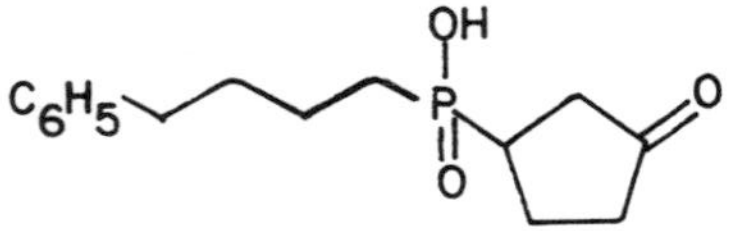	$C_6H_5(CH_2)_4P(OH)_2$	Ketone	D	77	87

$C_6H_5(CH_2)_4P(O)(OC_2H_5)CH_2CH_2CO_2H$	$C_6H_5(CH_2)_4P(OH)OC_2H_5$	Acid	D	61	72
$C_6H_5(CH_2)_4P(O)(OC_2H_5)CH_2CH_2CN$	$C_6H_5(CH_2)_4P(OH)OC_2H_5$	Nitrile	D	97, 63	72, 72
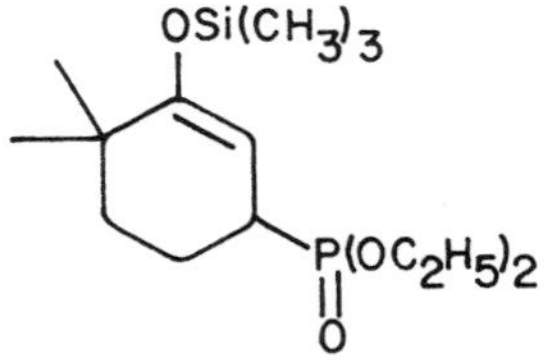	$(C_2H_5O)_2POSi(CH_3)_3$	Ketone	A	90	69
$C_6H_5(CH_2)_4P(O)(OC_2H_5)CH_2CH_2CO_2CH_3$	$C_6H_5(CH_2)_4P(OH)OC_2H_5$	Ester	D	87	87
$(n\text{-}C_4H_9O)_2P(O)CH_2CH(CH_3)CO_2C_4H_9\text{-}n$	$(n\text{-}C_4H_9O)_2POH$	Ester	E	65	43
$(n\text{-}C_4H_9O)_2P(O)CH(CH_3)CH_2CO_2C_4H_9\text{-}n$	$(n\text{-}C_4H_9O)_2POH$	Ester	E	81	43
$C_6H_5P(O)(OC_2H_5)CH(C_6H_5)CH_2NO_2$	$C_6H_5P(OC_2H_5)_2$	Nitro	A	32	31
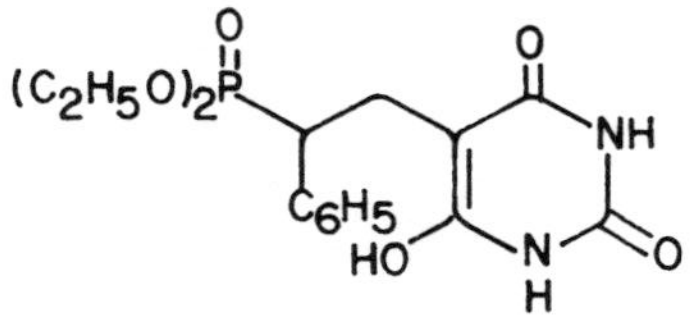	$(C_2H_5O)_2POSi(CH_3)_3$	Amide	B	60	66
$(C_6H_5)_2P(O)CH_2CH_2COC_2H_5$	$(C_6H_5)_2POH$	Ketone	E	55	57
$(C_6H_5)_2P(O)CH(CH_3)CH_2COCH_3$	$(C_6H_5)_2POH$	Ketone	E	58	57
$Ci6H_5(CH_2)_4P(O)(OC_2H_5)CH_2CH(CH_3)CO_2CH_3$	$C_6H_5(CH_2)_4P(OH)OC_2H_5$	Ester	D	93	72
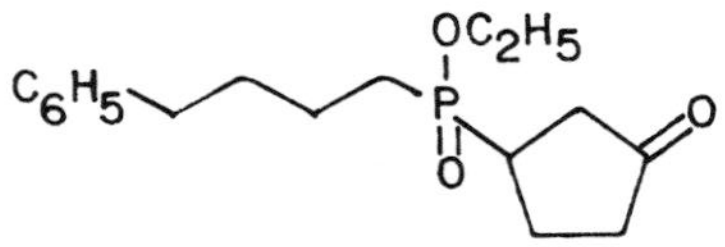	$C_6H_5(CH_2)_4P(OH)OC_2H_5$	Ketone	D	89	72

Table 1 (continued)
HYDROPHOSPHINYLATION REACTIONS

Product	Phosphorus reagent	Substrate type	Method	Yield (%)	Refs.
	(CH$_3$O)$_2$POH C$_6$H$_5$P(CH$_2$C$_6$H$_5$)OH	Ketone Ester	E E	13 45	38 56
	(CH$\leq$O)$_2$PONa	Ketoester	E	98	79
	(CH$_3$O)$_2$PNHC$_6$H$_5$	Amide	A	60	18
	(C$_2$H$_5$O)$_2$POH	Nitro	E	56	50

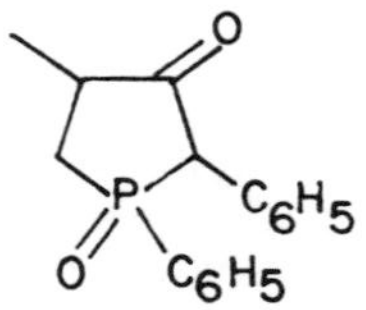

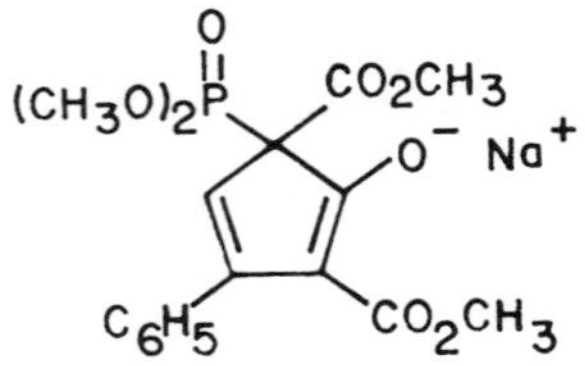

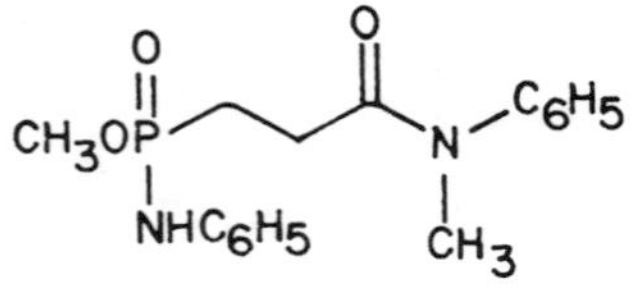

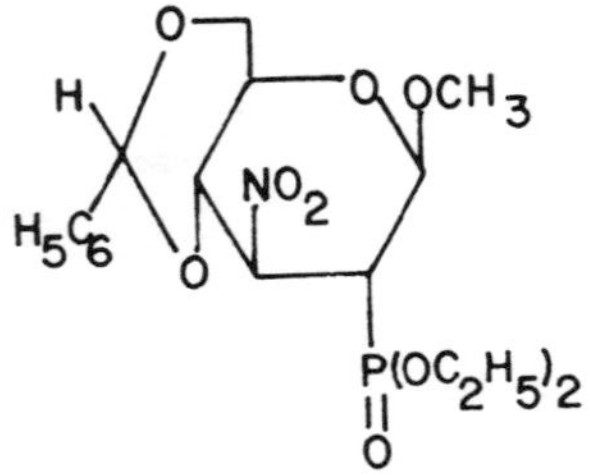

(CH$_3$O)$_2$P(O)CH(C$_6$H$_5$)CH$_2$COC$_6$H$_5$

Structure	Reagent	Type	Method	Yield	Yield
$(n\text{-}C_4H_9)_2P(O)C(CH_3)_2CH_2C(CH_3)=NC_6H_5$ (NHSO₂C₆H₅, P(OCH₃)₂, NHSO₂C₆H₅ structure)	$(n\text{-}C_4H_9)_2POH$ $(CH_3O)_2POH$	Imine Imine	A A	71 45	84 83
$(n\text{-}C_4H_9O)_2P(O)CH(C_6H_5)CH_2CO_2C_4H_9\text{-}n$ (cyclopentanone structure)	$(n\text{-}C_4H_9O)_2POH$ $(CH_3O)_3P$	Ester Ketone	E C	86 42	43 6
$[(CH_3O)_2P(O)CH(C_6H_5)CH_2]_2CO$ (steroid structure)	$(CH_3O)_2POH$ $(CH_3O)_3P$	Ketone Ketone	E C	41 63	39 3
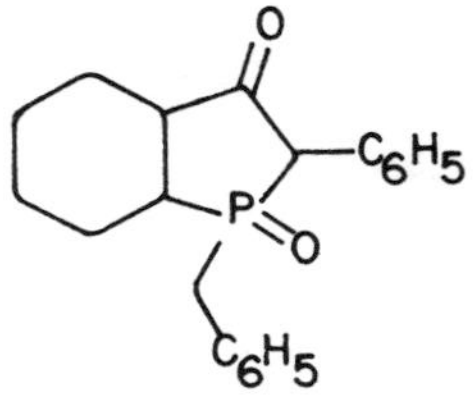 (bicyclic phosphorus structure)	$C_6H_5P(OH)CH_2C_6H_5$	Ester	E	50	56

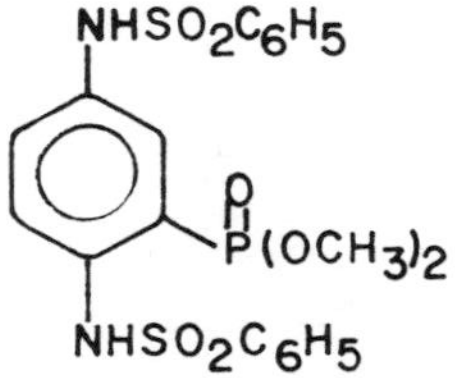

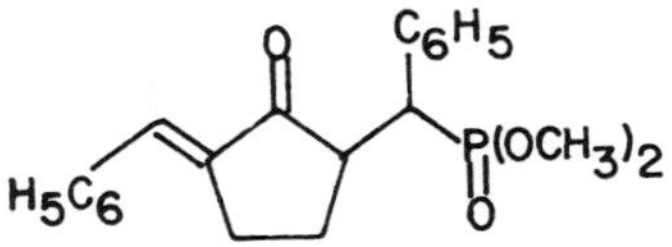

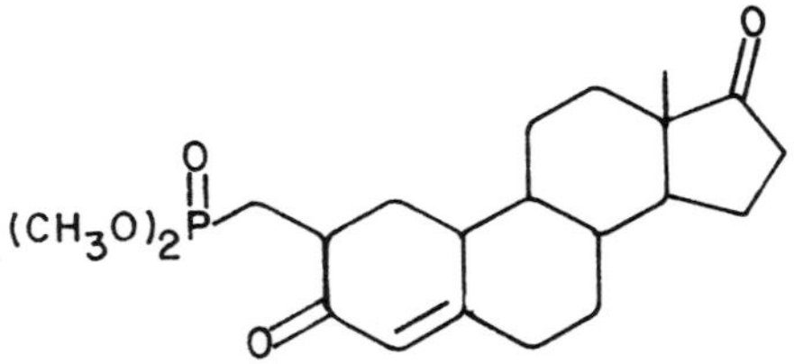

Table 1 (continued)
HYDROPHOSPHINYLATION REACTIONS

Product	Phosphorus reagent	Substrate type	Method	Yield (%)	Refs.
$(C_6H_5)_2P(O)CH_2CH=CHOSi(C_2H_5)_3$	$(C_2H_5)POCH_3$	Aldehyde	D	100	70
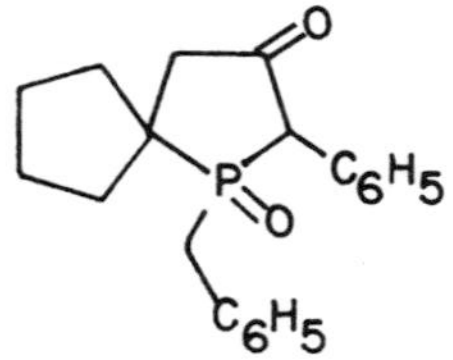	$(C_6H_5CH_2)_2POH$	Ester	E	65	56
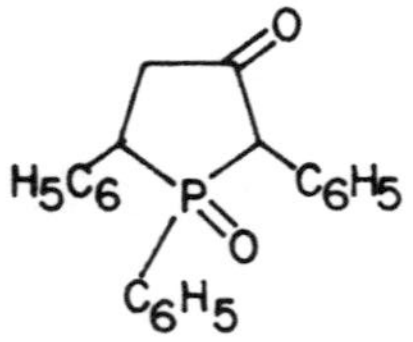	$C_6H_5P(OH)CH_2C_6H_5$	Ester	E	56	56
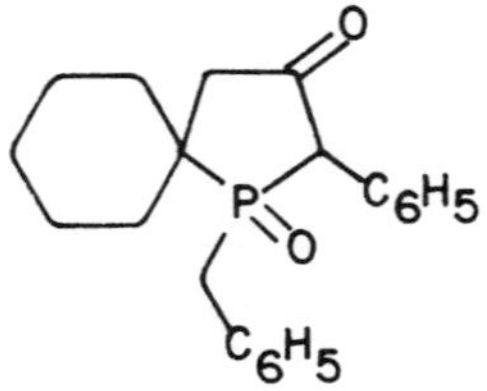	$(C_6H_5CH_2)_2POH$	Ester	E	61	56

Structure	Reagent	Type	Method	Yield	Yield
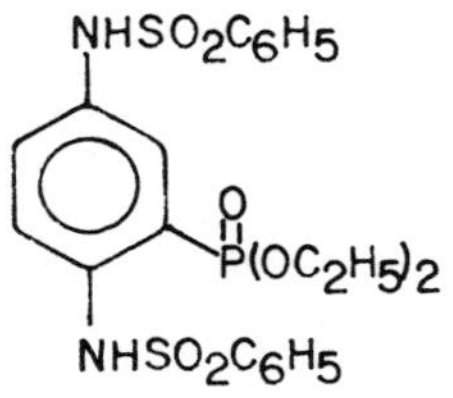 (NHSO₂C₆H₅ substituted benzene with P(OC₂H₅)₂)	$(C_2H_5O)_2POH$	Imine	A	45	83
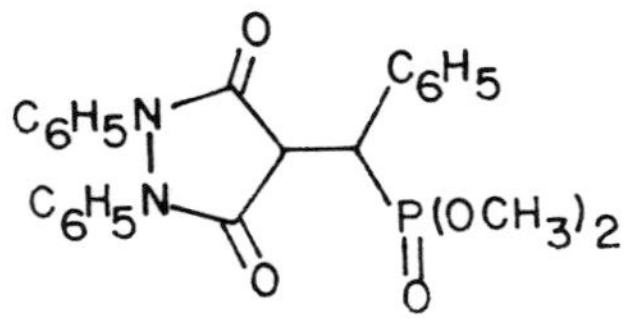 (phenyl pyrazolidinedione with P(OCH₃)₂)	$(CH_3O)_3P$	Amide	A	40	19
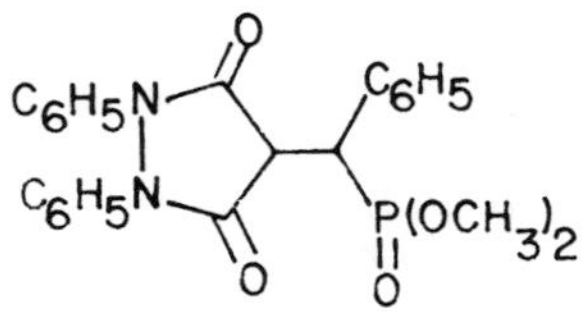 (phenyl pyrazolidinedione with P(OCH₃)₂)	$(CH_3O)_2POH$	Amide	A	60	46
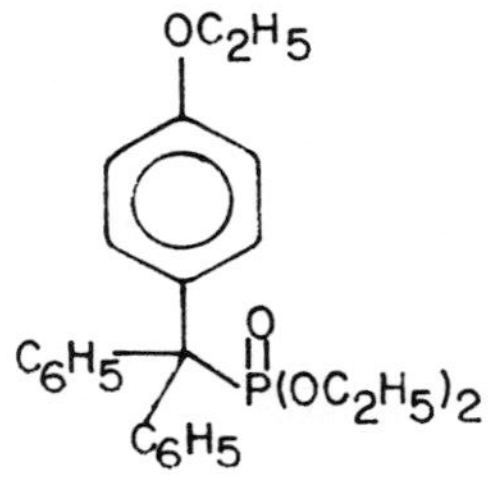 (p-ethoxyphenyl with P(OC₂H₅)₂, C₆H₅)	$(C_2H_5O)_3P$	Ketone	A	82	12

Table 1 (continued)
HYDROPHOSPHINYLATION REACTIONS

Product	Phosphorus reagent	Substrate type	Method	Yield (%)	Refs.
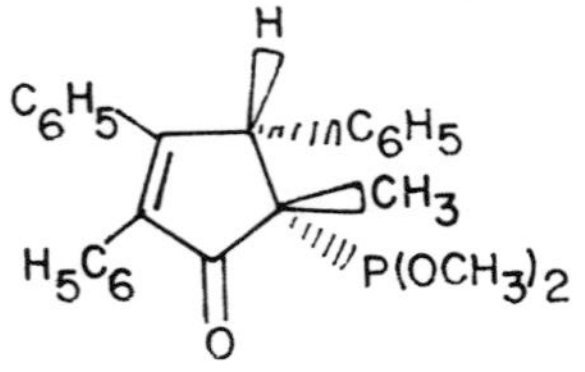	$(CH_3O)_2POH$	Ketone	A	87	82
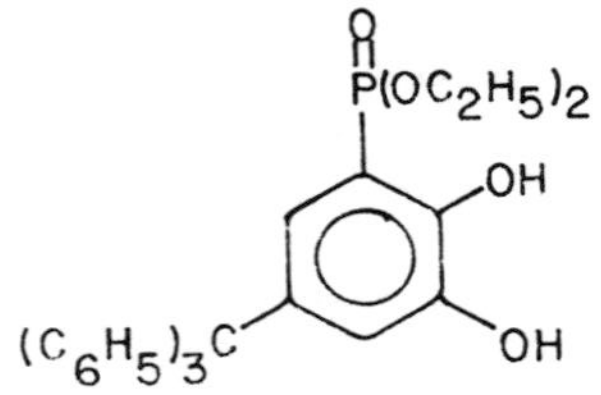	$(C_2H_5O)_2POH$	Ketone	A	92	8
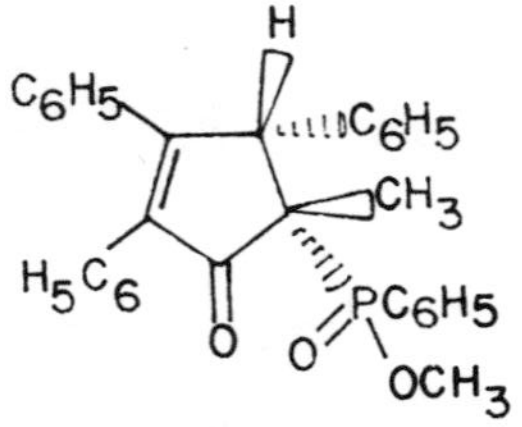	$C_6H_5P(OH)OCH_3$	Ketone	A	77	82

Note: A, reagents heated in the absence of base; B, hydrolytic work-up of reaction used; C, hydroxylic solvent used as reaction adjunct; D, silylating agent present in reaction mixture; E, base used with reactants; F, phosphorus-halide used as trapping agent; G, Mannich-condensation reaction conditions used.

REFERENCES

1. **Harvey, R. G. and Jensen, E. V.,** A novel reductive dimerization: the reaction of 1,2-dibenzoylethylene with P(III) esters, *Tetrahedron Lett.,* p. 1801, 1963.
2. **Harvey, R. G.,** Reactions of triethyl phosphite with activated olefins, *Tetrahedron,* 22, 2561, 1966.
3. **Bravet, J.-L., Benezra, C., and Weniger, J.-P.,** Steroylphosphonates. III. Dimethyl phosphonates in the estrone series, *Steroids,* 19, 101, 1972.
4. **Goldstein, S. L., Braksmayer, D., Tropp, B. E., and Engel, R.,** Isosteres of natural phosphates. 2. Synthesis of the monosidium salt of 4-hydroxy-3-oxobutyl-1-phosphonic acid, an isostere of dihydroxy-acetone phosphate, *J. Med. Chem.,* 17, 363, 1974.
5. **Tronchet, J. M. J., Neeser, J.-R., Gonzalez, L., and Charollais, E. J.,** Preparation of unsaturated sugars phosphonates using nucleophilic conjugate addition, *Helv. Chim. Acta,* 62, 2022, 1979.
6. **Arbuzov, B. A., Tudril, G. A., and Fuzhenkova, A. V.,** Reactions of several α-enones with trimethyl phosphite (transl.) *Bull. Acad. Sci. U.S.S.R.,* 29, 294, 1980.
7. **Arbuzov, B. A., Zoroastrova, V. M., Tudril, G. A., and Fuzhenkova, A. V.,** Reaction of 2,6-dibenzylidenecyclohexanone with trialkyl phosphites and dialkyl hydrogen phosphites, (transl.) *Bull. Acad. Sci. U.S.S.R.,* 21, 2545, 1972.
8. **Sifky, M. M. and Osman, F. H.,** The reaction of alkyl phosphites with 4-triphenylmethyl-1,2-benzoqui-none, *Tetrahedron,* 29, 1725, 1973.
9. **Cristau, H. J., Vors, J. P., Beziat, Y., Niangoran, C., and Cristol, H.,** Umpolung of α,β-ethylenic ketones and aldehydes by phosphorus groups, in *Phosphorus Chemistry, Proceedings of the 1981 International Conference,* Quin, L. D. and Verkade, J. G., Eds., American Chemical Society, Washington, D.C., 1981, 59.
10. **Okamoto, Y.,** A convenient preparation of dialkyl 3-(dialkoxyphosphinyl)-1-propenyl phosphate derivatives: addition of the mixed reagent trialkyl phosphite/dialkyl phosphorochloridate to α, β-unsaturated aldehyde, *Chem. Lett.,* p. 87, 1984.
11. **Okamoto, Y. and Azuhata, T.,** The reaction of diethyl acyl phosphites with α,β-unsaturated carbonyl compounds, *Bull. Chem. Soc. Jpn.,* 57, 2693, 1984.
12. **Shermolovich, Y. G., Markovskii, L. N., Kopeltsiv, Y. A., and Kolesnikov, V. T.,** Reactions of fuchsone with dialkyl hydrogen and trialkyl phosphites (transl.), *J. Gen. Chem. U.S.S.R.,* 50, 649, 1980.
13. **Stamm, H. and Gerster, G.,** Reactions with aziridines. XXI. The (Michaelis-)Arbuzov reaction with *N*-acyl aziridines and other amidoethylations at phosphorus, *Tetrahedron Lett.,* p. 1623, 1980.
14. **Okamoto, Y., Tone, T., and Saurai, H.,** The reaction of alkyl 3-alkoxy-2-bromopropionate with triethyl phosphite, *Bull. Chem. Soc. Jpn.,* 54, 303, 1981.
15. **Ohler, E., Haslinger, E., and Zbiral, E.,** Synthese und ¹H-NMR-Spektren von (3-Acylbicyclo[2.2.1]hept-5-en-2-yl)phosphonsaureestern, *Chem. Ber.,* 115, 1028, 1982.
16. **Ivanov, B. E., Kudryavtseva, L. A., Samurina, S. V., Ageeva, A. B., and Karpova, T. I.,** Reactions of phosphoramidites with malonic ester and paraformaldehyde (transl.), *J. Gen. Chem. U.S.S.R.,* 49, 1552, 1979.
17. **Pudovik, A. N., Batyeva, E. S., and Grifanova, Y. N.,** Reactions of phosphoramidous esters with maleimides (transl.), *J. Gen. Chem. U.S.S.R.,* 43, 1681, 1973.
18. **Pudovik, A. N., Batyeva, E. S., and Yastremskaya, N. V.,** Reactions of dialkyl phenyl phosphoramidites with α,β-unsaturated carboxamides (transl.), *J. Gen. Chem. U.S.S.R.,* 43, 2610, 1973.
19. **Arbuzov, B. A., Sorokina, T. D., Fuzhenkova, A. V., and Vinogradova, V. S.,** Reaction of 1,2-diphenyl-4-benzalpyrrolidine-3,5-dione with trimethyl phosphite and tris(dimethylamino)phosphine (transl.), *Bull. Acad. Sci. U.S.S.R.,* 22, 2577, 1973.
20. **Gusev, Y. K., Chistokletov, V. N., and Petrov, A. A.,** Reactions of some mixed phosphorous esters with acrylonitrile and methyl iodide (transl.), *J. Gen. Chem. U.S.S.R.,* 47, 39, 1977.
21. **Kamai, G. and Kukhtin, V. A.,** Reaction of addition of trialkyl phosphites to some unsaturated acids, *Dokl. Akad. Nauk S.S.S.R.,* 109, 91, 1956; *Chem. Abstr.,* 51, 1827 g, 1957.
22. **Chambers, J. R. and Isbell, A. F.,** A new synthesis of amino phosphonic acids, *J. Org. Chem.,* 29, 832, 1964.
23. **Rho, M. K. and Kim, Y. J.,** Synthesis of DL-aminoalkylphosphonic acids and their derivatives, *Daehan Hwahak Hwoejee,* 17, 135, 1973.
24. **Dixon, H. B. F. and Sparkes, M. J.,** Phosphonomethyl analogues of phosphate ester glycolytic intermediates, *Biochem. J.,* 141, 715, 1974.
25. **Horiguchi, M. and Rosenberg, H.,** Phosphonopyruvic acid: a probable precursor of phosphonic acids in cell-free preparations of *tetrahymena, Biochim. Biophys. Acta,* 404, 333, 1975.
26. **Soroka, M. and Mastalerz, P.,** The synthesis of phosphonic and phosphinic analogs of aspartic acid and asparagine, *Rocz. Chem.,* 50, 661, 1976.

27. **Krueger, W. E. and Maloney, J. R.,** Addition of trimethyl phosphite to β-nitrostyrene, *J. Org. Chem.,* 38, 4208, 1973.

28. **Ranganathan, D., Rao, C. B., and Ranganathan, S.,** Nitroethylene: synthesis of novel 2-nitroethyl-phosphonates, *Chem. Commun.,* p. 975, 1979.

29. **Ranganathan, D., Rao, C. B., Ranganathan, S., Mehrota, A. K., and Iyengar, R.,** Nitroethylene: a stable, clean, and reactive agent for organic synthesis, *J. Org. Chem.,* 45, 1185, 1980.

30. **Gareev, R. D. and Pudovik, A. N.,** Role of alcohols in reactions of phosphorous triesters with 1-nitropropene (transl.), *J. Gen. Chem. U.S.S.R.,* 48, 450, 1978.

31. **Teichmann, H., Thierfelder, W., and Weigt, A.,** P-functionalization of β-nitrostyrenes with trialkyl phosphites, *J. prakt. Chem.,* 319, 207, 1977.

32. **Teichmann, H., Thierfelder, W., Weight, A., and Schafer, E.,** Verfahren zur Herstellung von Substi-tuierten β-Nitroathanphosphonsaure und -phosphinsaureesters, East German Patent 130,354, 1978.

33. **Borisova, E. E., Gareev, R. D., and Shermergorn, I. M.,** Reactions of neutral phosphorus (III) esters with 3,3,3-trichloro-1-nitropropene (transl.), *J. Gen. Chem. U.S.S.R.,* 50, 1786, 1980.

34. **Gareev, R. D., Loginova, G. M., Abulkhanov, A. G., and Pudovik, A. N.,** Reaction of diethyl phenyl phosphite with 1-nitropropene (transl.), *J. Gen. Chem. U.S.S.R.,* 46, 2284, 1976.

35. **Pudovik, A. N. and Arbuzov, B. A.,** Addition of dialkyl phosphites to unsaturated ketones, nitriles, and esters, *Dokl. Akad. Nauk S.S.S.R.,* 73, 327, 1950; *Chem. Abstr.,* 45, 2853b, 1951.

36. **Pudovik, A. N. and Arbuzov, B. A.,** Addition of dialkyl phosphites to unsaturated compounds. I. Addition of dialkyl phosphites to 2,2-dimethylvinyl vinyl ketone, *Zh. Obshch. Khim. S.S.S.R.,* 21, 382, 1951; *Chem. Abstr.,* 45, 7517i, 1951.

37. **Paulsen, H., Greve, W., and Kuhne, H.,** Zuckerphosphonate durch Olefin-addition und Abramov-reak-tion, *Tetrahedron Lett.,* p. 2109, 1971.

38. **Arbuzov, B. A., Fuzhenkova, A. V., Tudril, G. A., and Zoroastrova, V. M.,** Reaction of dimethyl-phosphorous acid with some α,β-unsaturated ketones (transl.), *Bull. Acad. Sci. U.S.S.R.,* 24, 1285, 1975.

39. **Arbuzov, B. A., Tudril, G. A., and Fuzhenkova, A. V.,** Factors controlling the regioselectivity of the addition of dimethyl phosphite to dibenzylidene derivatives of ketones (transl.), *Bull. Acad. Sci. U.S.S.R.,* 28, 1466, 1979.

40. **Arbuzov, B. A., Zoroastrova, V. M., Tudril, G. A., and Fuzhenkova, A. V.,** Reaction of benzalacetone with dimethyl phosphonate (transl.), *Bull. Acad. Sci. U.S.S.R.,* 23, 2541, 1974.

41. **Tewari, R. S. and Shukla, R.,** Reactions of dialkyl phosphites with α, β-unsaturated ketones, *Ind. J. Chem.,* 10, 823, 1972.

42. **Kreutzkamp, N. and Schindler, H.,** Ungesattigte Phosphonsaure-ester aus Hydroxymethylen-athern, *Chem. Ber.,* 92, 1695, 1959.

43. **Johnston, F.,** Production of Diversified Phosphono Derivatives of Polyfunctional Organic Compounds, U.S. Patent 2,754,319, 1956.

44. **Linke, S., Kurz, J., Lipinski, D., and Gau, W.,** Annealation reactions of *N*-heterocycles to condensed pyridones with bridgehead nitrogen, *Ann. Chem.,* 542, 1980.

45. **Barycki, J., Mastalerz, P., and Soroka, M.,** Simple synthesis of 2-aminoethylphosphonic acid and related compounds, *Tetrahedron Lett.,* 3147, 1970.

46. **Sidky, M. M., Soliman, F. M., and Shabana, R.,** Organophosphorus compounds, XIV. Reaction of dialkyl phosphites and thiol acids with 4-benzylidene-1,2-diphenyl-3,5-pyrazolidinedione, *Egypt. J. Chem.,* 15, 79, 1972.

47. **Sharawat, B. S., Handa, I., and Bhatnagar, H. L.,** Reaction of acrylamide with bis(tetrahydrofurfyl) phosphite and diethyl phosphite, *Ind. J. Chem. Sect. A.,* 16A, 306, 1978.

48. **Isbell, A. F., Berry, J. P., and Tansey, L. W.,** The synthesis and properties of 2-aminoethylphosphonic and 3-aminopropylphosphonic acids, *J. Org. Chem.,* 37, 4399, 1972.

49. **Shokol, V. A., Gamaleya, V. F., and Molyavko, L. I.,** 2-(Dialkoxyphosphinyl)ethyl isocyanates (transl.), *J. Gen. Chem. U.S.S.R.,* 44, 87, 1974.

50. **Paulsen, H. and Greve, W.,** Synthese von Aminozuckerphosphonaten durch Addition von Dialkyl phos-phiten an Nitroolefin-Zucker, *Chem. Ber.,* 106, 2114, 1973.

51. **Baranov, G. M., Perekalin, V. V., Ponamarenko, M. V., and Orlovskii, I. A.,** Synthesis, Properties, and Structure of Nitro- and Aminoalkyl(alkene)phosphonates and Their Derivatives, *Khim Primen. Fos-fororg. Soldin., Tr. Konf., 4th,* p. 228, 1969; *Chem. Abstr.,* 78, 124686h, 1973.

52. **Yamamoto, H., Hanaya, T., Kawamoto, H., Inokawa, S., Yamashita, M., Armour, M.-A., and Nakashima, T. T.,** Synthesis and structural analysis of 5-deoxy-5-C-(hydroxyphosphinyl)-*D*-xylo- and glucopyranoses, *J. Org. Chem.,* 50, 3516, 1985.

53. **Thottathil, J. K.,** Phosphinic Acid Intermediate Products, West German Patent 3,434,124, 1985.

54. **Maier, L. and Rist, G.,** Organic phosphorus compounds. 77. Synthesis and properties of phosphinothricin homologs and analogs, *Phosphorus and Sulfur,* 17, 21, 1983.

55. **Cann, P. F., Warren, S., and Williams, M. R.,** Electrophilic substitutions at phosphorus: reactions of diphenylphosphinyl systems with carbonyl compounds, *J. Chem. Soc., Perkin 1,* p. 2377, 1972.

56. **Bodalski, R. and Pietrusiewicz, K.,** A new route to the phospholane ring-system, *Tetrahedron Lett.,* p. 4209, 1972.

57. **Bell, A., Davidson, A. H., Earnshaw, C., Norrish, H. K., Torr, R. S., and Warren, S.,** β-Diphenylphosphinoyl ketones (Ph₂POCH₂CH₂COR): stable reagents for β-ketocarbanions, *Chem. Commun.,* p. 988, 1978.

58. **Maier, L.,** The addition of hypophosphite esters to activated olefins, a new method for preparing 2-substituted ethyl phosphinates, *Helv. Chim. Acta,* 56, 489, 1973.

59. **Laskorin, B. N., Yakshin, V. V., and Bulgakova, V. B.,** Addition of salts of alkyl dihydrogen phosphites to unsaturated compounds (transl.), *J. Gen. Chem. U.S.S.R.,* 46, 2372, 1976.

60. **Cates, L. A. and Li, V.-S.,** Addition of hypophosphorous acid to α,β-unsaturated amides, *Phosphorus and Sulfur,* 21, 187, 1984.

61. Sumitomo Chemical Co., Ltd., α-Hydroxyethylphosphonic Acid, Japanese Patent 60 89,491, 1985.

62. **Hata, T. and Sekine, M.,** Silyl- and stannyl-esters of phosphorus oxyacids — intermediates for the synthesis of phosphate derivatives of biological interest, in *Phosphorus Chemistry Directed Toward Biology,* Stec, W. J., Ed., Pergamon, New York, 1980, 197.

63. **Novikova, Z. S. and Lutsenko, I. F.,** Reaction of trialkylsilyl dialkyl phosphites with unsaturated compounds (transl.), *J. Gen. Chem. U.S.S.R.,* 40, 2110, 1970.

64. **Novikova, Z. S., Mososhina, S. N., Sapozhnikova, T. A., and Lutsenko, I. F.,** Reactions of diethyl trimethylsilyl phosphite with carbonyl compounds (transl.), *J. Gen. Chem. U.S.S.R.,* 41, 2655, 1971.

65. **Okamoto, Y. and Sakurai, H.,** Preparation of (dialkoxyphosphinyl)-methyl-substituted ketone alkyl trimethylsilyl acetal derivatives, *Synthesis,* p. 497, 1982.

66. **Pudovik, A. N., Batyeva, E. S., and Zameletdinova, G. U.,** Reaction of diethyl trimethylsilyl phosphite with benzalbarbituric acid (transl.), *J. Gen. Chem. U.S.S.R.,* 43, 944, 1973.

67. **Pudovik, A. N., Batyeva, E. S., and Zameletdinova, G. U.,** Reactions of diethyl trimethylsilyl phosphite with maleimide (transl.), *J. Gen. Chem. U.S.S.R.,* 45, 922, 1975.

68. **Hata, T., Hashizume, A., Nakajima, M., and Sekine, M.,** A convenient method of ketone synthesis utilizing the reaction of diethyl trimethylsilyl phosphite with carbonyl compounds, *Tetrahedron Lett.,* p. 363, 1978.

69. **Liotta, D., Sunay, U., and Ginsberg, S.,** Phosphonosilylations of cyclic enones, *J. Org. Chem.,* 47, 2227, 1982.

70. **Evans, D. A., Hurst, K. M., and Takacs, J. M.,** New silicon-phosphorus reagents in organic synthesis. Carbonyl and conjugate addition reactions of silicon phosphite esters and related systems, *J. Am. Chem. Soc.,* 100, 3467, 1978.

71. **Evans, D. A., Hurst, K. M., Truesdale, L. K., and Takacs, J. M.,** The carbonyl insertion reactions of mixed tervalent phosphorus-organosilicon reagents, *Tetrahedron Lett.,* p. 2495, 1977.

72. **Thottathil, J. K., Ryono, D. E., Przybyla, C. A., Moniot, J. L., and Neubeck, R.,** Preparation of phosphinic acids: Michael additions of phosphorous acids/esters to conjugated systems, *Tetrahedron Lett.,* p. 4741, 1984.

73. **Evans, D. A., Takacs, J. M., and Hurst, K. M.,** Phosphoramide stabilized allylic carbanions. New homoenolate anion equivalents, *J. Am. Chem. Soc.,* 101, 371, 1979.

74. **Okamoto, Y., Azuhata, T., and Sakurai, H.,** Dialkyl 3-alkoxy-3-(trimethylsiloxy)-2-propenephosphonate; a one step preparation of (dialkoxyphosphinyl)methyl-substituted ketene alkyl trimethylsilyl acetal, *Chem. Lett.,* p. 1265, 1981.

75. **Nakano, M., Okamoto, Y., and Sakurai, H.,** Preparation of dialkyl 2-cyano-2-(trimethylsilyl)-ethenephosphonates, *Synthesis,* p. 915, 1982.

76. **Sekine, M., Yamamoto, I., Hashizume, A., and Hata, T.,** The reaction of tris(trimethylsilyl) phosphite with carbonyl compounds, *Chem. Lett.,* p. 485, 1977.

77. **Arbuzov, B. A., Dianova, E. N., and Vinogradova, V. S.,** Reaction of 3-methyl-4-benzylideneisoxazoline-5-one and 1,3-diphenyl-4-benzylidene-5-pyrazoline with tris(dimethylamino)phosphine (transl.), *J. Gen. Chem. U.S.S.R.,* 42, 742, 1972.

78. **Arbuzov, B. A., Fuzhenkova, A. V., and Rozhkova, R. A.,** Reaction of trimethyl phosphite with 2,5-bismethoxycarbonyl-3,4-diphenylcyclopentadienone (transl.), *J. Gen. Chem. U.S.S.R.,* 52, 10, 1982.

79. **Arbuzov, B. A., Fuzhenkova, A. V., Galyautdinov, N. I., and Saikhullina, R. F.,** Reaction of dimethylphosphorous acid with 2,5-bis(carbomethoxy)-3,4-diphenylcyclopentadienone (transl.), *Bull. Acad. Sci. U.S.S.R.,* 29, 826, 1980.

80. **Arbuzov, B. A., Fuzhenkova, A. V., and Galyautdinov, N. I.,** Reaction of dimethyl phosphite with tetracyclone (transl.), *Bull. Acad. Sci. U.S.S.R.,* 27, 381, 1978.

81. **Arbuzov, B. A., Fuzhenkova, A. V., and Galyutdinov, N. I.,** Reaction of tetracyclone with the ethyl ester of phenylphosphonous acid (transl.), *Bull. Acad. Sci. U.S.S.R.,* 28, 1258, 1979.

82. **Arbuzov, B. A., Fuzhenkova, A. V., and Galyutdinov, N. I.,** Reaction of dimethyl phosphite and monoethyl phenylphosphonite with 2-methyl-3,4,5-triphenylcyclopentadienone (transl.), *Bull. Acad. Sci. U.S.S.R.,* 29, 651, 1980.

83. **Sidky, M. M., Mahran, M. R., and Zayed, M. F.,** Organophosphorus compounds. XXIX. On the reaction of dialkyl phosphites with *p*-benzoquinonedibenzenesulfonamide, *Phosphorus and Sulfur,* 9, 337, 1981.

84. **Sobanov, A. A., Bakhtiyarov, I. V., Zimin, M. G., and Pudovik, A. N.,** Reaction of unsaturated ketimines with hydrophosphoryl compounds (transl.), *J. Gen. Chem. U.S.S.R.,* 55, 1059, 1985.

85. **Osipova, M. P., Mikhailova, L. A., and Kukhtin, V. A.,** Reaction of triethyl phosphite with cinnamaldehyde oxime (transl.), *J. Gen. Chem. U.S.S.R.,* 52, 394, 1982.

86. **Onysko, P. P., Gololobov, Y. G., Remennikov, G. Y., and Cherkasov, V. M.,** Anionic σ complexes of phosphorus compounds. VII. σ Complexes of dialkyl hydrogen phosphites with 5-nitropyrimidines (transl.), *J. Gen. Chem. U.S.S.R.,* 50, 995, 1980.

87. **Thottathil, J. K., Ryono, D. E., Przybyla, C. A., Moniot, J. L., and Neubeck, R.,** Preparation of phosphinic acids: Michael additions of phosphonous acids/esters to conjugated systems, *Tetrahedron Lett.,* p. 4741, 1984.

Chapter 5

FORMATION OF CARBON-PHOSPHORUS BONDS FROM PHOSPHORUS-HALIDES

I. INTRODUCTION

Several fundamental approaches are available for the generation of carbon-phosphorus bonds starting with phosphorus-halide compounds. These approaches, in general, are found to involve one of two types of reactivity of the phosphorus-halide compound. The phosphorus-halide compound can react via nucleophilic attack using an unshared electron pair on phosphorus [P(III) compounds only]. Alternatively, a carbon-phosphorus bond may be generated as a result of the high susceptibility of the acid halide for overall displacement by carbanionic species.

In either instance, some major difficulties stem from the high reactivity of the phosphorus-halide compound. Primarily, phosphorus polyhalides (as are often used) are capable of undergoing multiple reactions with nucleophilic agents. The particulars of this difficulty are addressed in the discussion of each reaction system. Also, the phosphorus-halide compounds are capable of undergoing extraneous reactions with a variety of functional groups which might be intended for incorporation in the target molecule.

Lesser difficulties arise in obtaining the particular phosphorus-halide compound most suitable for an efficient synthetic route. While phosphorus trichloride or phosphorus oxychloride are often suitable, allowing initial introduction of the carbon-phosphorus bond with halogen reactivities remaining on phosphorus, in other instances structurally more complex species will be found to be more suitable.

These structurally complex phosphorus-halide compounds, often already including one carbon-phosphorus bond, may be generated by several methods. Given this situation, a brief review here of phosphorus-halogen bond-forming processes is appropriate.

The phosphorodichloridates can be prepared by controlled reaction of the appropriate alcohol (or phenol) with a phosphorus polyhalide. For example, n-alcohols react with phosphorus oxychloride in inert solvent to generate the monoester-dichlorides (alkyl phosphorodichloridites) in moderate to good yield.[1,2]

$$\underline{n}\text{-}C_7H_{15}OH \quad + \quad POCl_3 \quad \longrightarrow \quad \underline{n}\text{-}C_7H_{15}OP(O)Cl_2 \quad + \quad HCl$$

$$67\%$$

Phosphorus pentachloride may be used as well, although an extra equivalent of alcohol is required. With the *in situ* generation of phosphorus oxychloride, the second equivalent of alcohol reacts to give the alkyl phosphorodichloridate.[3]

Aromatic esters can also be generated by the phosphorus oxychloride route, although there is usually added a Lewis acid catalyst to the reaction mixture. Aluminum chloride used with a twofold amount of phosphorus oxychloride and the phenol leads to yields that may be classed good or excellent.[4]

$$89\%$$

Similarly, the monoester-dichlorides of phosphorous acid (alkyl phosphorodichloridites) may be prepared by controlled reaction of alcohols or phenols with phosphorus trichloride.[5,6] It should be noted that significantly better yields are obtained with phenols than with alcohols, and that primary alcohols give better yields than secondary alcohols. Moreover, the use of bases (such as tertiary amines) is not advised in these reactions. The hydrogen chloride generated is simply vented from the reaction system. The alkyl phosphorodichloridites must be handled with caution. Not only are they highly susceptible to hydrolysis from moisture in the atmosphere, but many of these trivalent phosphorus dihalides are flammable in contact with air.

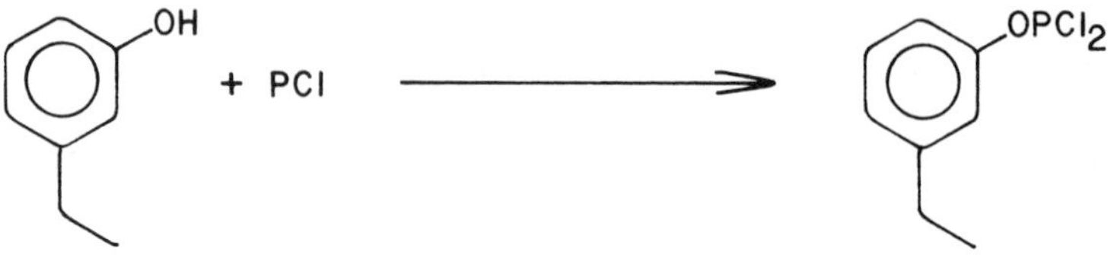

93%

While it is possible to perform controlled addition of alcohol to phosphorus polyhalides to generate the dialkyl phosphorochloridates, more often other more easily managed reaction systems are used. The two most facile approaches to these materials involve oxidative halogenation of P(III) esters.

Trialkyl phosphites react readily with halogens by a route analogous to the Michaelis-Arbuzov reaction. The reaction generally proceeds in good to excellent yield and generates an equivalent of alkyl halide. Either chlorine or bromine may be used without difficulty.[7]

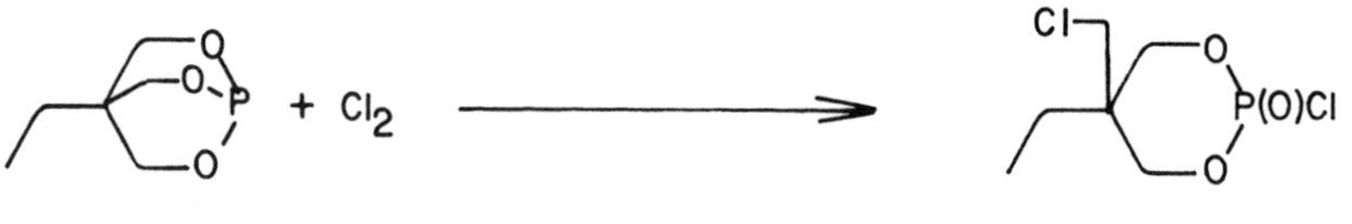

87%

An alternative route, commonly known as the Todd reaction, begins with the dialkyl phosphite and uses rather mild reaction conditions. The dialkyl phosphite is mixed with an excess of carbon tetrachloride, and a catalytic amount of a tertiary amine, generally triethyl amine, is added.[8-10] The reaction is recognized to proceed by reaction of the phosphorus anion with the carbon tetrachloride to give the dialkyl phosphorochloridate and trichloro-methide ion. The latter generates chloroform as an easily removed by-product.

$$CCl_4 \quad + \quad (C_2H_5O)_2POH \quad \xrightarrow{(C_2H_5)_3N} \quad (C_2H_5O)_2P(O)Cl \quad + \quad CHCl_3$$

78%

Other oxidative chlorinating agents have also been used successfully with dialkyl phosphites. These include cupric chloride,[11] sulfuryl chloride,[12] and chlorine.[13] All generate the dialkyl phosphorochloridate in good to excellent yield.

For the preparation of phosphonylchloridates, monoesters-monoacid chlorides of phosphonic acids, dialkyl phosphonates are generally used as the starting phosphorus reagent. These phosphonates are readily prepared via any of the numerous routes discussed in this work.

A single ester linkage of the phosphonate may be replaced by halogen in high yield using one of several reagents. Phosgene reacts cleanly with the phosphonate diesters to generate the target phosphonochloridate, an equivalent of alkyl halide, and carbon dioxdide.[14]

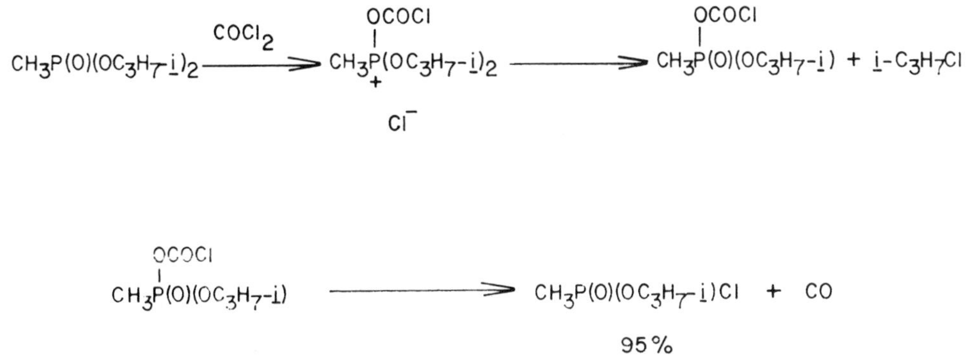

$$CH_3P(O)(OC_3H_7\text{-}\underline{i})_2 \xrightarrow{\ COCl_2\ } \overset{\displaystyle OCOCl}{\underset{\displaystyle +}{\overset{|}{CH_3\overset{}{P}(OC_3H_7\text{-}\underline{i})_2}}} \longrightarrow CH_3P(O)(OC_3H_7\text{-}\underline{i}) + \underline{i}\text{-}C_3H_7Cl$$

$$Cl^-$$

$$\overset{\displaystyle OCOCl}{\overset{|}{CH_3P(O)(OC_3H_7\text{-}\underline{i})}} \longrightarrow CH_3P(O)(OC_3H_7\text{-}\underline{i})Cl + CO$$

95%

An alternate method involves the treatment of the phosphonate diester with phosphorus pentachloride.[15] Also formed in this reaction are the alkyl halide and phosphorus oxychloride. The target phosphonochloridate may often be separated cleanly from the by-products by fractional distillation.

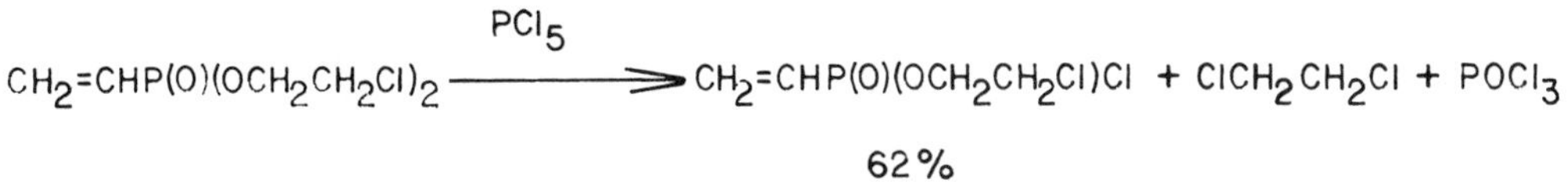

$$CH_2=CHP(O)(OCH_2CH_2Cl)_2 \xrightarrow{\ PCl_5\ } CH_2=CHP(O)(OCH_2CH_2Cl)Cl + ClCH_2CH_2Cl + POCl_3$$

62%

As only two of the chlorides of the phosphorus pentachloride are used in this conversion, triphenylphosphine dichloride might be anticipated to yield a similar result.[16,17] This has been observed in some recent work wherein the phosphonochloridate is generated in high yield with a by-product of triphenylphosphine oxide.[18] This last method has the advantage of providing minimal extraneous reactivity and allows relatively facile isolation of the phosphonochloridate.

$$CH_3P(O)(OCH_3)_2 \xrightarrow{\ (C_6H_5)_3PCl_2\ } CH_3P(O)(OCH_3)Cl + (C_6H_5)_3P(O) + CH_3Cl$$

99%

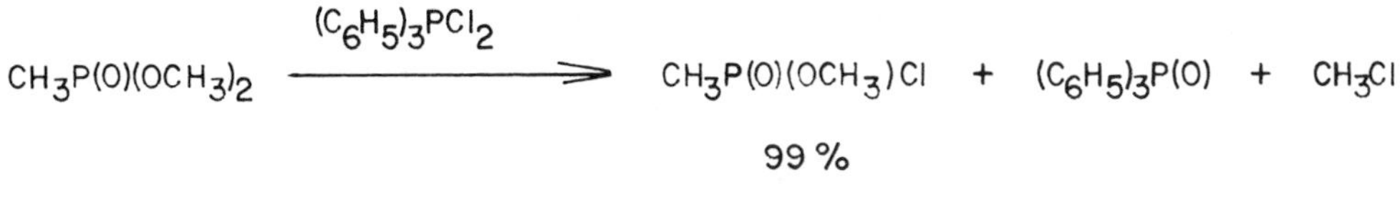

Thionophosphonic monoesters have also been reported to be converted into the posphonothiochloridates by reaction with either phosgene or phosphorus pentachloride. These reactions are notable in that they proceed both with high regio- and stereoselectivity. In reaction with chiral phosphonothioate monoesters, phosgene reacts selectively at the sulfur site generating the chiral phosphonochloridate of inverted configuration, with carbonyl sulfide as the by-product.[19] In contrast, the fully esterified species, the O,S-dialkyl alkylphosphonothioates, do not react with phosgene.[19]

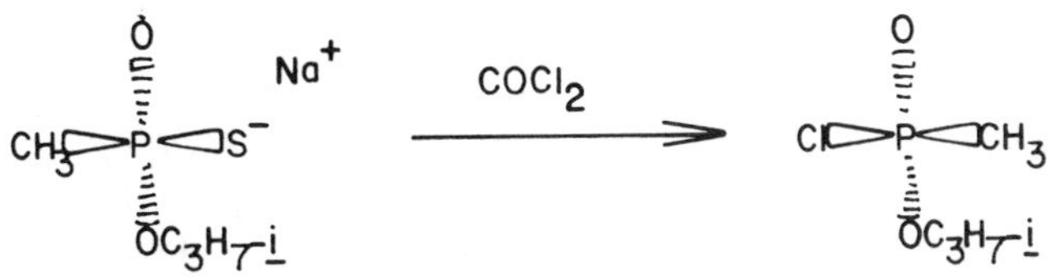

Phosphorus pentachloride, in contrast to phosgene, reacts with the same type of substrate to remove the oxygen atom only (as phosphorus oxychloride) and generate the phosphonothiochloridate with inversion of absolute configuration.[20]

$$C_2H_5O\cdots\overset{\displaystyle \overset{S}{\|}}{P}\diagdown_{OH} \quad \xrightarrow{\quad PCl_5 \quad} \quad Cl\diagup\overset{\displaystyle \overset{S}{\|}}{P}\cdots OC_2H_5$$

The methods for generation of phosphorus-halide compounds illustrated here are those of most general applicability which should be of greatest use in the preparation of materials for carbon-phosphorus bond formation. Other methods with their particular significance and utility are available.[21]

II. CARBANIONIC DISPLACEMENT REACTIONS AT PHOSPHORUS

As may be anticipated whenever a good (or moderate) leaving group is attached at a relatively electropositive center, carbanionic displacement of halide from phosphorus occurs readily to generate carbon-phosphorus bonds. These reactions are found to occur at both trivalent and quinquevalent phosphorus centers and are useful in a wide range of synthetic schemes.

Early efforts in the use of organometallics with phosphorus-halide compounds to generate carbon-phosphorus bonds involved organomercury reagents. These materials react with phosphorus trichloride to generate the monosubstitution products in reasonable yield.[22]

$$(n\text{-}C_4H_9)_2Hg \;+\; PCl_3 \quad \xrightarrow{\quad\quad} \quad n\text{-}C_4H_9PCl_2$$
$$61\%$$

The organomercurial route was particularly convenient for introducing olefinic functions directly on phosphorus.[23] However, the inherent toxicity of the organomercurials and their difficulty in preparation generally override their favorable aspects.

Relatively little has been done in recent years with these reagents, except for some special applications. For example, advantage was taken of the availability and specificity of divinylmercury in the preparation of vinyldifluorophosphine, involving the preferential displacement of bromide from difluorobromophosphine.[24] A review has appeared recently summarizing much of the early application of organomercurials to the generation of carbon-phosphorus bonds.[25]

An early alternative to the organomercurials, which still presents significant toxicological and reagent-preparation difficulties, is the use organolead compounds. Tetraethyllead, for example, can be used for the selective monoalkylation of phosphorus trichloride in high yield.[26]

$$(C_2H_5)_4Pb \;+\; PCl_3 \quad \xrightarrow{\quad\quad} \quad C_2H_5PCl_2$$
$$89\%$$

These two classes of reagents share the same advantages and disadvantages, the principal advantage being their ability to give controlled reaction. Reaction with phosphorus trichloride is easily controlled such that substitution for only one of the three available chlorides occurs.

Of course, very early it was known that Grignard reagents could be used for the alkylation of phosphorus by displacement of halogen. The preparation of tertiary phosphine oxides by the reaction of phosphorus oxychloride with excess of the Grignard reagent proceeds readily in good yield.[27,28] Similarly, tertiary phosphines may be prepared by reaction of an excess of the Grignard reagent with phosphorus trichloride.[29-32]

However, a major difficulty with the Grignard reagent is its high reactivity and subsequent low selectivity. Controlling the reaction when several identical or similar leaving groups are present is extremely difficult.[33]

Although ester linkages on phosphorus can be displaced by Grignard reagents,[34,35] a selectivity does exist for the more easily displaced halides when the amount of carbanionic reagent is limited. This selectivity, leaving ester functions untouched, is found with both quinquevalent and trivalent phosphorus centers.

For example, aromatic Grignard reagents react with diethyl phosphorochloridate to generate the diethyl arylphosphonate in moderate to excellent yield.[36]

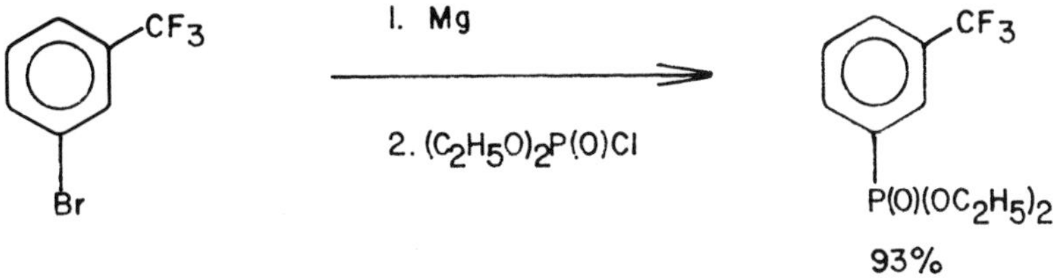

Similarly, acetylenic Grignard reagents are found to displace chloride selectively from phosphonochloridates.[37]

$$HC{\equiv}C\,MgCl \quad + \quad CH_3P(O)(OCH_3)Cl \quad \xrightarrow[-20°]{THF} \quad HC{\equiv}CP(O)(OCH_3)CH_3$$

50 %

It should be noted that the use of lithium acetylides in place of the Grignard reagent results in significantly lower yields of the same product, although aluminum acetylides give the desired materials in moderate to good yields.[38]

$$(RC{\equiv}C)_4\,Al\,Li \quad + \quad (R'O)_2P(O)Cl \quad \longrightarrow \quad RC{\equiv}CP(O)(OR')_2$$

Halogen displacement by Grignard reagent from dialkyl phosphorochloridites is found to proceed with high selectivity over ester displacement. The use of this reaction system for the synthesis of dialkyl alkylphosphonites in moderate to excellent yields has been noted in several recent patents.[39,40]

$$(C_2H_5O)_2PCl \quad + \quad CH_3MgBr \quad \xrightarrow[0°]{THF} \quad (C_2H_5O)_2PCH_3$$

82%

Chiral phosphonous esters and phosphines have also been synthesized using to good advantage a selectivity for displacement of aromatic esters preferentially to aliphatic esters by Grignard reagents.[41] Controlled reaction of phenyldichlorophosphine with the lithium alkoxide form of cinchonine yields a pair of diastereoisomers which are further esterified with phenol. Aromatic Grignard reagents react selectively with the mixed chiral ester displacing the aromatic ester function with inversion of configuration of phosphorus.[42] The chiral phosphine is then generated by further displacement reaction using a Grignard or alkyl lithium reagent with final purification by complexation with cuprous halide.

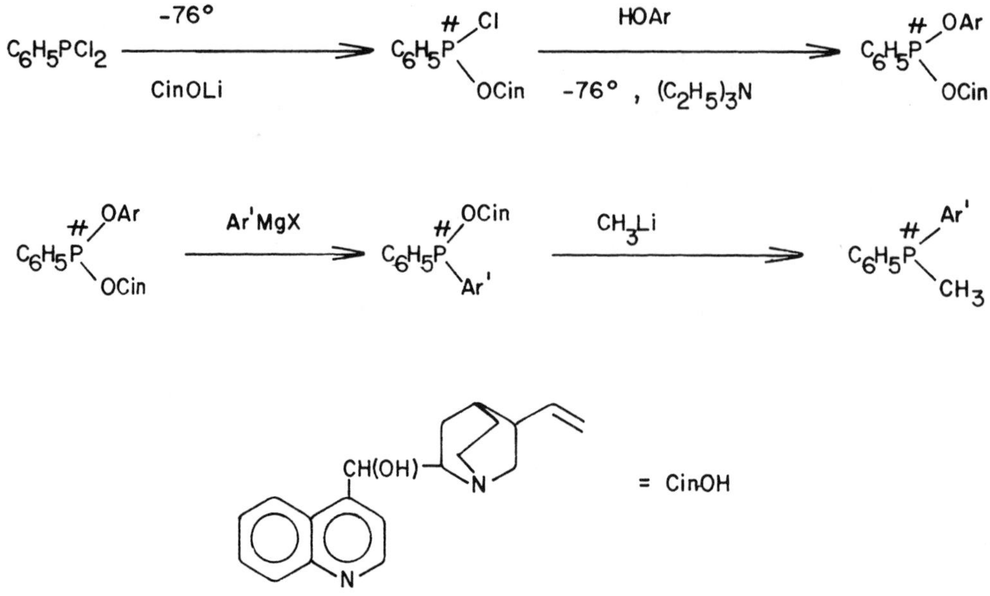

$$C_6H_5PCl_2 \xrightarrow[\text{CinOLi}]{-76°} C_6H_5P\overset{H}{\underset{\text{OCin}}{\diagup}}\overset{Cl}{} \xrightarrow[-76° , (C_2H_5)_3N]{\text{HOAr}} C_6H_5P\overset{H}{\underset{\text{OCin}}{\diagup}}\overset{OAr}{}$$

$$C_6H_5P\overset{H}{\underset{\text{OCin}}{\diagup}}\overset{OAr}{} \xrightarrow{\text{Ar'MgX}} C_6H_5P\overset{H}{\underset{\text{Ar'}}{\diagup}}\overset{OCin}{} \xrightarrow{CH_3Li} C_6H_5P\overset{H}{\underset{CH_3}{\diagup}}\overset{Ar'}{}$$

$$\text{CH(OH)} = \text{CinOH}$$

The high reactivity of the Grignard reagents may be used conveniently for exhaustive alkylation of other phosphorus halides. Alkyldichlorophosphines,[43,44] dialkylchlorophosphines,[45] diarylphosphinochloridates,[46] diarylphosphinothiochloridates,[47] and phosphonochloridates[48] react with excess of Grignard reagent for the replacement of each available halogen.

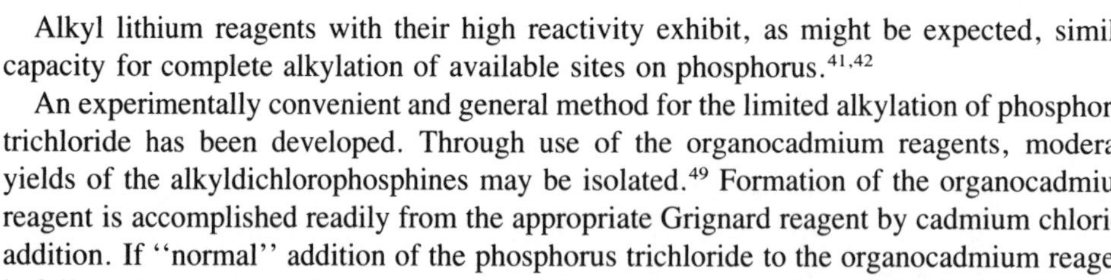

$$\overset{PCl_2}{} \xrightarrow{CH_3MgI} \overset{P(CH_3)_2}{}$$

80%

Alkyl lithium reagents with their high reactivity exhibit, as might be expected, similar capacity for complete alkylation of available sites on phosphorus.[41,42]

An experimentally convenient and general method for the limited alkylation of phosphorus trichloride has been developed. Through use of the organocadmium reagents, moderate yields of the alkyldichlorophosphines may be isolated.[49] Formation of the organocadmium reagent is accomplished readily from the appropriate Grignard reagent by cadmium chloride addition. If "normal" addition of the phosphorus trichloride to the organocadmium reagent is followed, a mixture of products representing mono-, di-, and trisubstitution is obtained. "Reverse" addition of the organocadmium reagent to an excess of the phosphorus trichloride results in formation of the monosubstitution product contaminated by only minimal amounts of higher boiling materials. This approach has been used in recent years for the introduction of the first alkyl group in the synthesis of unsymmetrical tertiary phosphines.[33,43]

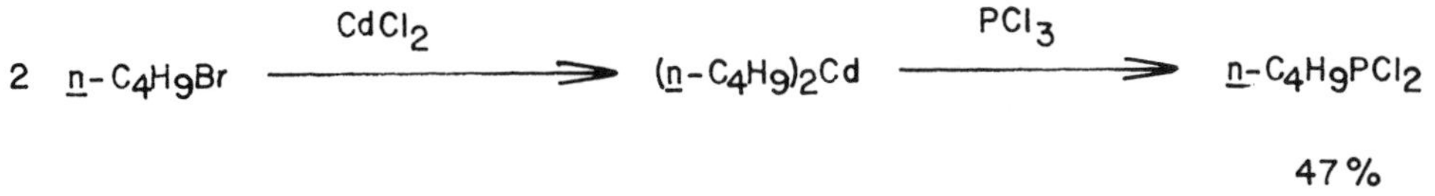

$$2 \quad \underline{n}-C_4H_9Br \xrightarrow{\quad CdCl_2 \quad} (\underline{n}-C_4H_9)_2Cd \xrightarrow{\quad PCl_3 \quad} \underline{n}-C_4H_9PCl_2$$

47%

Recently, organotin reagents have been employed with phosphorus polyhalides for the introduction of a single unsaturated carbon function.[50] For example, tetravinyltin reacts with phosphorus tribromide to give the vinyldibromophosphine in fair yield. Using ethynyl(tributyl)tin with phosphorus pentafluoride, a significantly better yield of the phosphorane is obtained, which is subsequently converted into the ethynylphosphonodifluoridate.

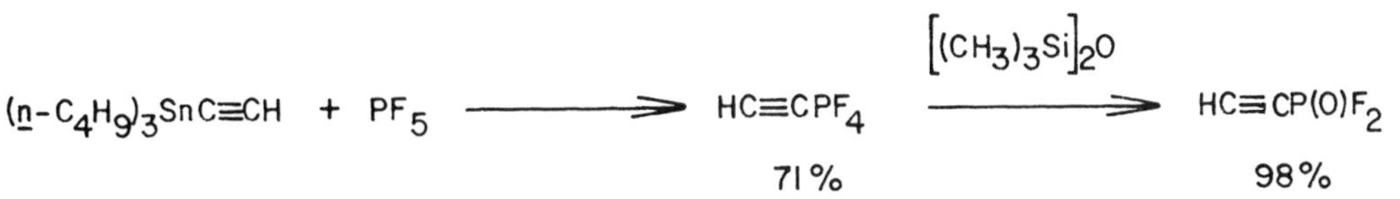

$$(\underline{n}-C_4H_9)_3SnC\equiv CH + PF_5 \longrightarrow HC\equiv CPF_4 \xrightarrow{\quad [(CH_3)_3Si]_2O \quad} HC\equiv CP(O)F_2$$

71%   98%

Although the treatment of phosphorus-halide compounds with enolate-type stabilized carbanionic reagents generally leads to the formation of hetero-atom-to-phosphorus bonds, there are certain exceptions. For example, in the reaction of the enolate anion of acetophenone with diethyl phosphorochloridate, the major product is the enol-phosphate. The phosphonate is formed only as a minor side-product.[51]

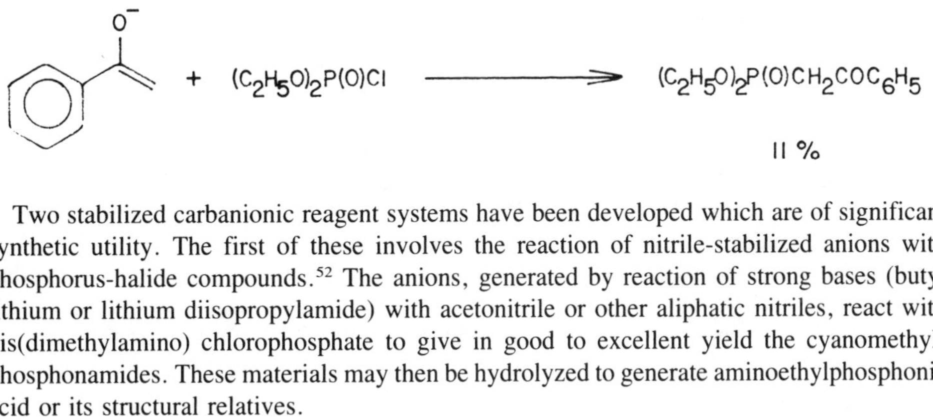

$$\text{(enolate anion of acetophenone)} + (C_2H_5O)_2P(O)Cl \longrightarrow (C_2H_5O)_2P(O)CH_2COC_6H_5$$

11 %

Two stabilized carbanionic reagent systems have been developed which are of significant synthetic utility. The first of these involves the reaction of nitrile-stabilized anions with phosphorus-halide compounds.[52] The anions, generated by reaction of strong bases (butyl lithium or lithium diisopropylamide) with acetonitrile or other aliphatic nitriles, react with bis(dimethylamino) chlorophosphate to give in good to excellent yield the cyanomethylphosphonamides. These materials may then be hydrolyzed to generate aminoethylphosphonic acid or its structural relatives.

$$[(CH_3)_2N]_2P(O)Cl + LiCH_2CN \longrightarrow [(CH_3)_2N]_2P(O)CH_2CN$$

95 %

A more recently developed reaction system involves the use of the stabilized anion of dithiane in reaction with diphenylchlorophosphine.[53] The resultant tertiary phosphine undergoes air oxidation spontaneously during the work-up procedure yielding the 2-[1,3]dithianyldiphenylphosphine oxide in moderate yield.

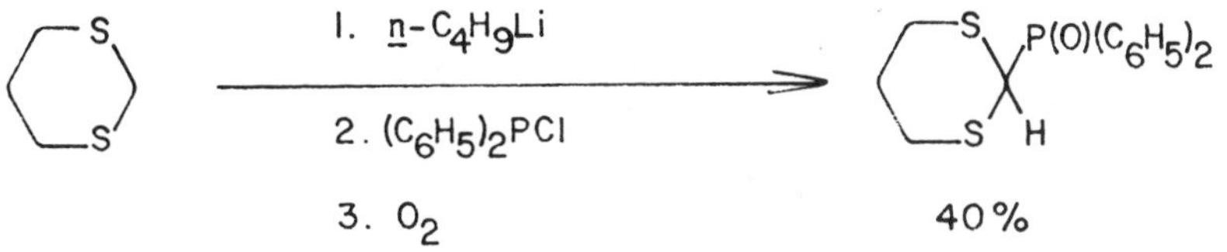

$$\text{(1,3-dithiane)} \xrightarrow[\substack{2.\ (C_6H_5)_2PCl \\ 3.\ O_2}]{1.\ \underline{n}-C_4H_9Li} \text{(2-[1,3]dithianyldiphenylphosphine oxide, } P(O)(C_6H_5)_2)$$

40%

The synthetic value of this material lies in its facile reaction with base and subsequent condensation with carbonyl compounds to generate ketene dithioketals. It would be anticipated that this primary product would also be capable of undergoing direct alkylation reactions. This would constitute a new approach to the synthesis of 1-ketophosphine oxides, materials which would have their own utility in Wittig-Horner type reactions.[54]

III. ALUMINUM HALIDE-MEDIATED REACTIONS

Aluminum chloride-mediated reaction of alkyl halides with phosphorus trichloride has also been used successfully for the introduction of a single alkyl function. Methyldichlorophosphine, a reagent of utility for the further syntheses of a wide variety of organophosphorus compounds, is prepared conveniently and in high yield by this approach,[55,56] as are other alkyldichlorophosphines.[57]

$$CH_3I + PCl_3 + AlCl_3 \longrightarrow \left[CH_3PCl_3\right]^+ \left[AlCl_3I\right]^- \xrightarrow{\text{KCl, Fe}} CH_3PCl_2$$

80%

High yield conversions have also been accomplished using phosphorus trichloride with benzene[58] and bromoform with phosphorus tribromide and aluminum bromide.[59]

Controlled alkylation of phosphorus oxychloride is also accomplished using a modification of this approach. Reaction of alkylaluminum dichloride with phosphorus oxychloride generates the aluminum chloride complex of the phosphonodichloridate.[60] These complexes can be used directly for further functionalization of the phosphorus center. For example, reaction with secondary amines gives the phosphonodiamides in fair to good yield.

Aryldichlorophosphines can be produced via a Friedel-Crafts-type reaction of phosphorus trichloride and aromatic hydrocarbons mediated by aluminum chloride. This general approach has been in use since the last century,[61] although significant improvements have recently been made in the experimental details. The major difficulty in the reaction arises from the formation of strong complexes between the primary product, the aryldichlorophosphine, and the aluminum chloride used. These complexes may be broken by conversion *in situ* to other derivatives of the phosphorus acids, for example, the phosphonic acids, by successive treatment with chlorine, alcohol, and hydrochloric acid.[62] Further efforts at breaking the Lewis-acid complex of the primary product involved the use of a transfer agent which would bind more strongly with the aluminum chloride. For this purpose phosphorus oxychloride has been used quite successfully,[63,64] as well as phosphonate diesters.[65]

Although it was early reported that decomposition of the primary product complexes with aluminum chloride were hydrolyzed only with great difficulty,[62] recent efforts have resulted in establishing conditions for hydrolysis which allow high yields of the free phosphonous acid to be isolated.[66,67]

Variation of the reaction conditions can be made which will result in changing the extent of substitution on the phosphorus center. Decreasing the amount of catalyst relative to the amount of aromatic hydrocarbon and increasing the reaction time results in a greater amount of disubstitution on phosphorus trichloride.[62]

IV. PHOSPHORUS ADDITION TO NEUTRAL UNSATURATED CARBON

A rather different approach for the generation of carbon-phosphorus bonds using phosphorus-halogen starting materials involves the addition of the trivalent phosphorus reagent to a carbonyl or imine carbon center. This, on the surface at least, would appear to represent a variation of the Abramov reaction as was discussed in Chapter 3.

The addition of trivalent phosphorus-halogen compounds to the carbonyl carbon of simple aldehydes has been known for quite some time. Phosphorus trichloride reaction with aldehydes, followed by water work-up, provides a convenient method for the synthesis of 1-hydroxyalkylphosphonic acids.[68-71] However, mechanistic investigations concerning this reaction, as well as the reaction system with added acetic anhydride, have provided relatively little structural information regarding intermediate species.[72,73] It is concluded that in either instance there is generated a polymeric phosphorus-halogen material which undergoes Arbuzov-type reactions with organic halides generated *in situ*.

In the absence of a third reagent, a *bis*-1-haloether is formed directly from the aldehyde and phosphorus trichloride, along with polymeric phosphorus-halogen material. Under the reaction conditions, the originally formed *bis*-1-haloether decomposes to the starting aromatic aldehyde and a benzal chloride. This resultant organic halide then undergoes an Arbuzov-type reaction with the polymeric phosphorus-halide. Hydrolysis of the extremely reactive 1-chloroalkylphosphonic dichloride yields the isolable 1-hydroxyalkylphosphonic acid.

$$CH_3CONH_2 + (C_6H_5)_2PCl + C_6H_5CHO \longrightarrow C_6H_5CH(NHCOCH_3)P(O)(C_6H_5)_2$$

63%

When acetic anhydride is used in the reaction system there is generated initially the aldehyde diacetate. This reacts with the phosphorus trichloride to generate acetyl chloride, the polymeric phosphorus-halogen material as mentioned previously, and the 1-chlorobenzylacetate. It is this last material which undergoes the Arbuzov-type reaction with the polymeric material forming the carbon-phosphorus bonded primary product.

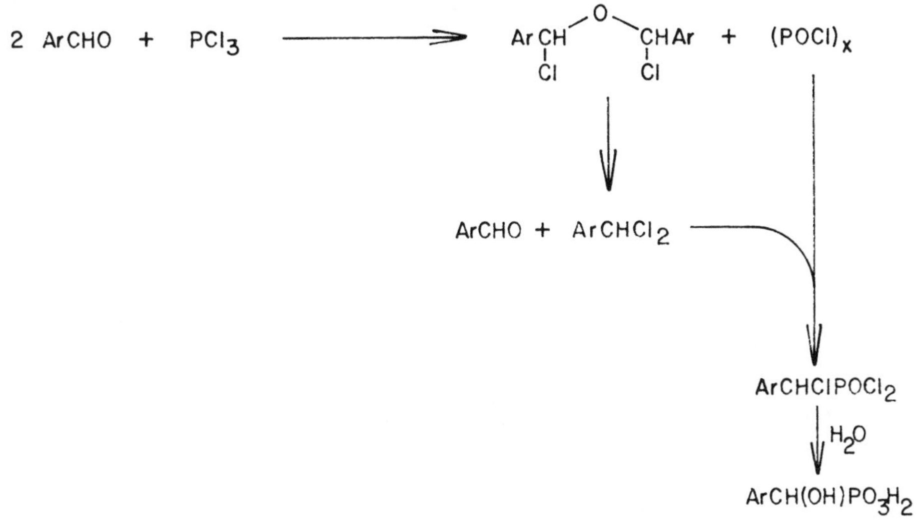

Several reports have been made of the use of this approach for the synthesis of vinylphosphonates[74-76] and tertiary vinylphosphine oxides[77] in moderate to good yield. A preparation of chloromethylphosphonic dichloride is also accomplished in this manner using formaldehyde.[78] A hazard in this method is presented by the formation of *bis*-chloromethyl ether, some of which may remain in the product or escape the reaction mixture.

The use of imine-type substrates for the formation of carbon-phosphorus bonds with phosphorus-halogen reagents has been quite common in recent years. Secondary amines have been used *in situ* with formaldehyde and phosphorus trichloride[79,80] as well as alkyldichlorophosphines[81] for the formation of *N,N*-disubstituted aminomethylphosphonates and -phosphinates. The reactions proceed under relatively mild conditions and in good yield.

Amides of simple aliphatic carboxylic acids have been used with aldehydes to generate Schiff bases *in situ* and to react further with diphenylchlorophosphine.[82] The target *N*-acylated 1-aminoalkyldiphenylphosphine oxides were obtained in moderate yield by this route.

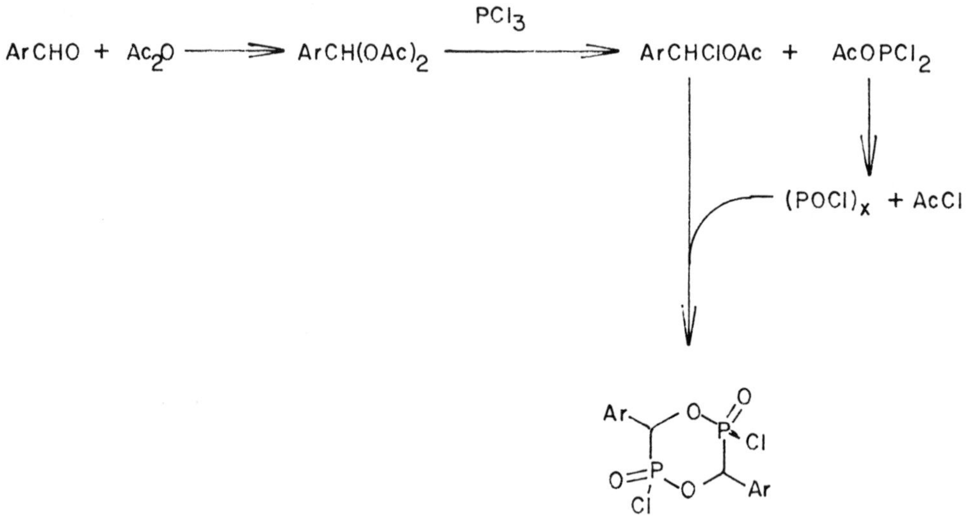

$$ArCHO + Ac_2O \longrightarrow ArCH(OAc)_2 \xrightarrow{PCl_3} ArCHClOAc + AcOPCl_2$$

$$(POCl)_x + AcCl$$

For the preparation of phosphorus analogues of naturally occurring α-amino acids, a similar procedure has been used. However, the amide component of the reaction most amenable is benzylcarbamate, as it allows facile generation of the free amino function. In addition to phosphorus trichloride, alkyl- and aryldichlorophosphines have been used to prepare analogues of alanine, phenylalanine, serine, proline, aspartic acid, and glutamic acid.[83-87] Those analogues generated using the alkyl- or aryldichlorophosphines are of greatest interest as the mateials are monobasic phosphorus-centered acids (phosphinic acids).

$$CH_3CHO + CH_3PCl_2 + C_6H_5CH_2OCONH_2 \xrightarrow[\text{2. H}_2\text{O , HCl}]{\text{1. CH}_3\text{COOH}} CH_3CH(NH_2)P(CH_3)O_2H$$

51%

A related approach involves the reaction of phosphorus-halogen species with acetals in the presence of Lewis acid catalysts.[88-90] Phosphorus trichloride and dialkyl phosphinous chlorides have been used to generate 1-alkoxyphosphonic dichlorides and dialkyl 1-alkoxy-phosphonates, respectively. Zinc chloride and iron(III) chloride have been used as catalysts. In other reports, good yields of adducts have been obtained by reaction at room temperature in the absence of any catalysts.[91,92]

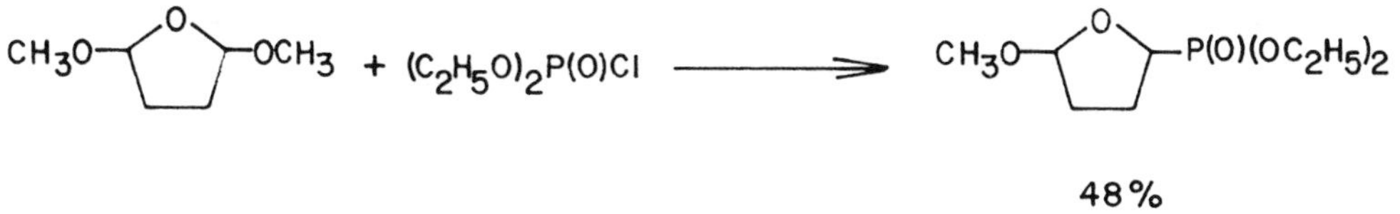

$$CH_3O-\overset{O}{\underset{}{\diagdown}}-OCH_3 + (C_2H_5O)_2P(O)Cl \longrightarrow CH_3O-\overset{O}{\underset{}{\diagdown}}-P(O)(OC_2H_5)_2$$

48%

Acylals[93] and amidals[94] have also been used in similar reaction systems but with rather poor yields of 1-substitutedphosphonyl compounds or the formation of mixtures. Orthoformates[95] and orthoacetates[96] have been used in reaction with trivalent phosphorus-halogen compounds to greater success. In one instance trialkyl phosphite has been used along with the phosphorus trichloride, although the necessity for this added reagent is not clear.[96]

$$(CH_3O)_3CCH_3 + 2 (CH_3O)_3P + PCl_3 \longrightarrow 3 (CH_3O)_2P(O)C(OCH_3)_2CH_3$$

94%

Phosphorus pentachloride has been found to undergo addition to olefinic linkages generating a new carbon-phosphorus bond.[97] The formation of the isolable phosphonic dichlorides requires the work-up of the reaction with an oxidizing agent. For this purpose a variety of materials has been used, including sulfur dioxide,[98] ethylene oxide,[99] and ketones.[100] The reaction has been reported to proceed by initial formation of a vinylic trichlorophosphonium salt which is then oxidized to the vinylicphosphonic dichloride.[98]

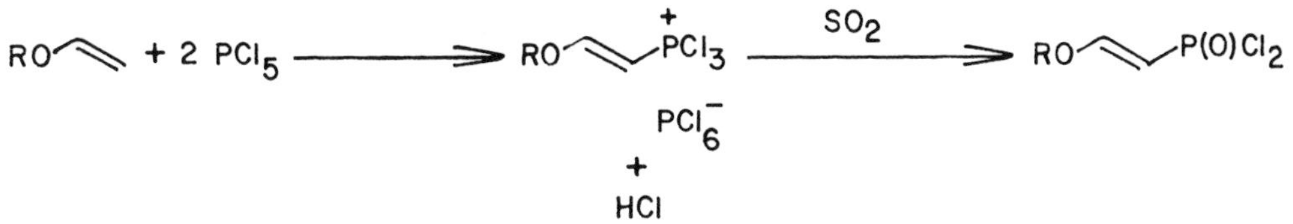

$$RO\diagdown\diagdown + 2 PCl_5 \longrightarrow RO\diagdown\diagup\overset{+}{PCl_3} \overset{SO_2}{\longrightarrow} RO\diagdown\diagup P(O)Cl_2$$

$$\underset{+}{PCl_6^-}$$

$$HCl$$

Depending on the reaction conditions, further addition of hydrogen chloride to the olefinic linkage may occur.[101] The reaction has been found to proceed most cleanly in solvents of low polarity which are weak electron acceptors. Yields in dichlorodimethylsilane are generally greater than 85%.[102]

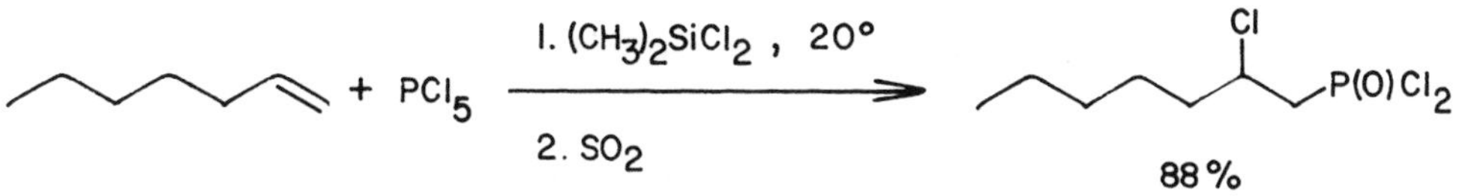

$$\diagup\diagdown\diagup\diagdown\diagdown + PCl_5 \xrightarrow[\text{2. } SO_2]{\text{1. } (CH_3)_2SiCl_2 , 20°} \diagup\diagdown\diagup\overset{Cl}{\underset{}{\diagdown}}\diagup P(O)Cl_2$$

88%

Several phosphorus-halogen systems have been observed to undergo facile addition to olefinic linkages conjugated with other unsaturated functions. Methyl phosphonic dichloride adds to the terminal position of acrylic acid in the presence of phosphorus trichloride (solvent) to give the 3-phosphonopropionyl chloride.[103]

$$CH_3P(O)Cl_2 + CH_2=CHCO_2H + PCl_3 \longrightarrow CH_3P(O)ClCH_2CH_2COCl$$

56%

Thiophosphonous chlorides undergo addition to acrylonitrile in the presence of carboxylic acids (acetic acid) generating the S-alkyl thiophosphinopropionitriles.[104] Reactions of mixtures of thiophosphoryl chloride and phosphorus pentasulfide have been noted to react with conjugated dienes,[105] simple olefins,[106] and aromatics[106] leading to the formation of thiophosphonyl compounds. Yields are found to be in the moderate to good range.

An intramolecular carbon-phosphorus bond-formation reaction has been reported upon decomposition of a diarylphosphinyl azide.[107] It has not been determined if such a reaction process could have application to intermolecular syntheses.

Oxidative chlorophosphonylation of olefins using phosphorus trichloride and oxygen has been reported to generate chloroalkylphosphonyl dichlorides.[108-113] Unlike other olefins, 1,2-dichloroethylene gives only *phosphate* product under typical reaction conditions.[114] Interestingly, reaction of the *saturated* hydrocarbon norbornane under the same conditions gives substitution product of relatively high positional specificity, but only in fair yield.[115]

O_2, PCl_3,   72 h

40%

Finally, there should be noted reactions of phosphorus trichloride with carboxylic acids and their derivatives. Several patents have recently appeared concerned with the use of mixtures of phosphorus trichloride and phosphorous acid with carboxylic acids for the preparation of geminal diphosphonic acids.[116-120] In each instance the substrate molecule has been an amino acid, the product aminoalkyl-1-hydroxy-1,1-*bis*-phosphonic acids being of utility as calcium complexing agents.

$$H_2NCH_2CH_2CO_2H \; + \; H_3PO_3 \; + \; PCl_3 \quad \xrightarrow[\text{2. } H_2O]{\text{1. } \Delta, C_6H_5Cl} \quad H_2NCH_2CH_2CH(OH)(PO_3H_2)_2$$

72%

Similar reaction conditions have been used in reactions using carboxylic acid amides and nitriles as substrates.[119,120] Here the primary products have been amidinoalkylidenediphosphonic acids. In reactions with simple carboxylic acids, phosphorus trichloride alone has been used.[121,122]

V. REARRANGEMENTS OF UNSATURATED PHOSPHITE ESTERS

The reaction of propargyl alcohols with phosphorus trichloride, under appropriate conditions, represents a standard method for the preparation of propargyl halides.[123] However, the formation of allenicphosphonates can be observed in instances when the reaction is performed in the presence of amines.[124,125] Moreover, dialkyl propargyl phosphites are found to generate dialkyl allenicphosphonates simply upon standing for extended periods of time.[127]

The route by which the transformation of propargylic phosphite esters into allenicphosphonate occurs has been described as intramolecular attack by the trivalent phosphorus site on the distant carbon of the acetylenic linkage.[127,128] This would be noted as an *Endo-5-dig* process.[129]

The simplest mode for performance of the reaction involves addition of the propargylic alcohol to a solution of the phosphorus halide. Rearrangement of the phosphorous ester proceeds at ambient temperature or with mild heating. If phosphorus trihalides are used, the product may be isolated as the phosphonic dichloride[130,131] or subjected to aqueous work-up for isolation of the phosphonic acid.[127]

In most instances, however, reaction has been performed using a dialkyl phosphorochloridite having only a single halogen on phosphorus available for reaction with alcohol.[128,132-136] A variety of alkyl esters linkages have been used, giving good to excellent yields of the dialkyl allenicphosphonates.

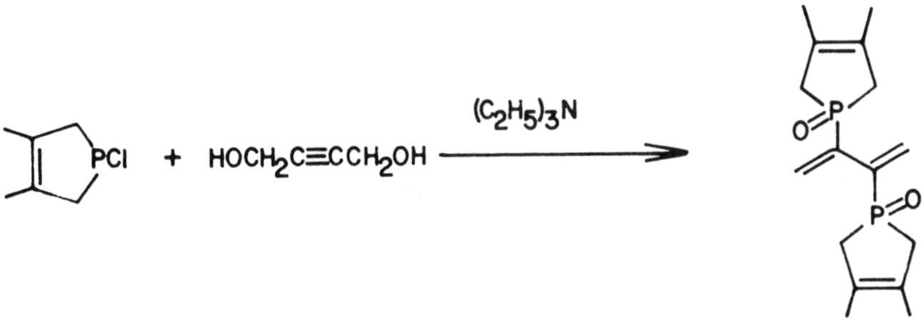

An interesting extension of this rearrangement is seen in the reaction of diethyl phosphorochloridite with 3-methylhex-4-yne-2-ene-1-ol.[137] Formation of the phosphite triester occurs readily (with pyridine present as base) followed by the vinylogous propargylic ester rearrangement.

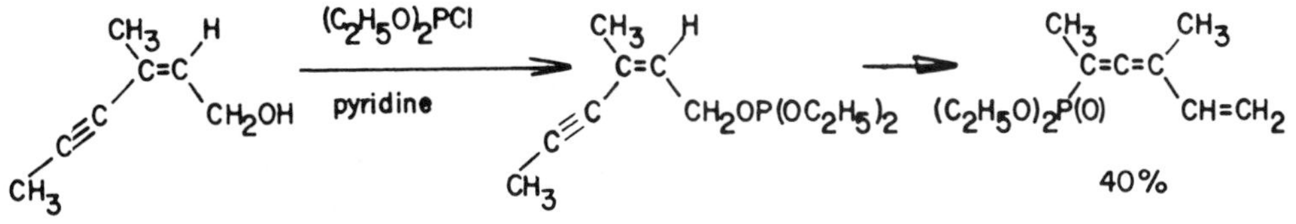

Variations of the rearrangement have been noted to occur with both thioallylic esters of phosphorus acid[138] and thiovinylic esters.[139] Yields in each instance were fair to good.

VI. COMPARISON OF METHODS

A. Formation of Phosphorus-Halides

Several approaches have been described for the preparation of phosphorus-halides from a variety of phosphorus esters. While all of these may reasonably be expected to generate a particular target material in acceptable yield, some methods allow isolation of the target species more easily than others.

For example, four reagents can be used for the replacement of a single ester linkage of a phosphonate by a chloride. Of the four, phosgene is the recommended reagent.[14] While this reagent is rather hazardous, particularly if used directly from a cylinder, it provides an extremely clean reaction with simple product isolation. If phosgene *is* used directly from a cylinder, extreme care should be taken in opening the valve. Corrosion can occur on standing, leading to difficulties in opening the valve and controlling the flow. Use of phosgene in solution (commercially available in toluene) is, however, relatively easy. Care must still be used, but a much safer procedure is allowed. The reaction can be performed using toluene as a solvent.

Although thionyl chloride may seem to be a good alternative to phosgene, impurities in the reagent and extraneous side-products often cause difficulties. Phosphorus pentachloride[15] also allows reasonable yields, but removal of the phosphorus oxychloride by-product is often a difficult problem. With triphenylphosphine dichloride,[16] a large quantity of triphenyl-phosphine oxide is generated as a by-product. Complete removal of this by-product from the target material is at times difficult.

Formation of dialkyl phosphorochloridates can also be accomplished from trivalent phosphorus esters. Chlorine treatment of trialkyl phosphites is the most facile method, allowing relatively simple product isolation and purification.[7] Use of dialkyl phosphites is also found to give the target materials in reasonable yield, but the formation of side-products and by-products introduces difficulties in purification. For example, in the Todd reaction,[8] complete separation of product from amine hydrochloride is at times difficult, and the formation of phosphorus anhydrides is a further complication.

The controlled reactions of alcohols with phosphorus polyhalides (e.g., phosphorus oxychloride and phosphorus trichloride) always present difficulties.[5] Isolation of target material from excess starting phosphorus-halide and polysubstituted side-products requires extremely careful separation techniques and the avoidance of moisture (and atmospheric oxygen in the instance of phosphorus trichloride).

B. Use of Organometallics with Phosphorus-Halides

Little difficulty is found in reactions for which all available halogens on phosphorus are to be replaced by carbon groups.[29] Even when the replacements are performed in the presence of ester linkages, formation of side-products can be minimized with relatively simple product purification procedures.[36]

Replacement of a single halogen on phosphorus trichloride, however, presents significant difficulties. Of the selective reagents available, the organocadmium reagents are experimentally the most desirable.[49] Although yields are only moderate with these reagents, the ease of generating, handling, and product isolation make this the reagent of choice.

The use of aluminum chloride and an alkyl halide with phosphorus trichloride presents excellent selectivity for monosubstitution in certain instances.[55] For reactions intended to produce a large quantity of the target material from simple alkyl halides, this approach may be preferred. However, with large or structurally complex alkyl halides, difficulties can be anticipated. If only a limited supply of an alkyl halide, obtained with difficulty, is available, use of the organocadmium route might be preferred. For the generation of phosphonodichloridates or aryldichlorophosphines, the use of the aluminum chloride routes is recommended.[60,61]

VII. EXPERIMENTAL PROCEDURES

Diallyl Phosphorochloridate — Preparation of a dialkyl phosphorochloridate from a dialkyl phosphite by the Todd reaction.[9]

To a stirred mixture of diallyl phosphite (32.4 g, 0.20 mol) and carbon tetrachloride (35 mℓ, 0.40 mol), cooled to 0°C, there was slowly added triethylamine (3.2 mℓ, 0.023 mol). The temperature was maintained at 0°C for 1 hr, during which time a vigorous reaction ensued. After this time the reaction was allowed to come to room temperature and was stirred overnight. The precipitate which had formed was removed by filtration, hydroquinone (0.5 g) was added, and the solvent evaporated under reduced pressure at room temperature. The residue was vacuum distilled (65°C/4 torr) to give pure diallyl phosphorochloridate (8.5 g, 22%).

2-Chloromethyl-2-ethyl-1,3-propanediol Phosphorochloridate — Preparation of a dialkyl phosphorochloridate by reaction of a trialkyl phosphite with chlorine.[7]

1-Ethyl-4-phospha-3,5,8-trioxabicyclo[2.2.2]octane (32.4 g, 0.20 mol) was dissolved in 50 mℓ of anhydrous diethyl ether. The solution was cooled to −20°C and chlorine gas was bubbled into the solution giving rise to a voluminous precipitate. The gas was added until the solution took on the characteristic green color of chlorine gas. The solution was then allowed to come to room temperature. When the solution reached 0°C, an exothermic reaction ensued which was controlled by means of an ice bath. The solution became completely clear and homogeneous upon completion of the reaction. The ether was evaporated from the solution under reduced pressure and the residue was vacuum distilled (135°C/0.3 torr) to give the pure 2-chloromethyl-2-ethyl-1,3-propanediol phosphorochloridate (40.5 g, 87%).

Isopropyl Methylphosphonochloridate — Preparation of a phosphonochloridate by reaction of a phosphonate diester with phosgene.[14]

A stream of phosgene was passed into diisopropyl methylphosphonate (270 g, 1.50 mol) for 10 hr with stirring at 20 to 30°C. After completion of the addition the mixture was allowed to stand overnight, and then the excess phosgene and isopropyl chloride was evaporated under reduced pressure. The residue was vacuum distilled (38°C/2 torr) to give isopropyl methylphosphonochloridate (222 g, 94%).

Diethyl 3-Trifluoromethylphenylphosphonate — Reaction of a Grignard reagent with a dialkyl phosphorochloridate.[36]

The Grignard reagent of 3-trifluoromethylphenyl bromide was prepared by the addition of the 3-trifluoromethylphenyl bromide (112.5 g, 0.50 mol) to magnesium (12.5 g, 0.515 mol) in anhydrous diethyl ether (280 mℓ). The dark green Grignard solution was transferred to a 500 mℓ constant-addition funnel under a nitrogen atmosphere through septa using a double-pointed needle. The Grignard reagent was then added dropwise to a stirred solution of freshly distilled diethyl phosphorochloridate (172.55 g, 1.0 mol) in anhydrous diethyl ether (500 mℓ) at −76°C. After the addition, which was complete in 1.5 hr, the mixture was stirred at ambient temperature for a further 1.5 hr and the reaction mixture was poured into 5% hydrochloric acid (300 mℓ). The ether layer was washed with water (3 × 500 mℓ), dried over magnesium sulfate, filtered, and concentrated under reduced pressure. Fractional distillation of the residue (105°C/0.9 torr) gave pure diethyl 3-trifluoromethylphenylphosphonate (131.15 g, 93%).

Vinyldichlorophosphine — Reaction of phosphorus trichloride with an organomercury compound.[23]

Freshly distilled phosphorus trichloride (420 g, 3.06 mol) together with dry degassed mineral oil (60 mℓ) were placed under a nitrogen atmosphere in a flat-bottomed flask equipped with magnetic stirrer, thermometer, dropping funnel, and reflux condenser. The flask was warmed to produce a gentle reflux of the phosphorus trichloride, and then divinylmercury (145 g, 0.57 mol) was added dropwise through a pressure-equalizing dropping funnel. The temperature of the reaction flask was maintained at 65 to 85°C during the addition and was continued at reflux for 1 hr after completion of the addition. On cooling, a distillation head was put in place of the reflux condenser and excess phosphorus trichloride (250 mℓ) was distilled at 200 torr. The remaining liquid was transferred under vacuum to a distillation flask cooled to −78°C, containing 25 mℓ of degassed mineral oil. It was necessary to heat the flask to 100°C to complete the transfer. The liquid material was fractionated (63.4°C/200 torr) to yield the vinyldichlorophosphine (36 g, 50%).

N-Butyldichlorophosphine — Monoalkylation of phosphorus trichloride using an organocadmium reagent.[49]

Di-*n*-butylcadmium was prepared by the rapid addition of cadmium chloride (50.4 g, 0.275 mol) to a cooled solution of *n*-butylmagnesium bromide formed from magnesium (12.15 g, 0.50 mol) and *n*-butyl bromide (75.4 g, 0.55 mol) in anhydrous diethyl ether (550 mℓ). Following the addition, the reaction mixture was stirred at 0°C for 2 hr. The di-*n*-butylcadmium reaction mixture, including the precipitate, was then added to a stirred solution of phosphorus trichloride (85.9 g, 0.625 mol) in anhydrous diethyl ether (100 mℓ) over a period of 25 min. The reaction mixture was maintained at −20°C. After the addition, anhydrous diethyl ether (100 mℓ) was flushed through the cadmium reagent vessel into the main reaction mixture. The reaction was allowed to come to room temperature and was then refluxed for 2.5 hr. The reaction mixture was allowed to stand overnight and the bulk of the ether solution was decanted. The residue was washed additionally with diethyl ether (100 mℓ) and the mixture filtered, and the solids further washed with diethyl ether (100 mℓ). (Care must be taken with the solution and the solids as the solution and the solids may catch fire upon hydrolysis of *n*-butyldichlorophosphine.) The combined ether washings and reaction mixture were evaporated under reduced pressure and the residue fractionated through a glass-helices packed column (56°C/22 torr) to give pure *n*-butyldichlorophosphine (37.4 g, 47%).

2-[1,3]Dithianyldiphenylphosphine Oxide — Reaction of a chlorophosphine with a stabilized carbanion reagent.[53]

Freshly sublimed 1,3-dithiane (1.0 g, 8.33 mmol) was placed in a flask equipped with a rubber septum. Tetrahydrofuran (20 mℓ) was added to the flask via syringe through the septum. The flask was cooled to −20°C and then a solution of *n*-butyl lithium in hexane (1.5 *M*, 6.67 mℓ, 10.0 mmol) was added, also via syringe through the septum maintaining a nitrogen atmosphere. The reaction mixture was stirred at −20°C for 1.5 hr and was then treated with chlorodiphenylphosphine (1.84 g, 8.33 mmol) in tetrahydrofuran (15 mℓ) with tetramethylethylenediamine (1.0 g, 8.5 mmol). The reaction mixture was stirred at −20°C for 1.5 hr and at room temperature for an additional 3 hr, and was then quenched by the addition of saturated ammonium chloride solution. The aqueous solution was extracted with chloroform, the solution dried, and solvent removed under reduced pressure. The residual solid was recrystallized from benzene to give pure 2-[1,3]dithianyldiphenylphosphine oxide (1.07 g, 40%) as white crystals of mp 242 to 243°C.

2,5-Dimethylbenzenephosphinic Acid — Aluminum chloride-mediated reaction of phosphorus trichloride with an aromatic hydrocarbon.[66]

To a flame-dried 3-neck 1-ℓ flask was added in order *p*-xylene (107 g, 1.0 mol), phosphorus trichloride (412 g, 3.0 mol), and anhydrous aluminum chloride (160 g, 1.2 mol). The reaction mixture was slowly heated to reflux with stirring. After 2.5 hr at reflux the reaction was allowed to cool to room temperature and the volatile components distilled at reduced pressure. The residual oil was slowly added to cold water (1 ℓ) with stirring and a white solid formed. The solid was removed by filtration, washed with water, and air dried. The solid was suspended in water (1 ℓ) to which was added 50% sodium hydroxide solution (90 mℓ) to cause dissolution. The solution was saturated with carbon dioxide and filtered through Celite. The basic solution was washed with methylene chloride (200 mℓ) and acidified with concentrated hydrochloric acid (200 mℓ). The white solid which separated was isolated by extraction with methylene chloride (3 × 250 mℓ). The extracts were dried over magnesium sulfate, filtered, and evaporated under reduced pressure to give the pure 2,5-dimethylbenzenephosphinic acid (99 g, 60%) as an oil which slowly crystallized to a solid of mp 77 to 79°C.

(1-Aminopropyl)phenylphosphinic Acid — Reaction of a carbonyl compound with benzyl carbamate and a dichlorophosphine.[84]

Freshly distilled propanal (4.4 g, 0.075 mol) is added at room temperature over a period of 20 min to a stirred mixture of benzyl carbamate (7.55 g, 0.05 mol), phenyldichlorophosphine (8.95 g, 0.05 mol), and glacial acetic acid (10 mℓ). The mixture was refluxed for 40 min, treated with 4 N hydrochloric acid (50 mℓ), and then refluxed again for 30 min. After cooling, the organic layer was removed and the aqueous layer was boiled with charcoal (2 g) and evaporated to dryness in vacuum. The residue was dissolved in methanol (40 mℓ) and treated with propylene oxide until a pH 6 to 7 was attained. The resultant precipitate was filtered, washed with acetone, and crystallized from methanol/water to give pure (1-aminopropyl)phenylphosphinic acid (4.08 g, 41%) of mp 256 to 258°C.

Diethyl 5-Methoxytetrahydrofuran-2-ylphosphonate — Reaction of an acetal with a phosphorochloridite.[91]

Diethyl phosphorochloridite (0.714 mℓ, 5 mmol) was added dropwise to a stirred solution of 2,5-dimethoxytetrahydrofuran (0.65 mℓ, 5 mmol) in dichloromethane at 0°C under an argon atmosphere. The solution was then stirred at room temperature for 12 hr after which the solution was concentrated under reduced pressure and the residue subjected to Kugelrohr distillation (120°C/0.03 torr). There was in this manner isolated the pure diethyl 5-methoxytetrahydrofuran-2-ylphosphonate (0.475 g, 48%) as a colorless oil.

Propadienylphosphonic Dichloride — Rearrangement of a proparglyic trivalent phosphorus ester.[130]

A 2-ℓ Morton flask, provided with motor stirrer, immersion thermometer, and a condenser topped by a drying tube, is charged with phosphorus trichloride (824 g, 6.0 mol) and heated in an oil bath to 74°C. With vigorous stirring, propargyl alcohol (11.2 g, 0.20 mol) is added subsurface through a thin delivery tube passing through the condenser. The addition is performed as quickly as possible (5 to 10 sec) under positive pressure. Gaseous hydrogen chloride is formed rapidly and passes out of the reaction system through the condenser. The solution is brought to reflux at 76°C and maintained at that temperature for 3 hr. Excess phosphorus trichloride is removed under reduced pressure and the residual oil is washed with toluene several times to remove the last traces of phosphorus trichloride. There is isolated by vacuum distillation (40°C/0.1 torr) pure propadienylphosphonic dichloride (25 g, 79%).

2-Chlorohept-1-ylphosphonic Dichloride — Reaction of an alkene with phosphorus pentachloride.[102]

To a suspension of phosphorus pentachloride (41.7 g, 0.2 mol) in dimethyldichlorosilane (100 mℓ) was added at 20°C 1-heptene (9.8 g, 0.1 mol) with stirring. The reaction mixture was stirred at 20°C for 3 hr cooled to −10°C, and sulfur dioxide dried over phosphorus pentoxide was bubbled into the reaction mixture until complete dissolution of the intermediate complex was observed. After evaporation of solvent, the residue was vacuum distilled to give pure 2-chlorohept-1-ylphosphonic dichloride (22.1 g, 88%).

6-Amino-1-hydroxyhexylidenediphosphonic Acid — Reaction of a carboxylic acid with phosphorous acid and phosphorus trichloride.[117]

A mixture of 6-aminocaproic acid (13 g, 0.1 mol) and phosphorous acid (12.7 g, 0.156 mol) in chlorobenzene (100 mℓ) was heated to 100°C with stirring. Phosphorus trichloride (22 g, 0.16 mol) was added dropwise to it within a period of 30 min. The solution was then heated with stirring for 3 hr. Insoluble material separated during this time. After cooling, the solvent was decanted and the residue was boiled with water (60 mℓ) for 30 min and subjected to hot filtration with activated charcoal through a layer of Supercel. The solution was concentrated under reduced pressure and the crystals formed were collected by filtration. Methanol was added to the mother liquors to complete the precipitation. There was in this way isolated pure 6-amino-1-hydroxyhexylidenediphosphonic acid (15 g, 55%) of mp 245°C.

Diethyl 1-(4-Pyridyl)-1,2-dihydropyridine-2-phosphonate — Reaction of phosphorus trichloride with a pyridylpyridinium chloride.[140]

Finely powdered 1-(4-pyridyl)-pyridinium chloride hydrochloride (11.45 g, 0.05 mol) and phosphorus trichloride (27.5 g, 0.2 mol) were placed in a flask protected from moisture. The mixture was heated at reflux for 20 hr. Excess phosphorus trichloride was evaporated under reduced pressure and ethanol (20 mℓ) was added to the reaction mixture with cooling by an ice bath. After 2 hr the excess ethanol was evaporated and a solution of potassium carbonate was added until the solution became basic. The oil which separated was extracted with diethyl ether (2 × 50 mℓ), the extracts were dried over magnesium sulfate, and the solvent evaporated under reduced pressure. The crude materials thus obtained were purified by column chromatography (silica gel; eluent: chloroform/ethanol, 8:1). There was thus isolated pure diethyl 1-(4-pyridyl)-1,2-dihydropyridine-2-phosphonate (8.1 g, 57%) which yielded a picrate salt of mp 163—164°C.

Table 1
FORMATION OF CARBON TO PHOSPHORUS BONDS FROM PHOSPHORUS-HALIDES

Product	Phosphorus reagent	Method	Yield (%)	Refs.
CH_3PCl_2	PCl_3	A	80	56
$ClCH_2P(O)Cl_2$	PCl_3	B	73	78
$Br_3CHP(O)Br_2$	PBr_3	A	88	59
$CH_2=CHPF_2$	PF_2Br	C	34	50
$CH_2=CHPCl_2$	PCl_3	C	50	23
$CH_2=CHPBr_2$	PBr_3	C	34	50
$CH_3CH_2PCl_2$	PCl_3	D	26	49
$CH_3OCH_2P(O)Cl_2$	PCl_3	E	58	89
$CH_3CH(NH_2)PO_3H_2$	PCl_3	B	52	84
$CH_2(OH)CH(NH_2)PO_3H_2$	PCl_3	B	48	87
$CH_3C(OH)(PO_3H_2)_2$	PCl_3	F	50	122
		F	97	123
$C_2H_5OCH_2P(O)Cl_2$	PCl_3	E	64	89
$H_2NCH_2CH_2C(OH)(PO_3H_2)_2$	H_3PO_3, PCl_3	F	72	118
$C_2H_5P(CH_3)_2$	$C_2H_5PCl_2$	G	78	23
$CH_2=CHP(OCH_3)CH_3$	$CH_3P(OCH_3)Cl$	G	60	48
$CH_3CH(NH_2)P(O)(OH)C_2H_5$	$C_2H_5PCl_2$	B	51	84
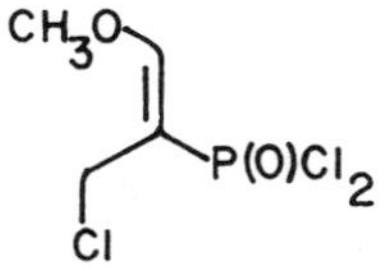 [structure: CH_3O–CH=C($P(O)Cl_2$)–CH_2Cl]	PCl_5	H	30	99
$(C_2H_5O)_2PCH_3$	$(C_2H_5O)_2PCl$	G	82	40
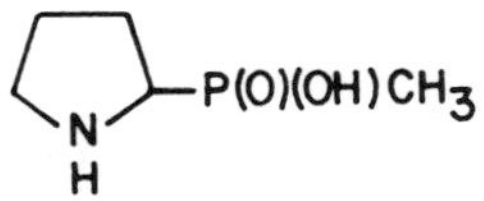 [structure: pyrrolidine-2-yl–$P(O)(OH)CH_3$]	CH_3PCl_2	B	35	85

Table 1 (continued)
FORMATION OF CARBON TO PHOSPHORUS BONDS FROM PHOSPHORUS-HALIDES

Product	Phosphorus reagent	Method	Yield (%)	Refs.
$(ClCH_2CH_2O)_2PCH_3$	$(ClCH_2CH_2O)_2PCl$	G	65	39
$C_6H_5PCl_2$	PCl_3	A	78	63
$n\text{-}C_6H_{13}PCl_2$	PCl_3	D	41	49
(structure)	PCl_3	B	58	84
$[(CH_3)_2N]_2P(O)CH_2CN$	$[(CH_3)_2N]_2P(O)Cl$	I	95	52
(structure)	$C_2H_5PCl_2$	B	42	83
$C_2H_5OCH_2CH_2CH(Cl)CH_2P(O)Cl_2$	PCl_5	H	28	102
$H_2N(CH_2)_5C(OH)(PO_3H_2)_2$	H_3PO_3, PCl_3	F	55	117
$H_2C=C=CHP(O)(OC_2H_5)_2$	$(C_2H_5O)_2PCl$	J	85	137
(structure)	PCl_3	D	32	43

Product	Reagent			
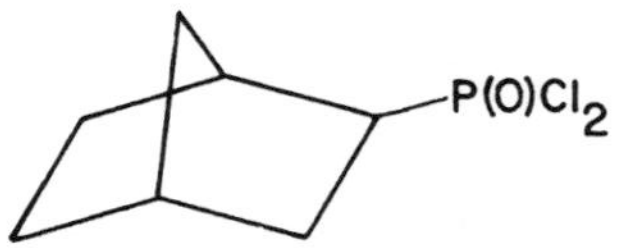 (P(O)Cl₂)	PCl_3	H	40	116
$C_6H_5CH(Cl)P(O)Cl_2$	PCl_3	B	32	72
$n\text{-}C_5H_{11}CH(Cl)CH_2P(O)Cl_2$	PCl_5	H	88	103
	$(C_2H_5)_2PCl$	J	43	140
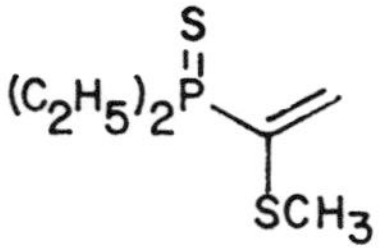 $(C_2H_5)_2\overset{\text{S}}{\underset{\|}{P}}\!-\!C(=CH_2)SCH_3$				
$CH_2=C=C(CH_3)_2P(O)(OC_2H_5)_2$	$(C_2H_5O)_2PCl$	J	78	137
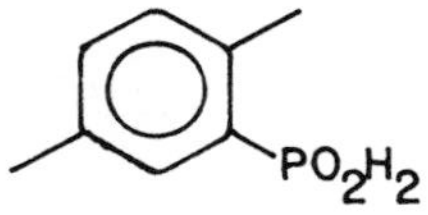 (PO₂H₂, dimethyl)	PCl_3	A	60	66
$4\text{-}C_2H_5C_6H_4PCl_2$	PCl_3	A	69	63
$n\text{-}C_8H_{17}PCl_2$	PCl_3	D	33	49
$C_6H_5CH(NH_2)P(O)(OH)CH_3$	CH_3PCl_2	B	65	84
(P(CH₃)₂)	(PCl₂)	G	80	43

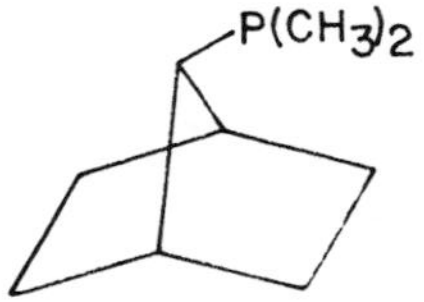

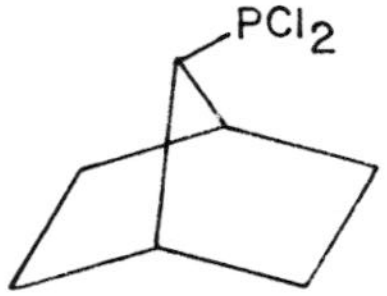

Table 1 (continued)

FORMATION OF CARBON TO PHOSPHORUS BONDS FROM PHOSPHORUS-HALIDES

Product	Phosphorus reagent	Method	Yield (%)	Refs.
$(CH_3)_2C=C=CHP(O)(OC_2H_5)_2$	$(C_2H_5O)_2PCl$	J	79	137
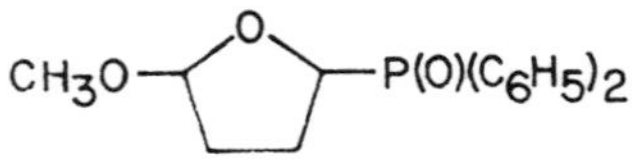	$(C_2H_5O)_2PCl$	E	67	91
(cyclic structure: CH_3O-furan-$P(O)(C_6H_5)_2$)				
$n\text{-}C_8H_{17}NHCH_2PO_3H_2$	PCl_3	B	96	80
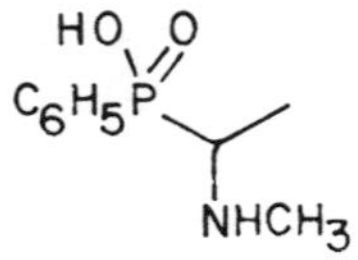 (structure: HO, O, C_6H_5P, $NHCH_3$)	$C_6H_5PCl_2$	B	55	86
$4\text{-}n\text{-}C_4H_9C_6H_4PCl_2$	PCl_3	A	39	63
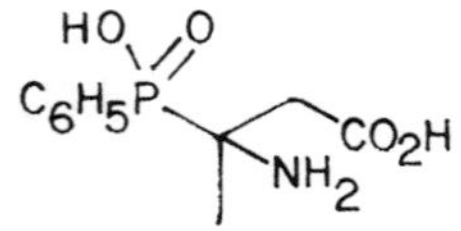 (structure: HO, O, C_6H_5P, CO_2H, NH_2)	$C_6H_5PCl_2$	B	55	83
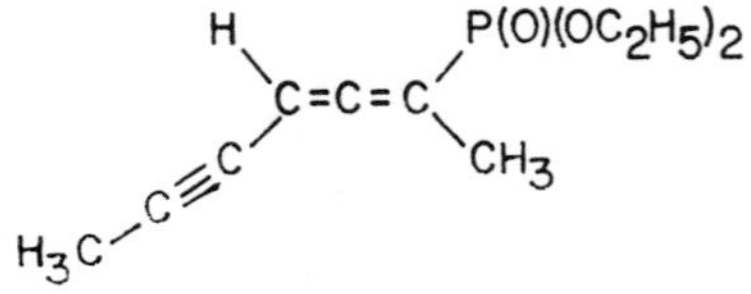 (structure: H_3C-C≡C-C=C=C< with H, CH_3, $P(O)(OC_2H_5)_2$)	$(C_2H_5O)_2PCl$	J	65	138

Structure	Reagent			
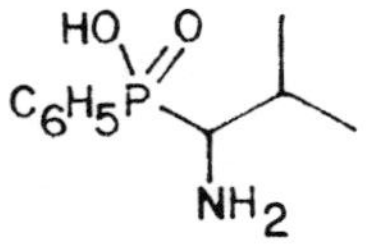 $C_6H_5P(O)(OH)$, $CH(NH_2)$	$C_6H_5PCl_2$	B	58	86
$4\text{-}CF_3C_6H_4P(O)(OC_2H_5)_2$	$(C_2H_5O)_2P(O)Cl$	G	93	36
	$[(C_2H_5)_2N]_2PCl$	J	26	140
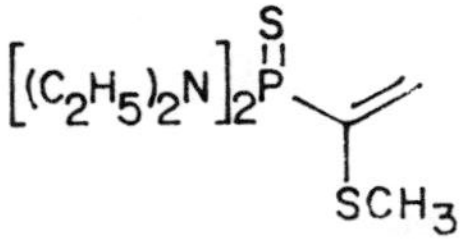 $[(C_2H_5)_2N]_2P(S)$, SCH_3				
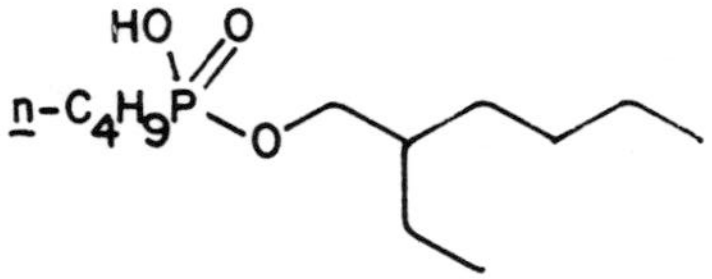 $n\text{-}C_4H_9$ phosphate	PCl_3	A	91	57
$(C_2H_5O)_2P(O)CH_2COC_6H_5$	$(C_2H_5O)_2P(O)Cl$	I	11	51
$4\text{-}FC_6H_4P(C_3F_7\text{-}n)_2$	$(n\text{-}C_3F_7)_2PCl$	G	39	45
$CH_2{=}CHP(OC_6H_5)C_6H_5$	$C_6H_5P(OC_6H_5)Cl$	G	55	48
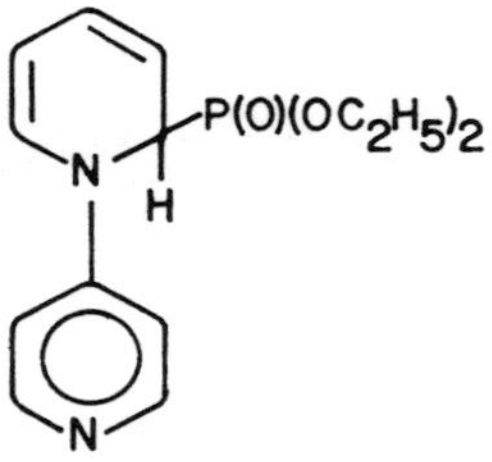	PCl_3	K	57	140

Table 1 (continued)
FORMATION OF CARBON TO PHOSPHORUS BONDS FROM PHOSPHORUS-HALIDES

Product	Phosphorus reagent	Method	Yield (%)	Refs.
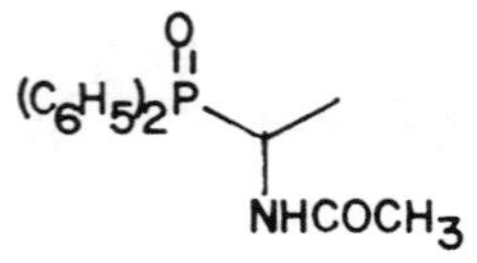	$(C_6H_5)_2PCl$	B	40	82
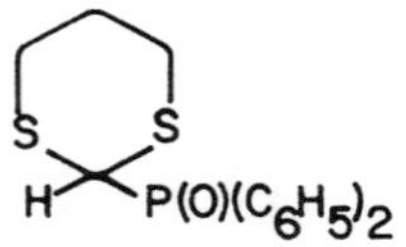	$(C_6H_5)_2PCl$	I	40	53
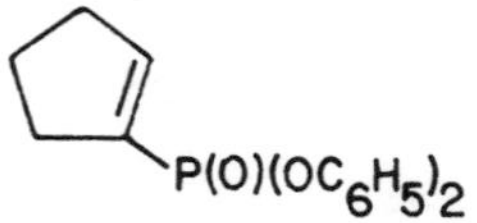	$(C_6H_5O)_2PCl$	B	57	76
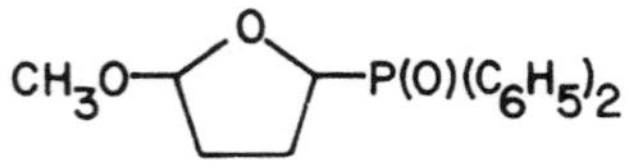	$(C_6H_5)_2PCl$	E	67	91
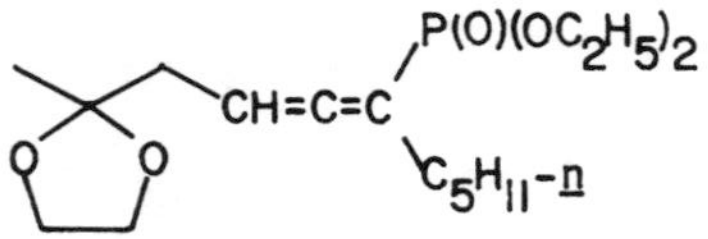	$(C_2H_5O)_2PCl$	J	88	136

189

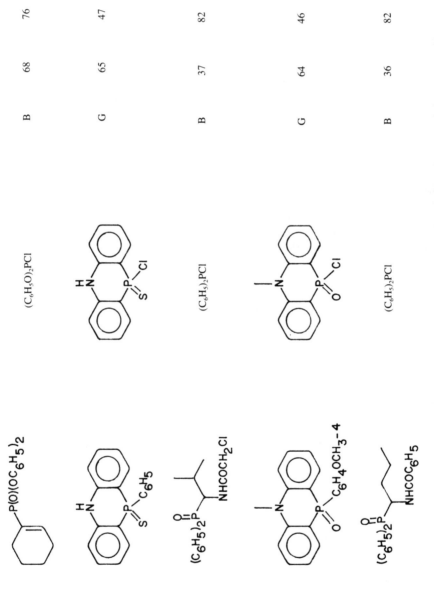

$P(O)(OC_6H_5)_2$	$(C_6H_5O)_2PCl$	B	68	76
		G	65	47
	$(C_6H_5)_2PCl$	B	37	82
		G	64	46
	$(C_6H_5)_2PCl$	B	36	82

Note: A, aluminum chloride used with organic halide; B, Abramov-type reaction of carbonyl compound; C, organometallic reagent other than Grignard or cadmium reagent used; D, cadmium reagent used; E, acetal used as substrate; F, carboxylic acid used as substrate; G, Grignard reagent used; H, alkene used as substrate; I, stabilized-carbanion used; J, Ynol used as substrate; K, substituted-pyridinium ion used as substrate.

REFERENCES

1. **Mizuma T., Minaki, Y., and Toyoshima, S.,** Synthesis and antiviral activity of alkyl *P,P*-diaziridino-phosphinate, *J. Pharm. Soc. (Japan),* 81, 51, 1961; *Chem. Abstr.,* 55, 12379g, 1961.
2. **Loew, B. and Massengale, J. T.,** Some unusual solubility properties of alkyl tetraalkylphosphorodiamidates, *J. Org. Chem.,* 22, 1186, 1957.
3. **Gerrard, W. and Phillips, R. J.,** Interaction of phosphorus pentachloride and alcohols, *Chem. Ind. (London),* p. 540, 1952.
4. **Orloff, H. D., Worrell, C. J., and Markley, F. X.,** The synthesis of alkyl aryl phosphates from aryl phosphorochloridates. I. The sodium alkoxide route, *J. Am. Chem. Soc.,* 80, 727, 1958.
5. **Malowan, J. E., Martin, D. R., and Pizzolato, P. J.,** Alkyl dichlorophosphites, *Inorg. Synth.,* 4, 63, 1953.
6. **Tolkmith, H.,** Aromatic phosphorodichloridites and phosphorodichloridothioates. I. Aryl phosphorodichloridites, *J. Org. Chem.,* 23, 1682, 1958.
7. **Wadsworth, W. S. and Emmons, W. D.,** Bicyclic phosphites, *J. Am. Chem. Soc.,* 84, 611, 1962.
8. **Atherton, F. R., Openshaw, H. T., and Todd, A. R.,** Phosphorylation. II. Reactions of dialkyl phosphites with polyhalogen compounds in the presence of bases — a method for the phosphorylation of amines, *J. Chem. Soc.,* p. 660, 1945.
9. **Steinberg, G. M.,** Reactions of dialkyl phosphites. Synthesis of dialkyl chlorophosphates, tetraalkyl pyrophosphates, and mixed orthophosphate esters, *J. Org. Chem.,* 15, 637, 1950.
10. **Kong, A. and Engel, R.,** A mechanistic investigation of the Todd reaction, *Bull. Chem. Soc. Jpn.,* 58, 3671, 1985.
11. **Smith, T. D.,** Reaction of dialkyl phosphites with cupric chloride, *J. Chem. Soc.,* p. 1122, 1962.
12. **McIvor, R. A., McCarthy, G. D., and Grant, C. A.,** Preparation and toxicity of alkyl thiopyrophosphates, *Can. J. Chem.,* 34, 1819, 1956.
13. **Walsh, E. N.,** Conversion of tertiary phosphites to secondary phosphonates. Diphenyl phosphonate, *J. Am. Chem. Soc.,* 81, 3023, 1959.
14. **Bryant, P. J. R., Ford-Moore, A. H., Perry, B. J., Wardrop, A. W. H., and Watkins, T. F.,** The preparation and physical properties of isopropyl methylphosphonofluoridate (sarin), *J. Chem. Soc.,* p. 1553, 1960.
15. **Welch, C. M., Gonzales, E. J., and Guthrie, J. D.,** Derivatives of unsaturated phosphonic acids, *J. Org. Chem.,* 26, 3270, 1961.
16. **Wiley, G. A., Hershkowitz, R. L., Rein, B. M., and Chung, B. C.,** Studies in organophosphorus chemistry. I. Conversion of alcohols and phenols to halides by tertiary phosphine dihalides, *J. Am. Chem. Soc.,* 86, 964, 1964.
17. **Wiley, G. A., Rein, R. M., and Hershkowitz, R. L.,** Studies in organophosphorus chemistry. II. Mechanism of the reaction of tertiary phosphine dihalides with alcohols, *Tetrahedron Lett.,* p. 2509, 1964.
18. **Ylagan, L., Benjamin, A., Gupta, A., and Engel, R.,** Organophosphorus chemistry. Ester-chloride conversion under mild conditions at phosphorus, *Synthetic Commun.,* in press.
19. **Aaron, H. S., Uyeda, R. T., Frack, H. F., and Miller, J. I.,** The stereochemistry of asymmetric phosphorus compounds. IV. The synthesis and stereochemistry of displacement reactions of optically active isopropyl methylphosphonochloridate, *J. Am. Chem. Soc.,* 84, 617, 1962.
20. **Michalski, J., Mikolajczyk, M., and Omelanczuk, J.,** Stereochemistry of nucleophilic displacement reaction at thiophosphoryl centre. An example of Walden cycle involving phosphorus, *Tetrahedron Lett.,* p. 1779, 1965.
21. **Sasse, K.,** Organische Phosphorverbindungen, in *Methoden der Organische Chemie,* Vol. 12, Parts 1 and 2, Muller, E., Ed., Georg Thieme Verlag, Stuttgart, 1963.
22. **Drake, L. R. and Marvel, C. S.,** Phosphonic acids and their alkyl esters from α, β-unsaturated ketones, *J. Org. Chem.,* 2, 387, 1938.
23. **Kaesz, H. D. and Stone, F. G. A.,** Preparation and characterization of vinyldichlorophosphine, vinyldimethylphosphine, and ethyldimethylphosphine, *J. Org. Chem.,* 24, 635, 1959.
24. **Lines, E. L. and Centofanti, L. F.,** Preparation and characterization of vinyldifluorophosphine, *Inorg. Chem.,* 13, 1517, 1974.
25. **Larock, R. C.,** Organomercurials in organic synthesis, *Tetrahedron,* 38, 1713, 1982.
26. **Kharasch, M. S., Jensen, E. V., and Weinhouse, S.,** Alkylation reactions of tetrathyllead. A new synthesis of ethyldichloroarsine and related compounds, *J. Org. Chem.,* 14, 429, 1949.
27. **Pickard, R. H. and Kenyon, J.,** Contributions to the chemistry of oxygen compounds. I. The compounds of tertiary phosphine oxides with acids and salts, *J. Chem. Soc.* 89, 262, 1906.
28. **Canavan, A. E. and Eaborn, C.,** Organosilicon compounds. XXI. Some compounds containing phosphorus, *J. Chem. Soc.,* p. 3751, 1959.

29. **Bodner, G. M., May, M. P., and McKinney, L. E.,** A Fourier transform carbon-13 NMR study of the electronic effects of phosphorus, arsenic, and antimony ligands in transition-metal carbonyl complexes, *Inorg. Chem.,* 19, 1951, 1980.

30. **Clark, P. W. and Mulraney, B. J.,** Synthesis and physical properties of chlorodi(o-tolyl)phosphine, lithium di(o-tolyl)phosphide and the diphosphine series (o-Tolyl)$_2$P(CH$_2$)$_n$P(o-tolyl) (n = 1—4, 6, 8), *J. Organomet. Chem.,* 217, 51, 1981.

31. **van Linthoudt, J.P., van den Berghe, E. V., and van der Kelen, G. P.,** NMR Study (^{1}H, ^{13}C, ^{29}Si, ^{31}P and ^{119}Sn) of the (C$_2$H$_5$)$_{3-n}$P{E^{IVB}(CH$_3$)$_3$}$_n$ Compounds (E^{IVB} = C, Si, Sn; n = 0, 1, 2, 3), *Spectrochim. Acta,* 36A, 17, 1980.

32. **Kosolapoff, G. M.,** Some variations of the Grignard synthesis of phosphinic acids, *J. Am. Chem. Soc.,* 72, 5508, 1950.

33. **van Linthoudt, J. P., van den Berghe, E. V., and van der Kelen, G. P.,** NMR Study (^{1}H, ^{13}C, ^{31}P) of the (C$_2$H$_5$)$_{3-n}$PX$_n$ Compounds (X = Cl, Br, I; n = 0, 1, 2, 3), *Spectrochim. Acta,* 35A, 1307, 1979.

34. **Burger, A. and Dawson, N. D.,** The reaction of dialkyl chlorophosphates with alkylmagnesium halides, *J. Org. Chem.,* 16, 1250, 1951.

35. **Golubski, Z. E.,** Alkylation of phosphinic acid salts in the presence of crown ethers, *Synthesis,* p. 632, 1980.

36. **Grabiak, R. C., Miles, J. A., and Schwenzer, G. M.,** Synthesis of phosphonic dichlorides and correlation of their P-31 chemical shifts, *Phosphorus and Sulfur,* 9, 197, 1980.

37. **Balthazor, T. M. and Flores, R. A.,** Dipolar cycloadditions of an acetylenic phosphinate, *J. Org. Chem.,* 45, 529, 1980.

38. **Yugadeev, T. A., Kushembaev, R. K., Nurgalieva, A. N., Zhumagaliev, S., Dzhakiyaev, G. M., and Godovikov, N. N.,** Synthesis of dialkyl [(1-chlorohexen-1-yl)ethynyl]-, [(3,6-dihydro-2,2-dimethyl-2H-pyran-4-yl)ethynyl]-, [3,6-dihydro-2,2-dimethyl-2H-thiopyran-4-yl)ethynyl]-, and [(1,2,3,6-tetrahydro-1,2,5-trimethyl-4-pyridyl)ethynyl]phosphonates, (transl.), *J. Gen. Chem. U.S.S.R.,* 50, 1804 (1980).

39. Nissan Chemical Industries, Ltd., Bis(-haloethyl) Methylphosphonites, Japanese Patent 80 62,096, 1980.

40. Nippon Kayaku Co., Ltd., Alkyl and Aryl Phosphinites, Japanese Patent 82 46,993, 1982.

41. **Chodkiewicz, W.,** One-pot synthesis of chiral phosphonous esters, conversion into asymmetric phosphines, *J. Organomet. Chem.,* 273, C55, 1984.

42. **Chodkiewicz, W., Jore, D., and Wodzki, W.,** Optically active phosphines: new synthetic approach, *Tetrahedron Lett.,* p. 1069, 1979.

43. **Quin, L. D. and Littlefield, L. B.,** Importance of the structure of the phosphorus functionality in allowing dihedral angle control of vicinal ^{13}C-^{31}P coupling. Carbon-13 NMR spectra of 7-substituted bicyclo[2.2.1]heptane derivatives, *J. Org. Chem.,* 43, 3508, 1978.

44. **Bodner, G. M., Gagnon, C., and Whittern, D. N.,** A Fourier transform carbon-13 NMR study of trivalent compounds of phosphorus, arsenic, antimony and bismuth, and their LNi(CO)$_3$ complexes, *J. Organomet. Chem.,* 243, 305, 1983.

45. **Yagupolskii, L. M., Pavlenko, N. V., Ignatev, N. V., Matyushecheva, G. I., and Semenii, V. Y.,** Aryl*bis*(heptafluoropropyl)phosphine Oxides. Electronic nature of the P(O)(C$_3$F$_7$)$_2$ group, (transl.), *J. Gen. Chem. U.S.S.R.,* 54, 297, 1984.

46. **Piskunova, O. G., Yagodina, L. A., Kordev. B. A., Bekanov, A. I., and Stepanov, B. I.,** Phenophosphazines. IV. Electronic effects in 5,10-dihydro-5-methylphenophosphazine 10-oxides, (transl), *J. Gen. Chem. U.S.S.R.,* 48, 1205, 1978.

47. **Bokanov, A. I., Gusev, A. I., Demidova, N I., Los, M. G., Segelman, I. R., and Stepanov, B. I.,** 5-Ethyl-5,10-dihydro-10-phenylphenophosphazine 10-sulfide, (transl.), *J. Gen. Chem. U.S.S.R.,* 51, 1216, 1981.

48. **Minowa, N., Fukatsu, S., Niida, T., and Mase, S.,** Phosphinic Acid Esters and Process for Preparing the Same, U.S. Patent 4,510,102, 1985.

49. **Fox, R. B.,** Organophosphorus compounds. Alkyldichlorophosphines, *J. Am. Chem. Soc.,* 72, 4147, 1950.

50. **Altoff, W., Fild, M., Rieck, H.-P., and Schmutzler, R.,** Synthesis and NMR spectroscopic studies of phosphorus compounds with vinyl and ethynyl groups, *Chem. Ber.,* 111, 1845, 1978.

51. **Borowitz, I. J., Firstenberg, S., Casper, E. W. R., and Crouch, R. K.,** The stereochemistry of vinyl phosphates from the Perkow reeaction and the phosphonylation of enolates, *J. Org. Chem.,* 36, 3282, 1971.

52. **Blanchard, J., Collignon, N., Savignac, P., and Normant, H.,** Preparation d'acides α-aminoethylphosphoniques, *Tetrahedron,* 32, 455, 1976.

53. **Juaristi, E., Valle, L., Valenzuela, B. A., and Aguilar, M. A.,** S-C-P Anomeric interactions. 4. Conformational analysis of 2-(diphenylphosphinoyl)-1,3-dithiane, *J. Am. Chem. Soc.,* 108, 2000, 1986.

54. **Mikolajczyk, M., Grzejszczak, A., Zatorski, A., and Mlotkowska, B.,** A new general synthesis of ketene acetals, *Tetrahedron Lett.,* p. 2731, 1976.

55. **Bayer, von E., Gugel, K. H., Hagele, K., Hagenmaier, H., Jessipow, S., Konig, W. A., and Zahner, H.,** Phosphinothricin and phosphinothricyl-alanyl-alanine, *Helv. Chim. Acta,* 55, 224, 1972.

56. **Soroka, M.,** A Simple preparation of methylphosphonous dichloride, *Synthesis,* p. 450, 1977.

57. Nippon Mining Co., Ltd., Butylphosphonic Acid Mono-2-ethylhexyl Ester, Japanese Patent 59,157,092, 1984.

58. **Kaegi, H. H. and Duncan, W. P.,** "Synthesis of phenyl-$^{14}C_6$-labeled *O*-(4-Bromo-2,5-dichlorophenyl) *O*-Methyl phenylphosphonothioate and *O*-(4-nitrophenyl) *O*-ethyl phenylphosphonothioate using a phase transfer catalyst, *J. Labelled Compd. Radiopharm.,* 18, 1831, 1981.

59. **Elkaim, J. C., Casabianca, F., and Riess, J. G.,** The direct synthesis of dibromomethylphosphonic dibromide and some of its derivatives, *Synth. React. Inorg. Met.-Org. Chem.,* 9, 479, 1979.

60. **Sonnek, G., Reinheckel, H., and Baumgarten, K. G.,** Aluminum alkyls with heteroatoms; preparation of phosphonic acid diamides, *Z. Chem.,* 21, 268, 1981.

61. **Michaelis, A.,** Ueber ein Homologes des Phosphenylchlorids, *Chem. Ber.,* 12, 1009, 1879.

62. **Kosolapoff, G. M. and Huber, W. F.,** Synthesis of aromatic phosphonic acids and their derivatives. I. The derivatives of benzene, toluene and chlorobenzenes, *J. Am. Chem. Soc.,* 69, 2020, 1947.

63. **Buchner, B. and Lockhart, L. B.,** An improved method of synthesis of aromatic dichlorophosphines, *J. Am. Chem. Soc.,* 73, 755, 1951.

64. **Neumaier, H.,** Process for Making Aryldichlorophosphines, U.S. Patent 4,536,357, 1985.

65. **Kormachev, V. V., Vasileva, T. V., and Karpova, R. D.,** Aryldichlorophosphines, Soviet Union Patent 1,151,540, 1985.

66. **Simmons, K. A.,** Preparation of Arylphosphonic Acids, U.S. Patent 4,316,858, 1982.

67. **Photis, J. M.,** Preparation of Arylphosphinic Acids, U.S. Patent 4,316,859, 1982.

68. **Fossek, W.,** Einwirkung von Phosphortrichlorid auf Aldehyde, *Monatsh.,* 7, 20 (1886).

69. **Page, H. J.,** Hydroxymethylphosphinic acid and some analogues, *J. Chem. Soc.,* 101, 423, 1912.

70. **Conant, J. B. and MacDonald, A. D.,** Addition reactions of phosphorus halides. I. The mechanism of the reaction of the trichloride with benzaldehyde, *J. Am. Chem. Soc.,* 42, 2337, 1920.

71. **Conant, J. B. and Wallingford, V. H.,** Addition reactions of phosphorus halides. VIII. Kinetic evidence in regard to the mechanism of the reaction, *J. Am. Chem. Soc.,* 46, 192, 1924.

72. **Miller, J. A. and Nunn, M. J.,** Aldehydes and phosphorus trichloride, *J. Chem. Soc. Perkin I,* p. 535, 1976.

73. **Michie, J. K. and Miller, J. A.,** A re-examination of the reactions of benzaldehyde with phosphorus trichloride in the presence of acetic anhydride, *J. Chem. Soc. Perkin I,* p. 785, 1981.

74. **Kenyon, G. L. and Westheimer, F. H.,** The stereochemistry of unsaturated phosphonic acids, *J. Am. Chem. Soc.,* 88, 3557, 1966.

75. **Fay, P. and Lankelma, H. P.,** The reaction of cyclohexene with phosphorus pentasulfide, *J. Am. Chem. Soc.,* 74, 4933, 1952.

76. **Nurtdinov, S. K., Ismagilova, N. M., Nazarov, V. S., Zykova, T. V., Salakhutdinao, R. A., Sultanova, R. B., and Tsivunin, V. S.,** Reactions of diphenyl phosphorochloridite and phenyl phosphorodichloridite with cyclic ketones, (transl), *J. Gen. Chem. U.S.S.R.,* 43, 1240, 1973.

77. **Nurtdinov, S. K., Sultanova, R. B., Nurtdinova, S. S., Zykova, T. V., Dorozhkina, G. M., and Tsivunin, V. S.,** Kinetics of the reaction of diethylphosphinous chloride with cyclohexanone, (transl.), *J. Gen. Chem. U.S.S.R.,* 50, 1592, 1980.

78. **Forstner, J., Peter, I., Gemes, I., Hevesi, J., Vasarhelyi, E., and Valvotis, E.,** Chloromethylphosphonic Acid Dichloride, Hungarian Patent 16,097, 1979.

79. **Elek, S., Fodor, I., Gulyas, I., Gyoker, I., Zoltai, A., and Zsupan, K.,** *N*-Phosphonomethyliminodiacetic Acid, Hungarian Patent 19,480, 1981.

80. Konishiroku Photo Industry Co., Ltd., Phosphonated Tertiary Amines, Japanese Patent 59 84,893, 16 May 1984.

81. **Maier, L.,** Zur Kenntnis der Umsetzung von Cyanomethyldichlorphosphin und 2-Chlorathyldichlorphosphin mit Benzylglycin und Formaldehyd in saurer Losung, *Phosphorus and Sulfur,* 11, 149, 1981.

82. **Oleksyszyn, J.,** Synthesis of *N*-acylated 1-aminoalkyldiphenylphosphine oxides by amidoalkylation of diphenylchlorophosphine, *Synthesis,* p. 444, 1981.

83. **Oleksyszyn, J., Gruszecka, E., Kafarski, P., and Mastalerz, P.,** New phosphonic analogs of aspartic and glutamic acid by aminoalkylation of trivalent phosphorus chlorides with ethyl acetoacetate of ethyl levulinate and benzyl carbamate, *Monatsh. Chemie,* 113, 59, 1982.

84. **Oleksyszyn, J., Tyka, R., and Mastalerz, P.,** Direct synthesis of 1-aminoalkanephosphonic and 1-aminoalkanephosphinic acids from phosphorus trichloride or dichlorophosphines, *Synthesis,* p. 479, 1978.

85. **Subotkowski, W., Tyka, R., and Mastalerz, P.,** Phosphonic analogs of proline, *Pol. J. Chem.,* 54, 503, 1980.

86. **Oleksyszyn, J.,** 1-*N*-Alkylaminoalkanephosphonic and 1-*N*-alkylaminoalkylphenylphosphinic acids, *Synthesis.* p. 722, 1980.

87. **Lejczak, B., Kafarski, P., Soroka, M., and Mastalerz, P.,** Synthesis of the phosphonic acid analog of serine, *Synthesis,* p. 577, 1984.

88. **Petrov, K. A., Chauzov, V. A., and Agafonov, S. V.,** Alkoxymethylphosphonyl Dichlorides, Soviet Union Patent 730,688, 1980.

89. **Petrov, K. A., Chauzov, V. A., and Agafonov, S. V.,** Alkoxymethylation of phosphorus trichloride and alkyl phosphorodichloridites with dialkoxymethanes (transl.), *J. Gen. Chem. U.S.S.R.,* 50, 628, 1980.

90. **Petrov, K. A., Chauzov, V. A., Agafonov, S. V., Pazhitnova, N. V., and Kostrova, S. M.,** Alkoxymethylation of dialkyl, ethylene, and diphenyl phosphorochloridites with dialkoxymethanes (transl.), *J. Gen. Chem. U.S.S.R.,* 50, 816, 1980.

91. **Ley, S. V., Lygo, B., Organ, H. M., and Wonnacott, A.,** Wittig and Horner-Wittig coupling reactions of 2-substituted cyclic ethers and their application to spiroketal synthesis, *Tetrahedron,* 41, 3825, 1985.

92. **Novikova, Z. S., Kabachnik, M. M., Snyatkova, E. V., and Lutsenko, I. F.,** α-Alkoxyalkylphosphonates, Soviet Union Patent 1,162,809, 1985.

93. **Gazizov, M. B., Sutanova, D. B., Razumov, A. I., Zykova, T. V., Pashinkin, A. P., and Salakhutdinov, R. A.,** Reactions of dialkyl phosphorochloridites with acylals of acetic acid (transl.), *J. Gen. Chem. U.S.S.R.,* 46, 1205, 1976.

94. **Petersen, H.,** Ureidomethyl Phosphonic Dihalides for Fireproofing Materials, West German Patent application 1,817,337, 1970.

95. **Morita, T., Okamoto, Y., and Sakurai,** The preparation of phosphonic acids having labile functional groups, *Bull. Chem. Soc. Jpn.,* 51, 2169, 1978.

96. **Golburn, P.,** Synthesis of dialkyl 1-alkoxyvinylphosphonates, *Synthesis,* p. 547, 1973.

97. **Andreae, S., and Seeboth, H.,** Synthesis of 2-furyl- and 5-nitro-2-furyl- substituted vinylphosphonic acids, *Collect. Lect. Int. Symp. Furan Chem.,* 3rd, Slovak Tech. Univ., Fac. Chem. Technol., Bratislava, Czechoslavakia 117, 1979.

98. **Rybkina, V. V., Rozinov, V. G., Glukhikh, V. I., and Kalabina, A. V.,** Causes of the formation of addition products in the phosphorylation of alkyl vinyl ethers with phosphorus trichloride (transl.), *J. Gen. Chem. U.S.S.R.,* 50, 2148, 1980.

99. **Kormachev, V. V., Mitrasov, Y. N., and Kurskii, Y. A.,** Reaction of adducts of phosphorus pentachloride and ethers of saturated and unsaturated alcohols with ethylene oxide (transl.,), *J. Gen. Chem. U.S.S.R.,* 51, 2107, 1981.

100. **Kormachev, V. V., Mitrasov, Y. N., Yokovlena, T. M., and Yaltseva, N. S.,** 2-Chloroalkylphosphonic Dichlorides, Soviet Union Patent 883,049, 1981.

101. **Kormachev, V. V., and Mitrasov, Y. N.,** Reaction of phosphorus pentachloride with 4-ethoxy-1-butene, *Izv. Vyssh. Uchebn. Zaved., Khim. Khim. Tekhnol.,* 24, 1570, 1981.

102. **Kolomiets, A. F., Fokin, A. V., Krolevets, A. A., Petrovskii, P. V., and Verenikin, O. V.,** Solvent effect on the reaction of alkenes with phosphorus pentachloride (transl.), *Bull. Acad. Sci. U.S.S.R.,* 1941, 1979.

103. **Ohorodnik, A., Neumaier, H., and Gehrmann, K.,** 2-Chloroformylethylmethylphosphonic Acid Chlorides, West German Patent application 2,936,609, 1979.

104. **Akamasin, V. D., Eliseenkova, R. M., and Rizpolozhenskii, N. I.,** Esters of ethyl(phenyl)-β-cyanoalkylthiolphosphonic and ethyl(phenyl)-γ-ketoalkylthiolphosphinic Acids (transl.), *Bull. Acad. Sci. U.S.S.R.,* 72, 1973.

105. **Uhing, E. H.,** Reaction Products of Unsaturated Hydrocarbons with P_4S_{10} and PSX_3, U.S. Patent 4,540,526, 1985.

106. **Uhing, E. H.,** Reaction Products of Unsaturated Hydrocarbons with P_4S_{10} and PSX_3, U.S. Patent 4,231,970, 1980.

107. **Baceiredo, A., Bertrand, G., and Mazerolles, P.,** Surprising reactivity of very crowded phosphinic derivative, *J. Chem. Soc., Chem. Commun.,* p. 1197, 1981.

108. **Zinovev, Y. M. and Soborovskii, L. Z.,** Synthesis of organophosphorus compounds from hydrocarbons and their derivatives. IX. Oxidative chlorophosphonation of 1-butene, 2-butene, and cyclohexene, *Zh. Obshch. Khim. C.C.C.P.,* 29, 611, 1959.

109. **Soborovskii, L. Z., Zinovev, Y. M., and Spiridonova, J. G.,** Synthesis of organophosphorus compounds from hydrocarbons and their derivatives. X. Oxidative chlorophosphonation of C_2H_4 Derivatives, *Zh. Obshch. Khim. C.C.C.P.,* 29, 1110, 1959.

110. **Daniewski, W. M., Gordon, M., and Griffin, C. E.,** Phosphonic acids and esters. XV. Preparation and proton magnetic resonance spectra of the diethyl α- and β-chlorovinylphosphonates, *J. Org. Chem.,* 31, 2083, 1966.

111. **Soborovskii, L. M., Zinovev, Y. M., and Muler, L. I.,** Synthesis of organophosphorus compounds from hydrocarbons and their derivatives. XIII. Synthesis and structure of derivatives of dichloroethyl- and dichloropropylphosphonic acids, *Zh. Obshch. Khim. C.C.C.P.,* 29, 3907, 1959.

112. **Rochlitz, F. and Vilcsek, H.,** β-Chloroethylphosphonic dichloride: its synthesis and use, *Angew. Chem., Int. Ed. Engl.,* 1, 652, 1962.

113. **Shvedova, Y. I., Kruglov, A. S., Dogadina, A. V., Ionin, B. I., and Petrov, A. A.,** Reactions of enynic hydrocarbons with phosphorus trihalides (transl.), *J. Gen. Chem. U.S.S.R.,* 54, 1486, 1984.

114. **Boyce, C. B. C. and Webb, S. B.,** Reaction of *trans*-1,2-dichloroethylene with phosphorus trichloride-oxygen: the direct formation of a phosphate derivative from an olefin, *J. Chem. Soc. C,* 1613, 1971.

115. **Hanstock, C. C., Tebby, J. C., and Coates, H.,** Oxidative chlorophosphonylation of 8,9,10-trinorborane, *J. Chem. Res.,* (S), p. 110, 1982.

116. **Jary, J., Rihakova, V., and Zobacova, A.,** 6-Amino-1-hydroxyhexylidene Diphosphonic Acid, Salts and a Process for Preparation Thereof, U.S. Patent 4,304,734, 1981.

117. **Blum, H. and Worms, K. H.,** 3-Amino-1-hydroxypropene-1,1-diphosphonic Acid, West German Patent application 2,934,498, 1981.

118. **Blum, H., Hempel, H. V., and Worms, K. H.,** Hydroxydiphosphonic Acids, West German Patent application 2,745,083, 1979.

119. **Worms, K. H. and Blum, H.,** Amidingruppenhaltige geminale Diphosphonsauren, *Liebig's Ann. Chem.,* p. 275, 1982.

120. **Sommer, K. and Raab, G.,** Acylaminomethanephosphonic Acids, Ger. Offen. 2,829,046, 1980.

121. **Pelayo, R. Z., and Salvador, D.,** 1-Hydroxy-1,1-ethylenediphosphonic Acid and Its Disodium Salt, Spanish Patent 471,259, 1979.

122. **Repasy, O., Bathory, J., Kerenyi, E., Belafi, L., and Szebenyi, N.,** 1-Hydroxyethylidene-1,1-bis(phosphonic acid), Hungarian Patent 34,759, 1985.

123. **Taylor, D. R.,** The chemistry of allenes, *Chem. Rev.,* 67, 317, 1967.

124. **Boisselle, A. P. and Meinhardt, N. A.,** Acetylene-allene rearrangement reactions of trivalent phosphorus chlorides with α-acetylenic alcohols and glycols, *J. Org. Chem.,* 27, 1828, 1962.

125. **Verny, M. and Vessiere, R.,** Transposition propargylique (5e Partie) action des halogenures de phosphor III sur l'hydroxy-2 methyl-2 butyne-3 oate d'ethyle, *Bull. Soc. Chim. Fr.,* p. 3004, 1968.

126. **Mark, V.,** A facile S_Ni' rearrangement: the formation of 1,2-alkadienylphosphonates from 2-alkynyl phosphites, *Tetrahedron Lett.,* p. 281, 1962.

127. **Elder, R. C., Florian, L. R., Kennedy, E. R., and Macomber, R. S.,** Phosphorus containing products from the reaction of propargyl alcohols with phosphorus trihalides. II. The crystal and molecular structure of 2-hydroxy-3,5-di-*tert*-butyl-1,2-oxaphosphol-3-ene 2-oxide, *J. Org. Chem.,* 38, 4177, 1973.

128. **Voskanyan, M. G., Gevorkyan, A. A., and Badanyan, S. O.,** Rearrangement during the reaction of vinylethynylcarbinols with phosphorus(III) compounds, *Arm. Khim. Zh.,* 23, 766, 1970.

129. **Baldwin, J. E.,** Rules for ring closure, *J. Chem. Soc., Chem. Commun.,* p. 734, 1976.

130. **Glamkowski, E. J., Rosas, C. B., Sletzinger, M., and Wantuck, J. A.,** Process for the Preparation of *cis*-1-Propenylphosphonic Acid, U.S. Patent 3,733,356, 1973.

131. **Angelov, K., Mikhailova, T. S., Ignatev, V. M., Dogadina, A. V., and Ionin, B. I.,** Synthesis of derivatives of 1,2-alkadienephosphonic acids, *Dokl. Bolg. Akad. Nauk,* 32, 619, 1980.

132. **Glamkowski, E. J., Gal, G., Purick, R., Davidson, A. J., and Sletzinger, M.,** A new synthesis of the antibiotic phosphonomycin, *J. Org. Chem.,* 35, 3510, 1970.

133. **Dangyan, Y. M., Voskanyan, M. G., Zurabyan, N. Z., and Badanyan, S. O.,** Reactions of unsaturated compounds. LVII. Vinylallenylphosphonates as diene fragments in the Diels-Alder reaction, *Arm. Khim. Zh.,* 32, 460, 1979.

134. **Arbuzov, B. A., Pudovik, A. N., Vizel, A. O., Shchukina, L. I., Muslinkin, A. A., Paramova, V. I., Kharitnov, V. V., Krupnov, V. K., and Vakulenko, O. V.,** Synthesis and structure of diphospholenebutadienes (transl.), *Bull. Acad. Sci., U.S.S.R.,* 904, 1981.

135. **Altenbach, H.-J., and Korff, R.,** β,ε-Dioxophosphonates by reductive nucleophilic acylation of 1,3-dioxo compounds: facile synthesis of jasmone, *Angew. Chem., Int. Ed. Engl.,* 21, 371, 1982.

136. **Altenbach, H.-J. and Korff, R.,** Einfache Synthese von β-Ketophosphonaten aus 1-Alkin-3-olen, *Tetrahedron Lett.,* p. 5175, 1981.

137. **Huche, M. and Cresson, P.,** Transpositions (2-3) et (2-5) de phosphites d'eynols, *Tetrahedron Lett.,* p. 4291, 1973.

138. **Rizpolozhenskii, N. I., Akamasin, V. D., and Eliseenkova, R. M.,** Reaction of acid chlorides of trivalent phosphorus acids with allyl mercaptan and allyl alcohol (transl.), *Bull. Acad. Sci., U.S.S.R.,* 77, 1973.

139. **Danchenko, M. N., Budilova, I. Y., and Sinitsa, A. D.,** Rearrangement of some *S*-vinyl esters of phosphorus(III) acids (transl.), *J. Gen. Chem. U.S.S.R.,* 55, 838, 1985.

140. **Boduszek, B. and Wieczorek, J. S.,** Synthesis of 1(4-pyridyl)-1,2-dihydropyridine-2-phosphonates and their derivatives, *Synthesis,* p. 454, 1979.

Chapter 6

DIRECT FORMATION OF AROMATIC AND VINYLIC CARBON-PHOSPHORUS BONDS

I. INTRODUCTION

This chapter differs from previous chapters in that it is concerned not with a particular type of reaction process for the generation of carbon-phosphorus bonds, but rather centers on the variety of methods available for the generation of a particular type of carbon-phosphorus bond. Specifically, the concern here is with the several methods for linking phosphorus with sp^2 hybridized carbon atoms of aromatic rings and olefinic linkages.

It is clearly seen that certain of the methods commonly available for the formation of aliphatic carbon-phosphorus bonds will be of only very limited utility for aromatic and olefinic systems. For example, it is well known that nucleophilic substitution type reactions occur at aromatic or olefinic carbon sites only under unusual circumstances. Thus, as expected, simple Michaelis-Arbuzov or Michaelis-Becker reactions are rarely involved in the generation of bonds between such sites and phosphorus.

In addition to these special instances in which a direct nucleophilic substitution reaction at an sp^2 carbon site is observed to occur, formation of aromatic carbon-phosphorus bonds has been accomplished via a wide variety of methods. Many of these methods are of use for the formation of olefinic carbon-phosphorus bonds as well, although certain variations are observed.

As noted previously, formation of aromatic or vinylic carbon-phosphorus bonds via elimination reactions or Wittig (Wadsworth-Emmons-Horner) procedures is not discussed here.

II. AROMATIC CARBON-PHOSPHORUS BOND FORMATION

A. Reactions and Mechanisms

Although a variety of aromatic halides had been reported to undergo direct displacement using Michaelis-Arbuzov or Michaelis-Becker conditions, most of these reactions subsequently have been found incapable of being reproduced. Generally, it may be anticipated that simple displacements at aromatic carbon involving a phosphorus-centered nucleophile will occur only under the same circumstances as will such reactions with other nucleophiles. That is, the presence on or within the aromatic ring of a suitable collection of strongly electron-withdrawing functionalities is required for these reactions to occur. These functionalities may be halides or other "-I" type linkages,[1-3] or simply a structural arrangement which results in the leaving group being attached to a suitably electron deficient carbon, as in 9-chloroacridine.[4]

The use of a photoinitiated form of the Michaelis-Arbuzov reaction has had considerably greater success for arylphosphonate ester syntheses than the classical thermal route. Photoinitiated homolysis of the aryl iodide and subsequent formation of the quasi-phosphonium ion allows an overall reaction appearing completely analogous to the thermal Michaelis-Arbuzov route.[5] The best results are obtained when the aryl iodide is photolyzed at low temperature in the presence of an excess of the trialkyl phosphite.[5] Arylphosphonate diesters are thus isolated in fair to excellent yield.[6-8] Problems associated with obtaining the appropriate aryl iodides pose the major difficulty in the use of this approach as a general synthetic method.

Trialkyl phosphites have also been arylated by anodic oxidation in acetonitrile,[9] proceeding via a route involving radical intermediates. This approach, however, leads to the desired arylphosphonate diesters in only poor to moderate yield. The method is also troubled by the formation of isomeric mixtures and by the generation of significant amounts of nonaromatic phosphonate esters as by-products.

An interesting report of direct nucleophilic substitution on aromatic rings involves displacement of the 3-nitro group from substituted, 3,4-dinitroaryloxybenzenes.[10] The attacking reagents used were phosphonite dialkyl esters which generated phosphinate diesters in moderate yield. The nitrite ion displaced in the initial attack is found to perform the usual second displacement step of the Michaelis-Arbuzov reaction giving alkyl nitrite as a by-product.

Direct generation of pyridinephosphonic acids can be accomplished by phosphorous acid attack on 2- or 4-(1-pyridyl)pyridinium salts.[11] However, yields of the phosphonic acids are generally low (<30%).

Systematic investigations concerned with the "direct" substitution of a trivalent phosphorus ester (a phosphite or other similar reagent) for a halogen on an aromatic ring have focused on the use of metal species as catalysts. These metal species are involved initially with the phosphorus reagent for the generation of complexes which proceed further in reaction with the aromatic halide to generate Michaelis-Arbuzov or Michaelis-Becker type products. These efforts have explored in detail the utility of several oxidation states of nickel, copper, and palladium for the facilitation of the substitution reactions.

The earliest work in this direction was that of Tavs[12] which involved heating Ni(II) salts (halides) with a mixture of the aryl halide and the trialkyl phosphite. Temperatures in the range commonly employed for the performance of Michaelis-Arbuzov reactions were used. Yields greater than 65% were generally obtained using this technique, although bulky substituents *ortho* to the site of the displaced halide tended to lower the yield.

$$(\underline{i}\text{-}C_3H_7O)_3P \;+\; C_6H_5Br \xrightarrow[160°]{NiCl_2} C_6H_5P(O)(OC_3H_7\text{-}\underline{i})$$

The absence of free radical intermediates in this process was clearly demonstrated.[12] It was also demonstrated that an initial reduction of the Ni(II) to a Ni(0) state occurred.[13] That is, there is initially formed tetrakis(trialkyl phosphite)Ni(0) which further reacts with the aryl halide. Once the Ni(0) species has been generated and the appropriate reaction temperature has been attained, there is no induction period for the generation of product. Moreover, Ni(0) species need be used in only catalytic amounts with the aryl halide and phosphite. The analogous substitution reaction has also been accomplished using triaryl phosphites,[14] a variety of phosphonite esters,[15-17] and thiophosphinite esters.[18]

$$(C_6H_5O)_3P \;+\; \underset{1\%}{C_6H_5I} \xrightarrow[320°,\; 8.5\ h]{0.2\ \%\ NiCl_2} \underset{85\%}{(C_6H_5O)_2P(O)C_6H_5}$$

Other preformed Ni(0) species added directly to the reaction system provide suitable catalysis for the reaction as well. Nickel in its elemental state (Raney nickel) and nickel tetracarbonyl have both been found to be of use in the substitution of aromatic halogen with triphenyl phosphite.[19] Nickel tetracabonyl has also been used successfully in substitution reactions involving silyl phosphite esters.[20]

Both cuprous and cupric salts have been used successfully for these aromatic substitution reactions. Copper(II) acetate with trialkyl phosphites forms reactive complexes which further undergo facile reaction with aromatic halides to generate the arylphosphonate diesters in good yield. This reaction proceeds well even when highly sensitive functionalities are present in the aromatic halide.[21,22] Copper(I) chloride has also been used with triethyl phosphite and aryl iodides for these substitution reactions.[23]

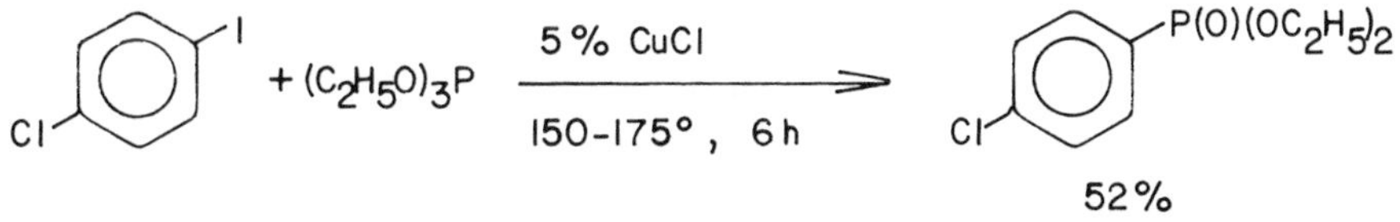

Palladium(0) complexes have also been used as catalytic agents for substitution reaction on aromatic halides by dialkyl phosphites in the presence of a tertiary amine.[24,25] Substituents *ortho* to the displacement site decrease yields in this reaction, which otherwise are good to excellent.

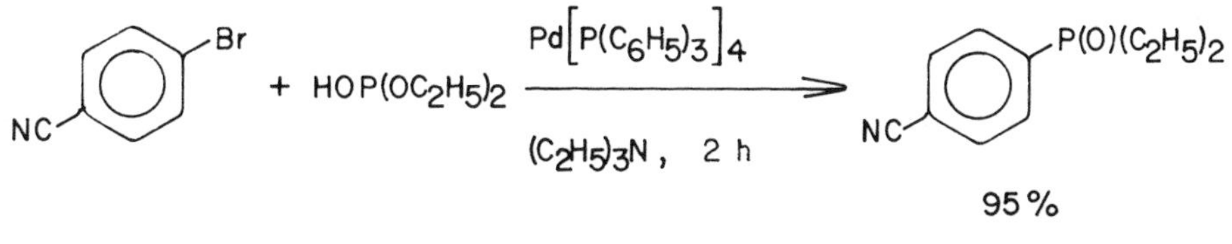

The reaction proceeds via initial oxidative addition of the aryl halide to the Pd(0) species, followed by phosphorus atom displacement of the metal from the aromatic ring.

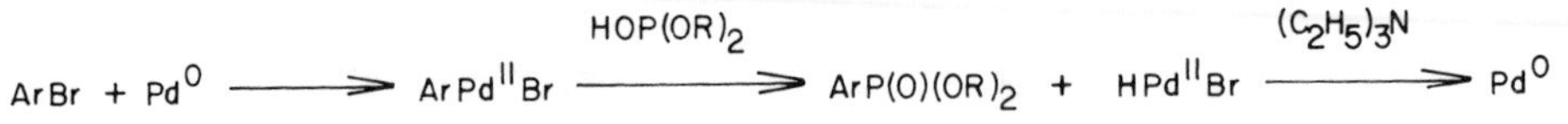

$$\text{ArBr} + \text{Pd}^0 \longrightarrow \text{ArPd}^{II}\text{Br} \xrightarrow{\text{HOP(OR)}_2} \text{ArP(O)(OR)}_2 + \text{HPd}^{II}\text{Br} \xrightarrow{(C_2H_5)_3N} \text{Pd}^0$$

The role of the amine is to regenerate the palladium in the original Pd(0) form. The reaction system has been used successfully with monoalkyl phosphonites[26,27] and secondary phosphine oxides[28] as well as the dialkyl phosphites.

A rather different approach has been of use in the attachment of a phosphorus ester or acid function to a pyridine ring system. With pyridinium ions bearing alkoxy groups on the nitrogen, an addition-elimination reaction occurs upon treatment with salts of dialkyl phosphites.[29-31]

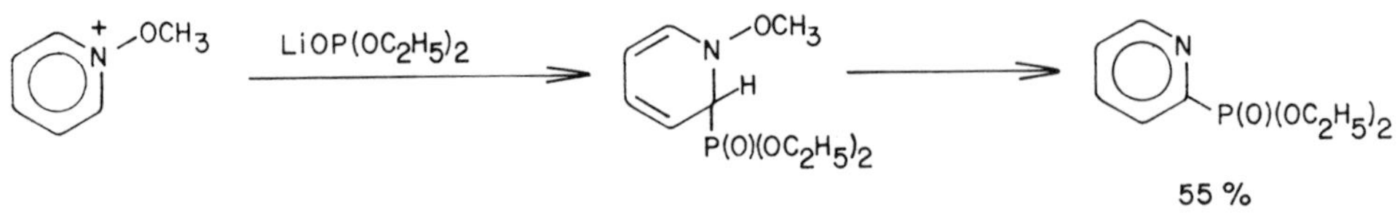

55 %

Reaction proceeds to introduce the phosphorus function at the 2-position in moderate yield. If the 2-position is blocked by a carbon functionality, substitution occurs at the 4-position. When the pyridine nitrogen is quaternized with a triphenylmethyl group, substitution occurs specifically at the 4-position.[32,33]

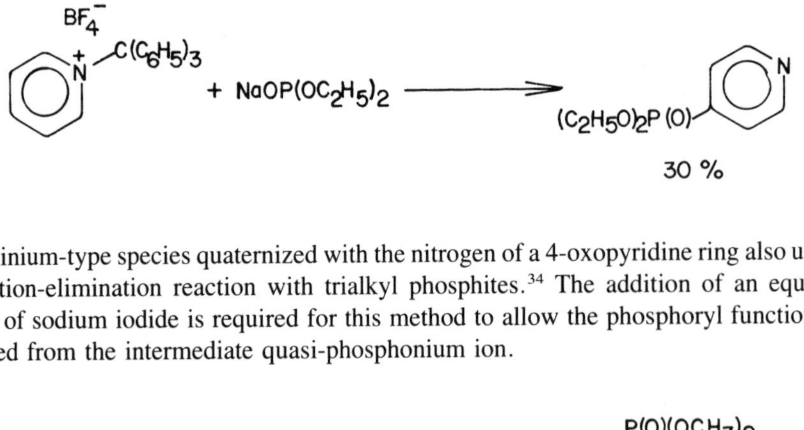

30 %

Pyridinium-type species quaternized with the nitrogen of a 4-oxopyridine ring also undergo an addition-elimination reaction with trialkyl phosphites.[34] The addition of an equivalent amount of sodium iodide is required for this method to allow the phosphoryl function to be generated from the intermediate quasi-phosphonium ion.

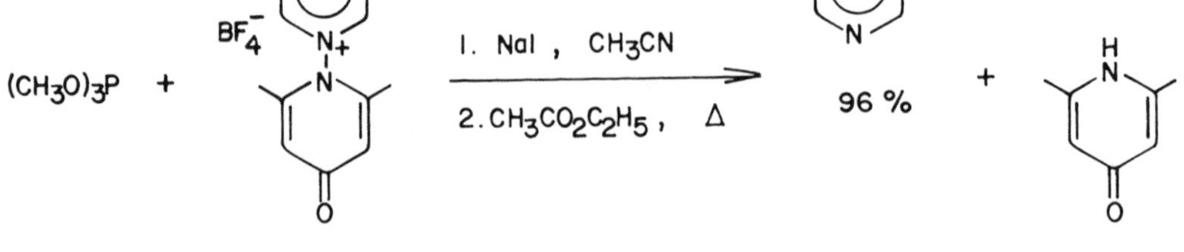

96 %

Treatment of N-carboalkyoxypyridinium salts with trialkyl phosphites also results in a "Michaelis-Arbuzov addition" process of the phosphorus atom to the aromatic ring. In contrast to the reaction systems mentioned previously, the resultant 1,4-dihydropyridine-phosphonate diesters are isolable and may be purified by distillation.[35]

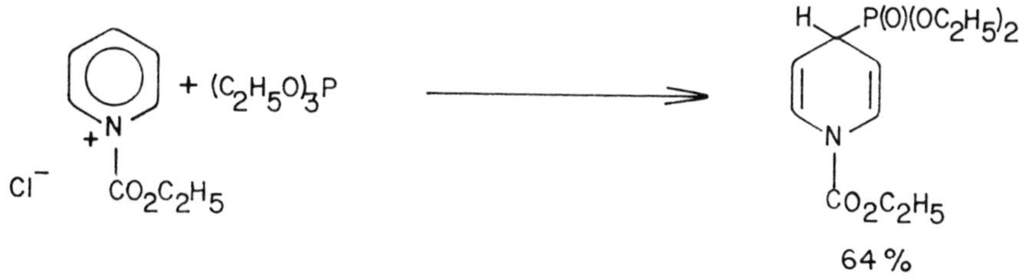

64%

Addition without elimination occurs upon treatment of 1-(4-pyridyl)pyridinium salts with phosphorus trichloride.[36] Upon addition of alcohol, the 1,2-dihydropyridinephosphonate diesters are obtained in moderate to good yield. Regeneration of the aromatic ring is accomplished by bromine oxidation and the free pyridine-2-phosphonic acid by subsequent acid cleavage.

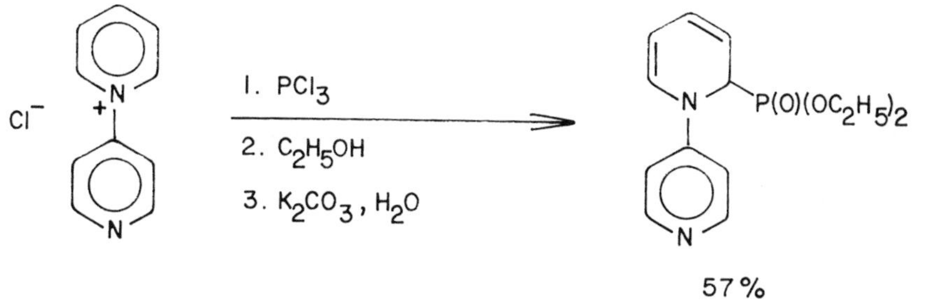

57%

Several reports have appeared of addition reactions of phosphites to π-complexes of aromatic hydrocarbons and transition metals.[37-40] While interesting, these reactions have relatively little synthetic utility.

Finally, there has been a report of a salt of a dialkyl phosphite undergoing addition to an aromatic sulfonium ion.[41] The resultant adduct is isolable and can be used in further synthetic procedures.

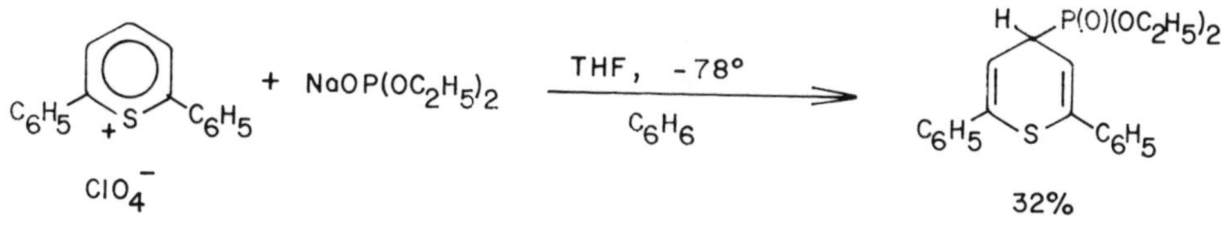

32%

For many years the Friedel-Crafts reaction has been of interest for the direct attachment of phosphorus to an aromatic ring. The limited success and the difficulties associated with the classical approaches have been chronicled.[42] However, improvements in the yield and purification procedures have been reported recently.

In addition to recent improvements in the techniques using aluminum chloride as the catalyst,[43-47] use of stannic chloride as the Lewis-acid catalyst has been reported to give superior yields with minimization of side-reactions involving sensitive functionalities which may be present on the aromatic starting compound.[48]

It is of interest that use of this organic-soluble catalyst minimizes multiple reaction on a single phosphorus center, even with aromatic rings activated by alkoxy groups.

An improved yield of the double Friedel-Crafts reaction product with phosphorus trichloride via aluminum chloride catalysis has been reported.[49] Isolated yields of diarylchlorophosphines from the aromatic hydrocarbon are reported to be increased as the result of using tris(2-chloroethyl) phosphite for decomposition of the adduct-Lewis acid complex.

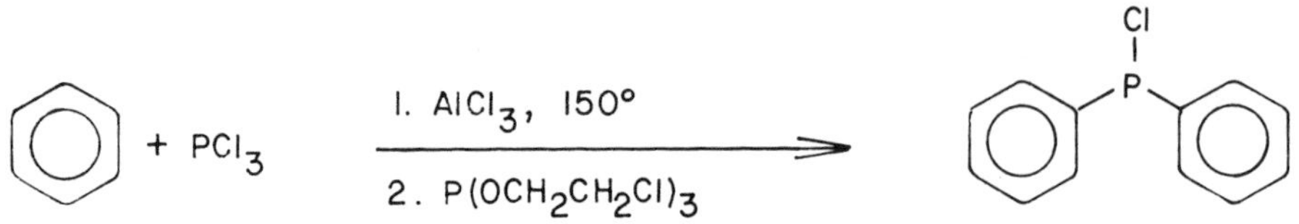

The use of aromatic organometallics for substitution of halogen on phosphorus has been used in several instances in recent work. The yields of phosphonate esters[50,51] and phosphines[52] generated in this manner are highly variable, but in certain instances are greater than 90%.

Attempts to synthesize arylphosphonic acids via aryldiazonium ions, while successful, have produced the materials in only mediocre yield.[53] The reaction involves treatment of the diazonium tetrafluoborate with phosphorus trichloride. In addition to the relatively low yields of the desired products, varying amounts of diarylphosphinic acids were also produced.

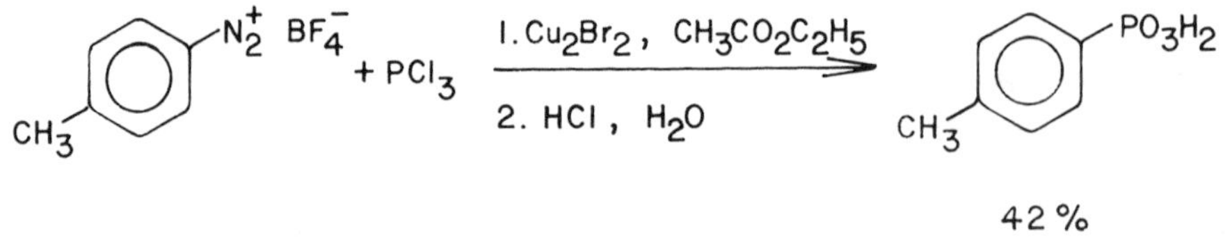

Other attempts using phosphites with diazonium species failed to result in carbon-phosphorus bond formation.

Finally, 2-hydroxyarylphosphonates can be generated from the corresponding phosphates.[54] Treatment of the aryl phosphate ester with strong base results in the abstraction of a proton *ortho* to the ester linkage, followed by O-to-C migration of the phosphorus.

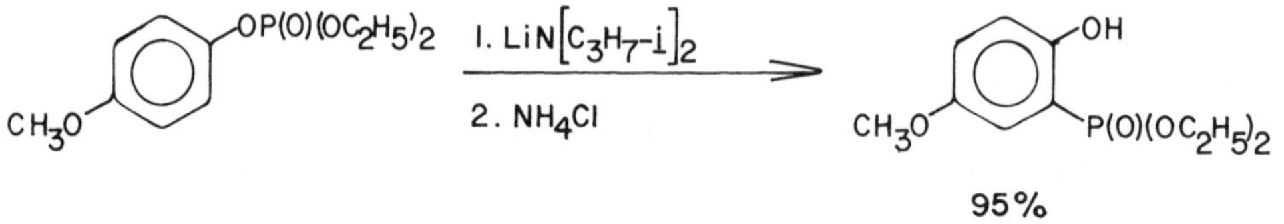

B. Comparison of Methods

Several approaches have been demonstrated in recent years to have general utility for the generation in high yield of aryl-carbon to phosphorus bonds. The most promising of these appears to involve the use of transition metal complexes to catalyze the overall Michaelis-Arbuzov and Michaelis-Becker type reactions using aryl halides. A major source of difficulty in the use of this approach arises when substituents (either bulky or polar) are located *ortho* relative to the site of substitution.

Photochemicaly assisted Michaelis-Arbuzov reaction is often found to proceed in quite good yield. However, the major drawback for this method in the design and performance of a synthetic scheme is the requirement for using the aryl iodide in the reaction. Although there is a decrease in yield when substituents are located *ortho* to the site of substitution, the reaction is still viable for synthetic purposes.

The use of the Friedel-Crafts approach is much more variable with regard to yield. In designing a synthetic scheme involving preparation of a new arylphosphonate ester, exploratory investigations of solvent and catalyst would be required.

In the generation of carbon-phosphorus bonds to nitrogen-containing aromatic rings, several approaches are available. Although fewer examples have been reported, the use of the pyridine ring system with *N*-pyridone substitution appears to provide the best yields. The major difficulty in the use of this approach synthetically would appear to be the preparation of the appropriate *N*-pyridone species.

C. Experimental Procedures

Diethyl Phenylphosphonate — Reaction of an aryl halide with a dialkyl phosphite in the presence of a Pd(0) catalyst and a tertiary amine.[25]

To a stirred mixture of diethyl phosphite (0.61 g, 4.4 mmol), triethylamine (0.44 g, 4.4 mmol), and tetrakis(triphenylphosphine)palladium (0.23 g, 0.2 mmol) there was added bromobenzene (0.63 g, 4.0 mmol) under a nitrogen atmosphere. The mixture was stirred for 2.5 hr at 90°C. At the end of this time, ether (50 mℓ) was added and the resultant precipitate of triethylamine hydrobromide was removed by filtration. After evaporation of the solvent, the residue was vacuum distilled (Kugelrohr) to give pure diethyl phenylphosphonate (0.79 g, 92%) which exhibited IR and ^{1}H NMR spectra in accord with the assigned structure.

Dimethyl 2-Methylphenylphosphonate — Photoinitiated reaction of an aryl iodide with a trialkyl phosphite.[6]

A mixture of 2-iodotoluene (8.78 g, 0.04 mol) and trimethyl phosphite (24.8 g, 0.20 mol) was placed in a 45-mℓ double-jacketed silica reaction vessel. The mixture was degassed by flushing with dry nitrogen for 5 min and irradiated with a 450 watt Hanovia (Model 679A-10) high pressure quartz mercury vapor lamp fitted with an aluminum reflector head. The lamp was placed 5 cm from the inner portion of the reaction vessel. The reaction temperature was maintained at 0°C by the circulation of coolant from a thermostatted refrigeration unit. Irradiation was continued at this temperature for 24 hr. At the end of this time the volatile materials were removed with a water aspirator and the residue was vacuum distilled (96 to 97°C/0.25 torr) to give the pure dimethyl 2-methylphenylphosphonate (7.28 g, 91%).

4-Methoxyphenylphosphonous Dichloride — Friedel-Crafts reaction of a substituted benzene with phosphorus trichloride.[48]

A solution of anisole (10.8 g, 0.1 mol), phosphorus trichloride (41 g, 0.30 mol), and anhydrous stannic chloride (2 mℓ) was refluxed under dry nitrogen for 76 hr. An additional 2 mℓ of stannic chloride was added every 12 hr. At the end of this time the reaction mixture was concentrated under reduced pressure and the residue was vacuum distilled (74 to 78°C/

0.05 torr) through a 10-in. Vigreux column to give pure 4-methoxyphenylphosphonous dichloride (19.8 g, 91%).

Diethyl 4-Acetylphenylphosphonate — Reaction of an aryl halide with a trialkyl phosphite catalyzed by Ni(II).[12]

To a suspension of nickel(II) chloride (2.6 g, 0.2 mol) in 4-bromoacetophenone (40.0 g, 0.2 mol) heated to a temperature of 160°C is added dropwise over a period of 1 hr triethyl phosphite (50.0 g, 0.3 mol). Over a further period of heating for 1 hr ethyl bromide is distilled. The residue is vacuum distilled (155—158°C/0.06 torr) to give pure diethyl 4-acetylphenylphosphonate (39.5 g, 78%).

Dimethyl Pyridin-4-ylphosphonate — Reaction of an *N*-pyridonepyridinium salt with a trialkyl phosphite.[34]

To a stirred suspension of *N*-(2,6-dimethyl-4-oxypyridin-1-yl)pyridinium tetrafluoborate (0.58 g, 2 mmol) in dry acetonitrile (20 mℓ) under nitrogen was added trimethyl phosphite (0.25 g, 2 mmol) followed by finely divided sodium iodide (0.30 g, 2 mmol). After 1 hr at 25°C the solvent was removed under reduced pressure and water (20 mℓ) was added. The mixture was extracted with methylene chloride (3 × 15 mℓ) and the extracts were dried over magnesium sulfate, filtered, and evaporated under reduced pressure. The residue was dissolved in ethyl acetate (40 mℓ), heated at reflux for 4 hr, evaporated under reduced pressure, and eluted on an alumina column (grade I, neutral) with chloroform to yield pure dimethyl pyridin-4-ylphosphonate (0.36 g, 96%) of mp 139 to 140°C.

Table 1
FORMATION OF AROMATIC-CARBON TO PHOSPHORUS BONDS

Product	Phosphorus reagent	Method	Yield (%)	Refs.
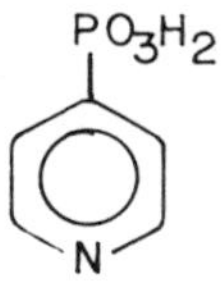 PO_3H_2 — pyridine	H_3PO_3	A	25	11
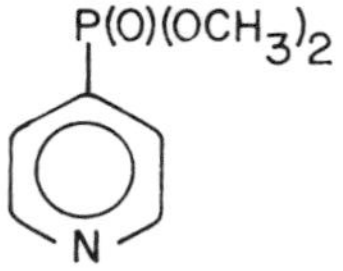 $P(O)(OCH_3)_2$ — pyridine	$(CH_3O)_3P$	B	96	34
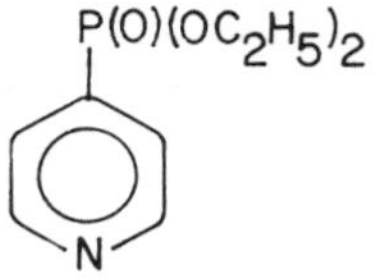 $P(O)(OC_2H_5)_2$ — pyridine	$NaOP(OC_2H_5)_2$	C	39	33
$P(O)(OC_3H_7-i)_2$ — pyridine	$(C_2H_5O)_3P$ $NaOP(OC_2H_5)_2$ $NaOP(OC_3H_7-i)_2$	B D C	82 39 30	34 32 33

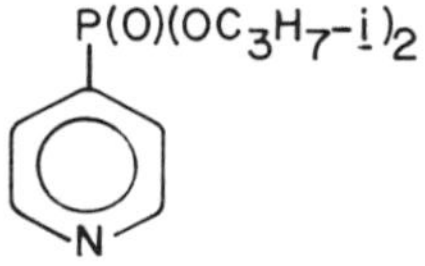

Table 1 (continued)
FORMATION OF AROMATIC-CARBON TO PHOSPHORUS BONDS

Product	Phosphorus reagent	Method	Yield (%)	Refs.
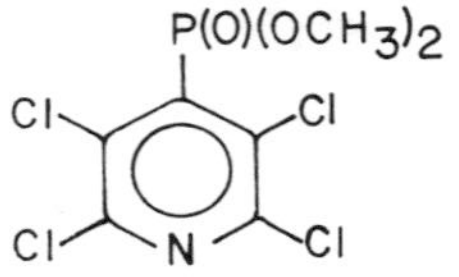 tetrachloropyridine–$P(O)(OCH_3)_2$	$(CH_3O)_3P$	D	30	32
		E	40	1
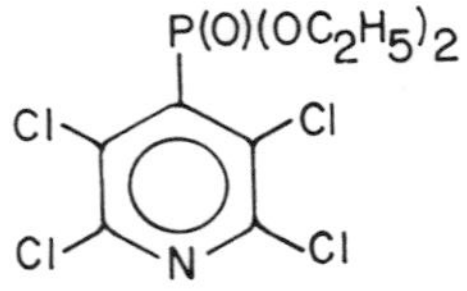 tetrachloropyridine–$P(O)(OC_2H_5)_2$	$(C_2H_5O)_3P$	E	50	1
$C_6H_5PO_3H_2$	PCl_3	F	17	53
$C_6H_5P(O)(OCH_3)_2$	$(CH_3O)_3P$	G	16	9
		H	12	12
$C_6H_5P(O)(OC_2H_5)_2$	$(C_2H_5O)_3P$	G	59	9
		H	91	12
		H	76	23
	$HOP(OC_2H_5)_2$		92	25
$C_6H_5P(O)(OC_3H_7\text{-}i)$	$(i\text{-}C_3H_7O)_3P$	G	41	9
		H	86	12
	$HOP(OC_3H_7\text{-}i)_2$	H	90	25
$C_6H_5P(O)(OC_6H_5)_2$	$(C_6H_5O)_3P$	H	96	19
$C_6H_5P(O)(C_6H_5)C_4H_9\text{-}n$	$HOP(C_6H_5)C_4H_9\text{-}n$	H	86	28
$C_6H_5P(O)(CH_3)OC_4H_9\text{-}n$	$HOP(CH_3)OC_4H_9\text{-}n$	H	89	27
$C_6H_5P(O)(C_6H_5)OC_2H_5$	$HOP(C_6H_5)OC_2H_5$	H	79	26
$C_6H_5P(O)(C_6H_5)OC_4H_9\text{-}n$	$HOP(C_6H_5)OC_4H_9\text{-}n$	H	97	26
	H_3PO_3	A	28	11

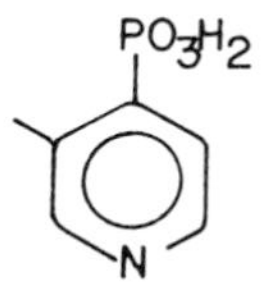

Product	Reagent	Ref.	Yield	Yield
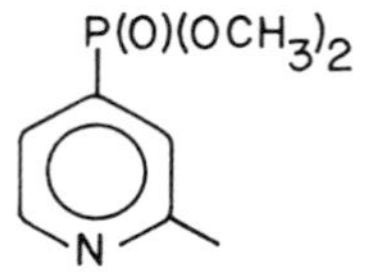 (methylpyridinyl)PO_3H_2	$(CH_3O)_3P$	B	60	34
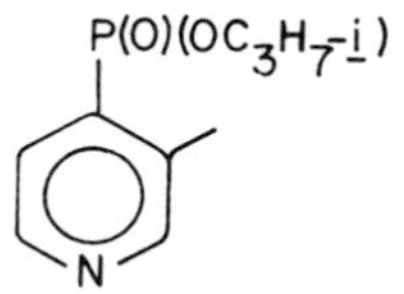 (methylpyridinyl)$P(O)(OCH_3)_2$	$NaOP(OC_3H_7\text{-}i)_2$	C	53	33
(dimethylpyridinyl)$P(O)(OC_3H_7\text{-}i)$	$C_6H_5OP(O)(OC_2H_5)_2$	D	53	32
		I	93	54
$2\text{-}HOC_6H_4P(O)(OC_2H_5)_2$	$(C_2H_5O)_3P$	J	66	6
$4\text{-}HOC_6H_4P(O)(OC_2H_5)_2$	$(C_2H_5O)_3P$	J	53	6
$4\text{-}NO_2C_6H_4P(O)(OC_2H_5)_2$	PCl_3	F	44	53
$2\text{-}ClC_6H_4PO_3H_2$	PCl_3	F	50	53
$4\text{-}ClC_6H_4PO_3H_2$	$HOP(CH_3)OC_4H_9\text{-}n$	H	85	27
$4\text{-}ClC_6H_4P(O)(CH_3)OC_4H_9\text{-}n$	$HOP(C_6H_5)OC_4H_9\text{-}n$	H	76	26
$4\text{-}ClC_6H_4P(O)(C_6H_5)OC_4H_9\text{-}n$	$(C_2H_5O)_3P$	H	41	12
$2\text{-}ClC_6H_4P(O)(OC_2H_5)_2$	$(C_2H_5O)_3P$	H	78	12
$4\text{-}ClC_6H_4P(O)(OC_2H_5)_2$		H	52	23

Table 1 (continued)
FORMATION OF AROMATIC-CARBON TO PHOSPHORUS BONDS

Product	Phosphorus reagent	Method	Yield (%)	Refs.
4-NO$_2$C$_6$H$_4$P(O)(CH$_3$)OC$_4$H$_9$-n	HOP(OC$_2$H$_5$)$_2$	H	85	23
4-NO$_2$C$_6$H$_4$P(O)(C$_6$H$_5$)OC$_4$H$_9$-n	HOP(CH$_3$)OC$_4$H$_9$-n	H	93	27
4-NO$_2$C$_6$H$_4$P(O)(OC$_2$H$_5$)$_2$	HOP(C$_6$H$_5$)OC$_4$H$_9$-n	H	86	26
2,3,5,6-F$_4$C$_6$H–P(O)(OC$_2$H$_5$)$_2$ (tetrafluorophenyl, structure)	HOP(OC$_2$H$_5$)$_2$	H	73	25
	NaOP(OC$_2$H$_5$)$_2$	E	10	2
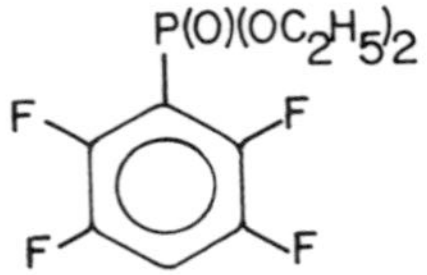 4-CH$_3$C$_6$H$_4$PO$_3$H$_2$	PCl$_3$	F	42	53
2-CH$_3$C$_6$H$_4$P(O)(C$_6$H$_5$)C$_4$H$_9$-n	HOP(C$_6$H$_5$)C$_4$H$_9$-n	H	32	28
4-CH$_3$C$_6$H$_4$P(O)(C$_6$H$_5$)C$_4$H$_9$-n	HOP(C$_6$H$_5$)C$_4$H$_9$-n	H	37	28
2-CH$_3$C$_6$H$_4$P(O)(CH$_3$)OC$_4$H$_9$-n	HOP(CH$_3$)OC$_4$H$_9$-n	H	80	27
4-CH$_3$C$_6$H$_4$P(O)(CH$_3$)OC$_4$H$_9$-n	HOP(CH$_3$)OC$_4$H$_9$-n	H	81	27
2-CH$_3$C$_6$H$_4$P(O)(C$_6$H$_5$)OC$_4$H$_9$-n	HOP(C$_6$H$_5$)OC$_4$H$_9$-n	H	72	26
4-CH$_3$C$_6$H$_4$P(O)(C$_6$H$_5$)OC$_4$H$_9$-n	HOP(C$_6$H$_5$)OC$_4$H$_9$-n	H	69	26
2-CH$_3$C$_6$H$_4$P(O)(OC$_2$H$_5$)$_2$	(C$_2$H$_5$O)$_3$P	H	90	12
	(C$_2$H$_5$O)$_3$P	J	73	6
4-CH$_3$C$_6$H$_5$P(O)(OC$_2$H$_5$)$_2$	HOP(OC$_2$H$_5$)$_2$	H	91	25
	(C$_2$H$_5$O)$_3$P	H	83	12
		J	54	23
		J	95	6
4-CH$_3$C$_6$H$_4$P(O)(OC$_6$H$_4$CH$_3$-4)$_2$	HOP(OC$_2$H$_5$)$_2$	H	94	25
	(4-CH$_3$C$_6$H$_4$O)$_3$P	H	94	19
	H$_3$PO$_3$	A	25	11

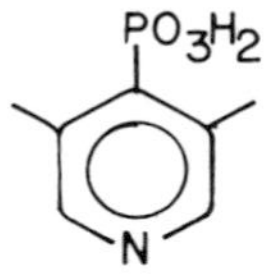

Structure (3,5-dimethylpyridin-4-yl) bearing $-PO_3H_2$

Structure (3,5-dimethylpyridin-4-yl) bearing $-P(O)(OC_3H_7\text{-}i)$

Structure (2,3,5,6-tetrafluoro-4-cyanophenyl) bearing $-P(O)(OCH_3)_2$

Product	Reagent	C	28	33
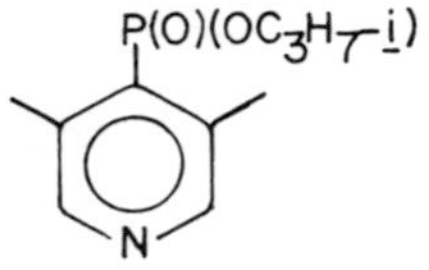	$NaOP(OC_3H_7\text{-}i)_2$			
$4\text{-}CH_3\text{-}2\text{-}HOC_6H_3P(O)(OC_2H_5)_2$	$3\text{-}CH_3C_6H_4OP(O)(OC_2H_5)_2$	I	87	54
$4\text{-}CH_3OC_6H_4PCl_2$	PCl_3	K	91	48
$4\text{-}CH_3OC_6H_4P(O)(C_4H_9\text{-}n)OC_2H_5$	$HOP(C_4H_9\text{-}n)OC_2H_5$	H	80	27
$4\text{-}CH_3OC_6H_4P(O)(CH_3)OC_4H_9\text{-}n$	$HOP(CH_3)OC_4H_9\text{-}n$	H	74	27
$4\text{-}CH_3OC_6H_4P(O)(C_6H_5)OC_4H_9\text{-}n$	$HOP(C_6H_5)OC_4H_9\text{-}n$	H	60	26
$4\text{-}CH_3OC_6H_4P(O)(OC_2H_5)_2$	$(C_2H_5O)_3P$	J	72	12
		H	70	6
$4\text{-}OCHC_6H_4P(O)(OC_2H_5)_2$	$HOP(OC_2H_5)_2$	J	95	25
$2\text{-}CH_3O\text{-}5\text{-}HOC_6H_3P(O)(OC_2H_5)_2$	$(C_2H_5O)_3P$	I	34	6
$4\text{-}HO_2CC_6H_4PO_3H_2$	$3\text{-}CH_3OC_6H_4OP(O)(OC_2H_5)_2$	F	95	54
$2\text{-}HO_2CC_6H_4P(O)(OC_2H_5)_2$	PCl_3	H	39	53
	$HOP(OC_2H_5)_2$	E	24	25
	$NaOP(OCH_3)_2$		63	2

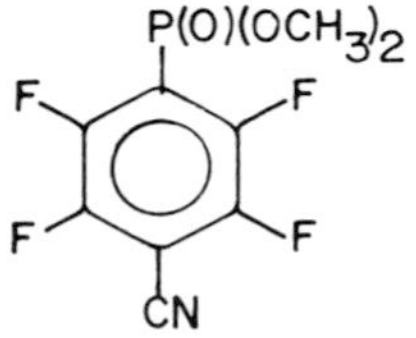

Table 1 (continued)
FORMATION OF AROMATIC-CARBON TO PHOSPHORUS BONDS

Product	Phosphorus reagent	Method	Yield (%)	Refs.
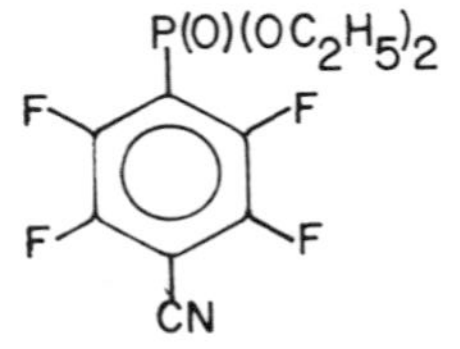 P(O)(OC$_2$H$_5$)$_2$ / fluorophenyl-CN	(C$_2$H$_5$O)$_3$P	E	26	3
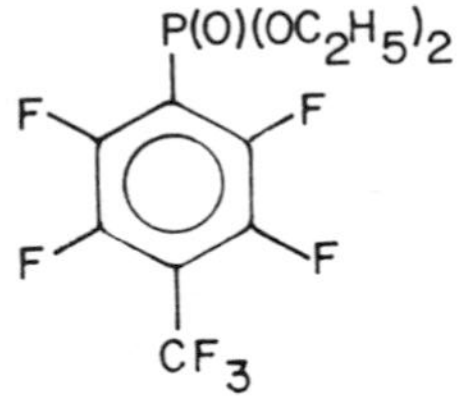 P(O)(OC$_2$H$_5$)$_2$ / fluorophenyl-CF$_3$	NaOP(OC$_2$H$_5$)$_2$	E	57	2
	NaOP(OC$_2$H$_5$)$_2$	E	65	2
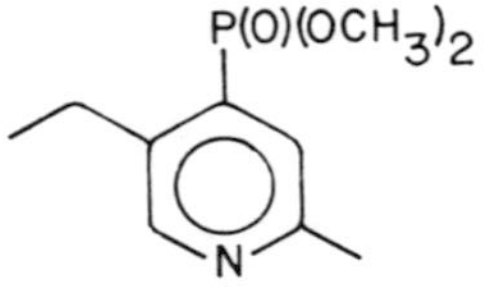 P(O)(OCH$_3$)$_2$ / pyridine	(CH$_3$O)$_3$P	B	78	34
4-(CH$_3$)$_2$NC$_6$H$_4$P(O)(C$_6$H$_5$)C$_4$H$_9$-n	HOP(C$_6$H$_5$)C$_4$H$_9$-n	H	51	28
4-(CH$_3$)$_2$NC$_6$H$_4$P(O)(CH$_3$)OC$_4$H$_9$-n	HOP(CH$_3$)OC$_4$H$_9$-n	H	86	27
4-CH$_3$O-3-CH$_3$C$_6$H$_3$PCl$_2$	PCl$_3$	K	68	48
	(C$_2$H$_5$O)$_3$P	H	66	12
4-CH$_3$COC$_6$H$_4$P(O)(OC$_2$H$_5$)$_2$	CH$_3$P(OC$_2$H$_5$)$_2$	J	50	8

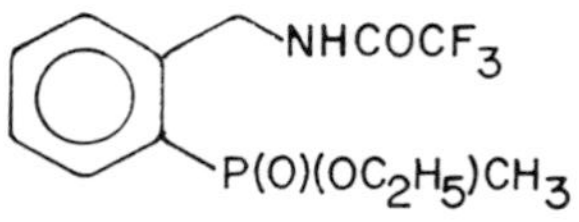

Product	Reagent	Method		
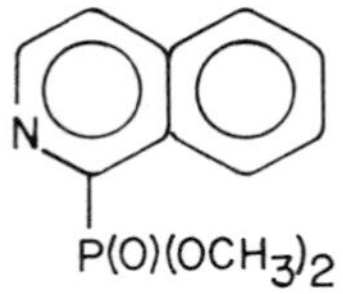 benzene: —NHCOCF$_3$, —P(O)(OC$_2$H$_5$)CH$_3$	(CH$_3$O)$_3$P	B	53	34
isoquinoline: P(O)(OCH$_3$)$_2$	PCl$_3$	K	29	48
	HOP(OC$_2$H$_5$)$_2$	H	43	25
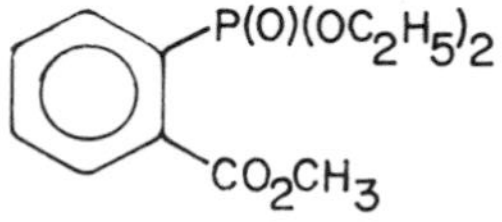 2,4-(CH$_3$O)$_2$C$_6$H$_3$PCl$_2$				
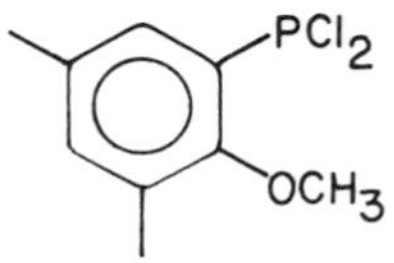 benzene: —P(O)(OC$_2$H$_5$)$_2$, —CO$_2$CH$_3$	PCl$_3$	K	50	48
benzene: —PCl$_2$, —OCH$_3$, CH$_3$				
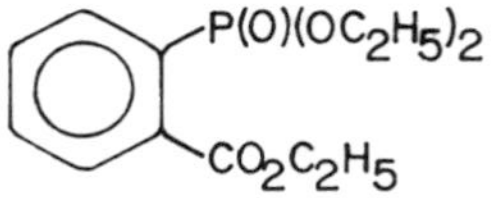 benzene: —P(O)(OC$_2$H$_5$)$_2$, —CO$_2$C$_2$H$_5$	(C$_2$H$_5$O)$_3$P	H	69	12

Table 1 (continued)
FORMATION OF AROMATIC-CARBON TO PHOSPHORUS BONDS

Product	Phosphorus reagent	Method	Yield (%)	Refs.
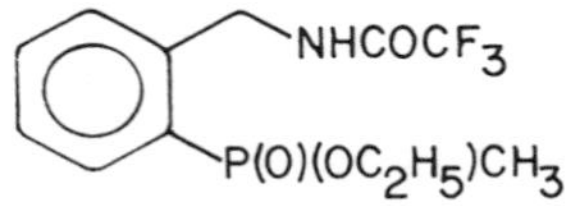 (structure: NHCOCF$_3$, P(O)(OC$_2$H$_5$)CH$_3$)	CH$_3$P(OC$_2$H$_5$)$_2$	J	90	8
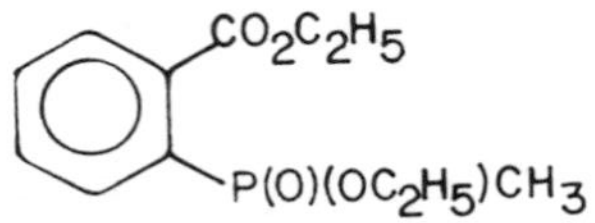 (structure: CO$_2$C$_2$H$_5$, P(O)(OC$_2$H$_5$)CH$_3$)	CH$_3$P(OC$_2$H$_5$)$_2$	H	95	15
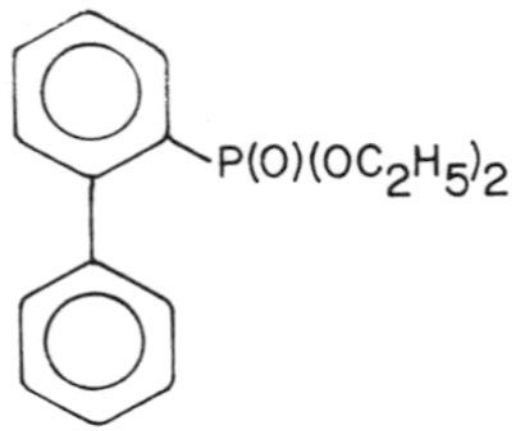 (structure: biphenyl, P(O)(OC$_2$H$_5$)$_2$)	(C$_2$H$_5$O)$_3$P	H	14	12
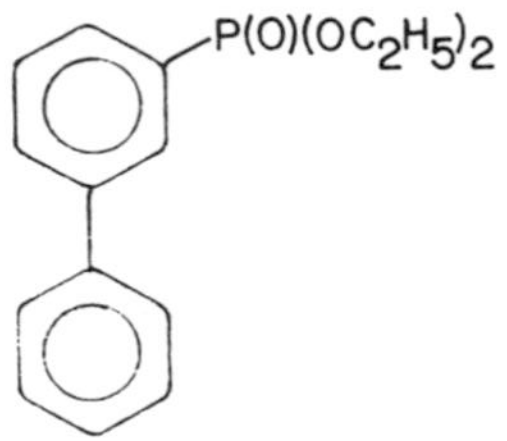 (structure: biphenyl, P(O)(OC$_2$H$_5$)$_2$)	(C$_2$H$_5$O)$_3$P	H	73	12

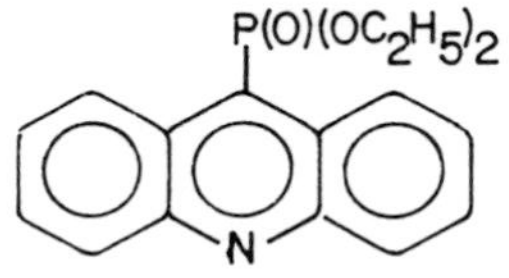P(O)(OC2H5)2 N	$(C_2H_5O)_3P$	E	62	4
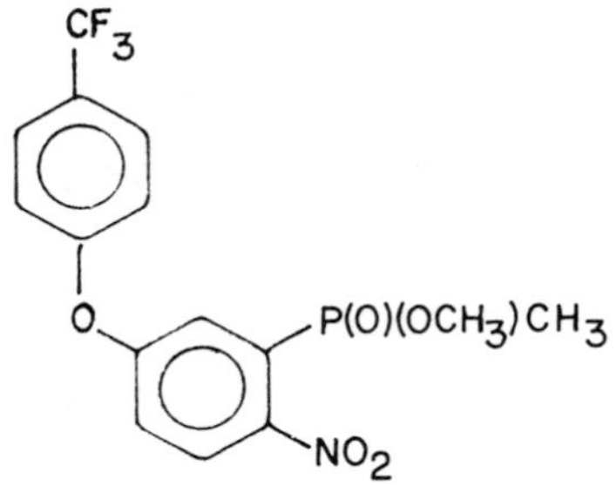CF3 O P(O)(OCH3)CH3 NO2	$CH_3P(OCH_3)_2$	L	50	10
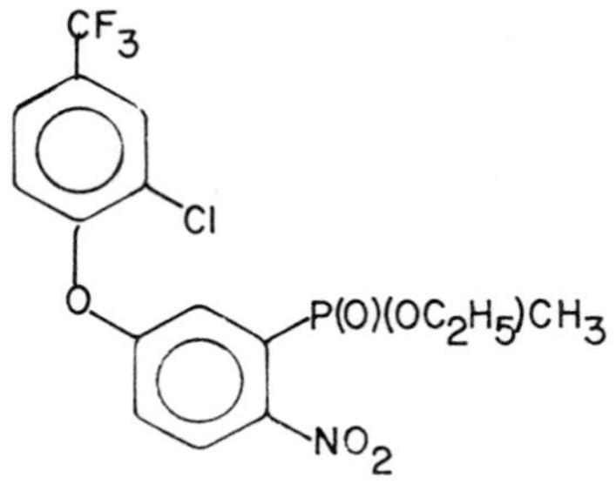CF3 Cl O P(O)(OC2H5)CH3 NO2	$CH_3P(OC_2H_5)_2$	L	52	10

Table 1 (continued)
FORMATION OF AROMATIC-CARBON TO PHOSPHORUS BONDS

Product	Phosphorus reagent	Method	Yield (%)	Refs.
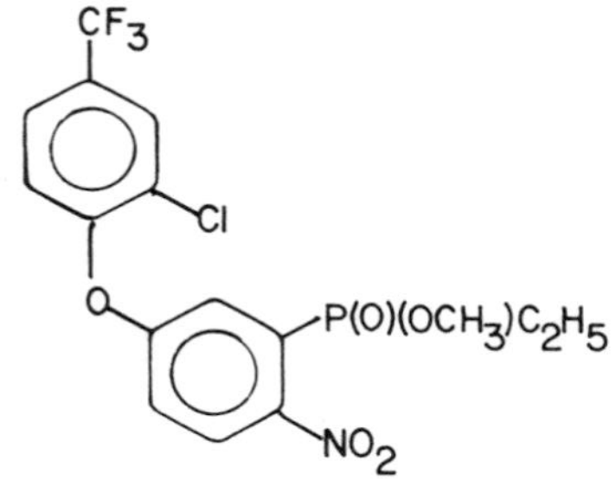 (structure: CF_3-substituted chlorophenyl ether linked to nitrophenyl bearing $P(O)(OCH_3)C_2H_5$)	$C_2H_5P(OCH_3)_2$	L	31	10
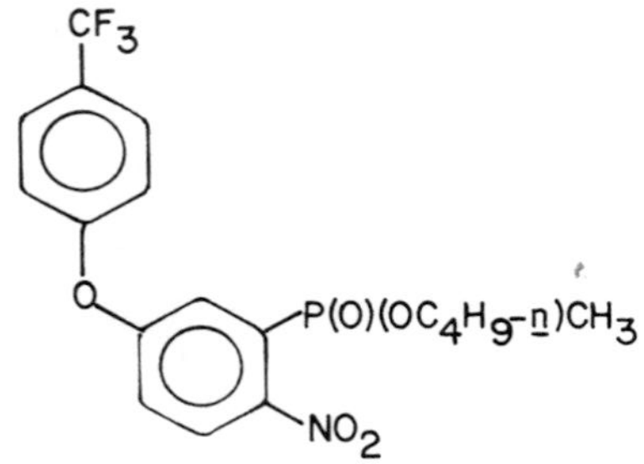 (structure: CF_3-substituted phenyl ether linked to nitrophenyl bearing $P(O)(OC_4H_9\text{-}n)CH_3$)	$CH_3P(OC_4H_9\text{-}n)_2$	L	24	10

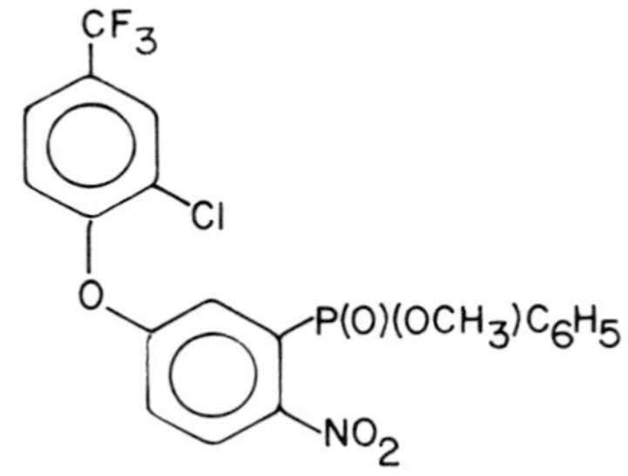

$C_6H_5P(OCH_3)_2$

L   54   10

Note: A, pyridinium ion used as substrate; B, pyridone used as substrate; C, *N*-alkoxypyridinium ion used as substrate; D, *N*-tritylpyridinium ion used as substrate; E, direct thermal reaction of aryl halide; F, diazonium species used as substrate; G, anodic oxidation; H, metal catalyzed "Michaelis-Arbuzov-type" reaction of aryl halide; I, LDA facilitated reaction; J, photoinitiated reaction of aryl halide; K, Friedel-Crafts-type reaction.

III. VINYLIC CARBON-PHOSPHORUS BOND FORMATION

A. Reactions and Mechanisms

Investigations of the direct formation of vinylic-carbon to phosphorus bonds have been concentrated on fewer general methods than for aromatic-carbon to phosphorus bond formation. Two major approaches have been the subject of the bulk of the recent reports. Several other techniques have been noted occasionally and will be given brief treatment here.

As with aromatic halides, the reaction of vinylic halides with trialkyl or dialkyl phosphites facilitated by transition metal species has been rigorously explored. The first work in this direction involved heating the vinyl halide and trialkyl phosphite in the presence of nickel(II) chloride in benzene solution with hydroquinone present.[55] Moderate yields of the vinyl-phosphonate diester were obtained directly. Interestingly, in the one instance where a direct comparison of halide effect could be made, chloride provided a significantly higher yield of phosphonate product than did bromide.

This procedure was later utilized in the preparation of several β-bisphosphonates to be used in Wittig-Horner reactions.[56]

The use of copper(I) halide complexes of the trialkyl phosphites in 30% excess of the vinyl halide resulted in improved yields of the vinylphosphonate diesters.[57]

The reaction was demonstrated to involve the intermediacy of the quasi-phosphonium ion similar to the normal Michaelis-Arbuzov reaction. This is generated by an initial addition of the copper complex to the olefinic carbon and is decomposed to the product by normal displacement reaction involving the halide ion.[58] It is noted that vinyl bromides provide higher yields of phosphonate diesters than do the vinyl chlorides, and the copper(I) bromide complexes likewise provide higher product yields than the chloride complexes. A series of cyclic vinylicphosphonate diesters have been synthesized via this method as well as the simple acyclic species.[59]

The use of triaryl phosphite complexes of copper(I) with vinyl halides fails to give phosphonate products. However, there does occur a halide exchange reaction wherein chloride from the complex is exchanged with bromide of the vinyl bromide.[60] The exchange reaction does not proceed in the opposite direction. Presumably the initial interaction of the complex with the olefinic carbon occurs in such systems, but completion of the reaction is superceded by the halide attack.

Similarly, as with the previously discussed aromatic halides, dialkyl phosphites may be used for phosphonate diester formation with vinyl halides in the presence of palladium(0) catalysts. Good to excellent yields of vinylphosphonate diesters are obtained using catalytic amounts of tetrakis(triphenylphosphine)palladium in the presence of triethyl amine.[25,61]

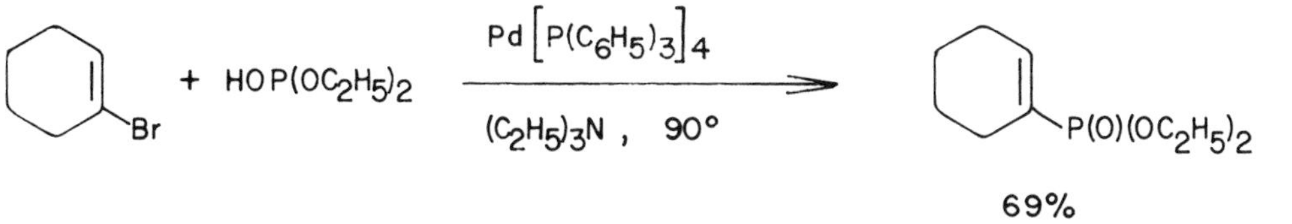

It should be noted that both of these metal-catalyzed methods provide the vinylphosphonate diesters with retention of the original stereochemistry about the olefinic linkage.

The second approach to the preparation of vinylphosphonate diesters, which has been investigated extensively, involves addition of the phosphorus reagent (either a trialkyl phosphite or a dialkyl phosphite anion) to the β-carbon atom site of an α,β-unsaturated carbonyl compound. When the β-carbon atom site also bears a potential leaving group, its departure generates a quasi-phosphonium ion which appears as though a simple displacement reaction had occurred. With further displacement of an alkyl ester function, generation of the vinylphosphonate diester is accomplished. Early work demonstrated the capability of this reaction with either halide or alkoxy leaving groups.[62] Moderate yields of the products were obtained using either trialkyl phosphites or sodium salts of dialkyl phosphites.

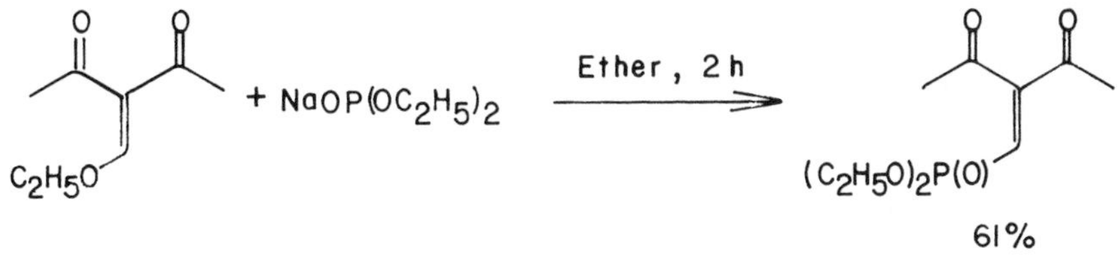

Further efforts have concentrated on the use of trialkyl phosphites with the vinyl halides.[63-68] Vinylogous ketones have also been found to proceed well in this type of reaction.[69]

In a similar vein, diphenyl(β-styryl)phosphine had been prepared by reaction of β-bromostyrene with the lithium salt of diphenylphosphine.[70] It is unlikely that simple direct displacement of the halide has occurred, but rather the phenyl ring of the β-bromostyrene acted as a charge-stabilizing function for an initial adduct, which then lost halide generating the product vinylphosphine.

Direct displacement of vinylic halides by phosphorus reagents for the formation of carbon to phosphorus bonds have been reported in several instances. The use of sodium salts of dialkyl phosphites at low temperature in tetrahydrofuran was found to provide the vinylphosphonate diesters in good yield.[56]

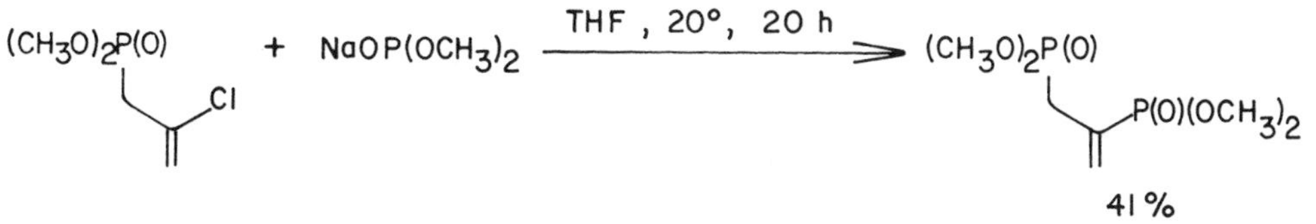

Interestingly, the isolated yields were found to be as good or better than those from the use of trialkyl phosphites and nickel(II) chloride.

Triisopropyl phosphite was found to react with trifluoroiodoethene to generate the vinylphosphonate diester in moderate yield.[71]

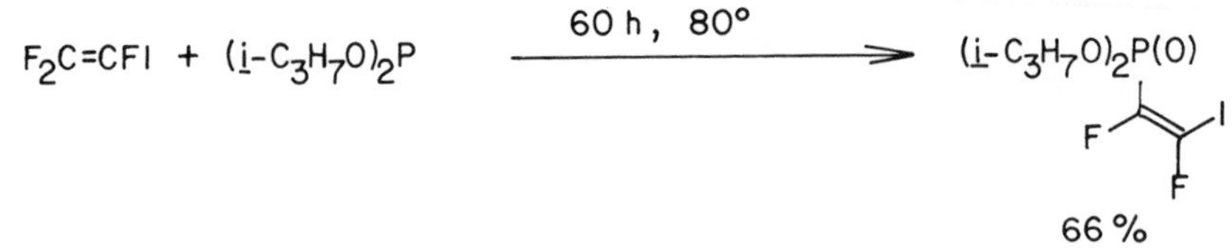

$$F_2C=CFI + (i-C_3H_7O)_2P \xrightarrow{\text{60 h, 80°}} (i-C_3H_7O)_2P(O)$$

66 %

In the initial reaction fluoride was displaced rather than the iodide, and only the phosphonate of the (Z) configuration was formed. In further reaction the iodide was replaced.

A photoinitiated reaction of trialkyl phosphite with another electron deficient vinylic halide, where the olefinic carbon is covalently bound to a metal center, has been reported.[72] However, only low yields of the phosphonate diester product are obtained. A similar result is found if the lithium salt of the dialkyl phosphite is used.

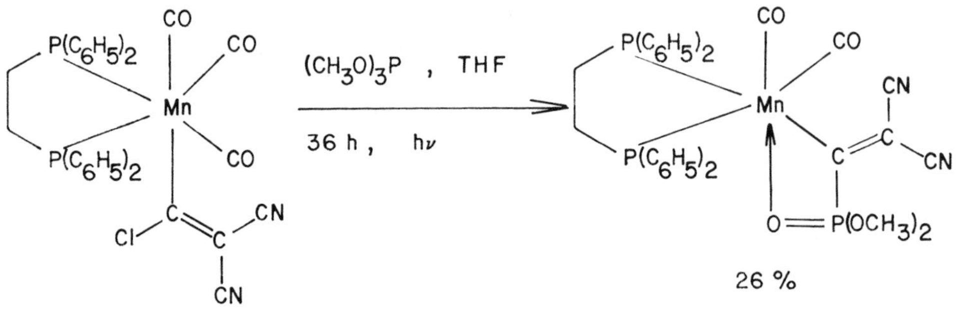

$\xrightarrow[\text{36 h, } h\nu]{(CH_3O)_3P \text{ , THF}}$

26 %

In another report, an acetylenic carbon covalently bound to an iron center has been found to undergo thermal reaction with triethyl phosphite generating a carbon to phosphorus bond at a vinylic site.[73] The zwitterionic quasi-phosphonium species is internally stabilized by coordination of the anionic center to a second iron atom.

Two reports have been made recently of mixed vinyl phosphite esters which undergo rearrangement to phosphonate diesters in moderate yield.[74,75] In both instances the vinyl phosphite esters were prepared by reaction of the dialkyl phosphorous chloride with highly enolized carbonyl compounds. The mixed ester products undergo thermal rearrangement to the phosphonate diesters.

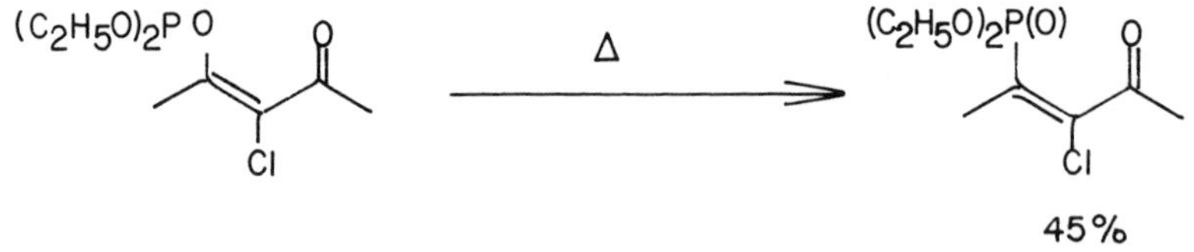

$\xrightarrow{\Delta}$

45%

Recently, vinylphosphonic acids were reported to be formed in good yield by the reaction of phosphorous acid and phosphorus trichloride with methyl aromatic ketones.[76]

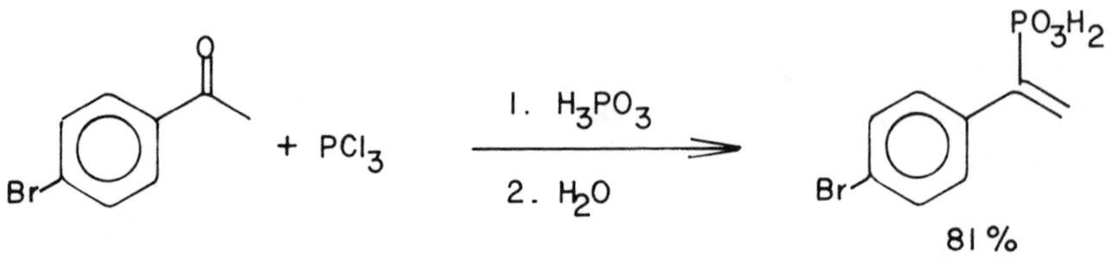

1. H_3PO_3

2. H_2O

81%

Vinylmercurials have been reported in two instances to be of use for the generation of vinylic-carbon to phosphorus bonds. The thermal reaction of divinylmercury with phosphorus trichloride results in the substitution of only one of the available chlorides to give vinyldichlorophosphine in 50% yield.[77] The remaining chlorides may be substituted in standard ways.

$$(CH_2=CH)_2Hg \ + \ PCl_3 \quad \longrightarrow \quad CH_2=CHPCl_2$$

$$50\%$$

The 1-alkenylmercury halides undergo photoinitiated free-radical reactions with salts of dialkyl phosphites and alkyl phosphonites.[78] When irradiated in a Pyrex vessel with dimethyl sulfoxide as solvent, the salts undergo addition of the vinylic function to phosphorus in moderate to good yield, maintaining the original stereochemistry about the olefinic linkage.

B. Comparison of Methods

Of the variety of methods used for generating vinylic-carbon to phosporus bonds, several have received sufficient study to be considered of general value. The addition-elimination reactions of β-halovinyl ketones are useful for the preparation of a limited range of compounds, specifically the β-phosphonovinylketones. While only moderate yields are obtained in most instances, this remains the fundamental direct method for the preparation for this class of compounds.

The several methods involving transition metal facilitation of the overall Michaelis-Arbuzov or Michaelis-Becker type reactions all provide the desired phosphorus products. There are some differences among them, however.

The approach using the copper(I) halide complex of the trialkyl phosphite provides the target vinylphosphonate diesters in excellent yield. There are, however, associated experimental difficulties. As it is necessary to use an equivalent of the phosphite complex, an equivalent of copper salt is left at the end of the reaction. Simple chromatographic or distillation techniques are found insufficient to remove these and a further treatment with ethylenediamine is required prior to final purification.

This difficulty is avoided using the Pd(0) catalyzed process, where excellent yields are obtained as well. Both methods have the advantage of providing product with the original stereochemistry of the olefinic linkage maintained.

The use of Ni(II) salts as catalysts for these reactions has the advantage of experimental simplicity. However, significantly lower yields have been noted. In fact, where direct comparisons are available, direct Michaelis-Becker reaction conditions provide comparable or slightly superior yields.

C. Experimental Procedures

Diethyl (Z)-1-Propenylphosphonate — Reaction of a vinylic bromide with a dialkyl phosphite catalyzed by Pd(0)[25]

There was mixed (Z)-1-bromo-1-propene (0.58 g, 4.8 mmol) with diethyl phosphite (0.55 g, 4.0 mmol), triethyl amine (0.41 g, 4.0 mmol), and tetrakis(triphenylphosphine)palladium (0.23 g, 0.2 mmol) in toluene (1 mℓ). The mixture was heated at 90°C for 1.5 hr. At the end of this time ether (50 mℓ) was added and the solution was filtered. The filtrate was evaporated under reduced pressure and the residue was chromatographed on a silica gel column to give pure diethyl (Z)-1-propenylphosphonate (0.70 g, 98%).

Diisopropyl 2,2-Diphenylvinylphosphonate — Reaction of a vinyl bromide with a Cu(I) complex of a trialkyl phosphite[58]

A mixture of (triisopropyl phosphito)copper(I) bromide (17.6 g, 0.05 mol) and 1-bromo-2,2-diphenylethene (9.1 g, 0.035 mol) were heated at 200°C for 1 hr under a nitrogen atmosphere in a flask equipped with a Vigreaux column topped by a Dean-Stark trap. The alkyl halide produced in the reaction was collected in the trap. After cooling, the reaction mixture was poured into toluene (60 mℓ) and ethylenediamine was added (5 mℓ). After filtering and washing the precipitate with toluene, the combined toluene solutions were washed with 10% hydrochloric acid (10 mℓ) and water (10 mℓ), dried over magnesium sulfate, and evaporated under reduced pressure. The residue was chromatographed on silica gel to give the pure diisopropyl 2,2-diphenylvinylphosphonate (12.1 g, 96%).

Diisopropyl (E)-2-Benzoylvinylphosphonate — Reaction of a trialkyl phosphite with a β-halovinylketone[63]

(E)-2-Chlorovinyl phenyl ketone (1.83 g, 11 mmol) and triisopropyl phosphite (2.08 g, 10 mmol) were heated under an argon atmosphere for 1 hr at 120 to 130°C. When all of the isopropyl chloride formed had distilled, the residue was chromatographed on a column of silica gel (18 g) being eluted with a 1:1 mixture of methylene chloride-ethyl acetate. The eluent was evaporated of solvent and the residue vacuum distilled to give pure diisopropyl (E)-2-benzoylvinylphosphonate (1.33 g, 45%).

Table 2
FORMATION OF VINYLIC-CARBON TO PHOSPHORUS BONDS

Product	Phosphorus reagent	Method	Yield (%)	Refs.
$CH_2=CHP(O)(OC_2H_5)_2$	$(C_2H_5O)_3P$	A	70	55
$CH_2=CHP(O)(OC_3H_7\text{-}i)_2$	$(i\text{-}C_3H_7O)_3P$	A	65	55
$(C_2H_5O)_2P(O)CH=CHP(O)(OC_2H_5)_2$	$(C_2H_5O)_3P$	A	94	55
$(E)\text{-}CHF=CFP(O)(OC_3H_7\text{-}i)_2$	$(i\text{-}C_3H_7O)_3P$	B	66	71
$CH_3COCH=CHP(O)(OC_3H_7\text{-}i)_2$	$(i\text{-}C_3H_7O)_3P$	B	48	63
$CH_3OCOCH=CHP(O)(OC_3H_7\text{-}i)_2$	$(i\text{-}C_3H_7O)_3P$	B	44	64
$C_2H_5COCH=CHP(O)(OC_3H_7\text{-}i)_2$	$(i\text{-}C_3H_7O)_3P$	B	44	66
$CH_3COCCl=C(CH_3)P(O)(OC_2H_5)_2$	$(C_2H_5O)_2POC(CH_3)=CClCOCH_3$	C	45	74
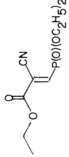	$HOP(OC_2H_5)_2$	A	69	61
$(CH_3)_2CHCOCH=CHP(O)(OC_3H_7\text{-}i)_2$	$(i\text{-}C_3H_7O)_3P$	B	57	68
$(CH_3CO)_2C=CHP(O)(OC_2H_5)_2$	$NaOP(OC_2H_5)_2$	B	61	62
$(CH_3CO)_2C=CHP(O)(OC_4H_9\text{-}n)_2$	$NaOP(OC_4H_9\text{-}n)_2$	B	33	62
(vinylic CN/ester phosphonate structure)	$NaOP(OC_2H_5)_2$	B	54	62

Table 2 (continued)
FORMATION OF VINYLIC-CARBON TO PHOSPHORUS BONDS

Product	Phosphorus reagent	Method	Yield (%)	Refs.
$(CH_3)_2CHCH_2COCH=CHP(O)(OC_3H_7\text{-}i)_2$	$(i\text{-}C_3H_7O)_3P$	B	42	68
$(C_2H_5O)_2P(O)CH_2CH=CHP(O)(OCH_3)_2$	$NaOP(OCH_3)_2$	B	40	56
$(C_2H_5O)_2P(O)CH_2CH=CHP(O)(OC_2H_5)_2$	$NaOP(OC_2H_5)_2$	B	77	56
cyclooctenyl–$P(O)(OC_2H_5)_2$	$(C_2H_5)_3PCuBr$	A	71	59
$C_6H_5CH=CHP(O)(OC_2H_5)_2$	$(C_2H_5O)_3PCuCl$	A	86	57
	$HOP(OC_2H_5)_2$	A	93	61
$C_6H_5CH=CHP(O)(OC_3H_7\text{-}i)_2$	$(i\text{-}C_3H_7O)_3PCuBr$	A	76	57
$(Z)\text{-}C_6H_5CH=CHP(O)(OC_4H_9\text{-}n)_2$	$HOP(OC_4H_9\text{-}n)_2$	A	92	61
$CH_2=C(C_6H_5)P(O)(OC_2H_5)_2$	$HOP(OC_2H_5)_2$	A	66	61
$CH_2=C(C_6H_5)P(O)(OC_3H_7\text{-}i)_2$	$(i\text{-}C_3H_7O)_3P$	A	49	57
$C_6H_5COCH=CHP(O)(OC_3H_7\text{-}i)_2$	$(i\text{-}C_3H_7O)_3P$	B	50	68
$(i\text{-}C_3H_7O)_2P(O)CH_2CH=CHP(O)(OC_3H_7\text{-}i)_2$	$NaOP(OC_3H_7\text{-}i)_2$	B	64	56
	$(C_2H_5O)_3P$	A	50	56
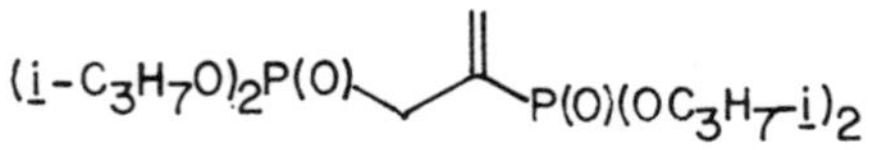 $(i\text{-}C_3H_7O)_2P(O)\!-\!CH_2\,C(=CH_2)\!-\!P(O)(OC_2H_5)_2$	$NaOP(OC_2H_5)_2$	B	70	56
$(i\text{-}C_3H_7O)_2P(O)\!-\!CH_2\,C(=CH_2)\!-\!P(O)(OC_3H_7\text{-}i)_2$	$(i\text{-}C_3H_7O)_3P$	A	57	56

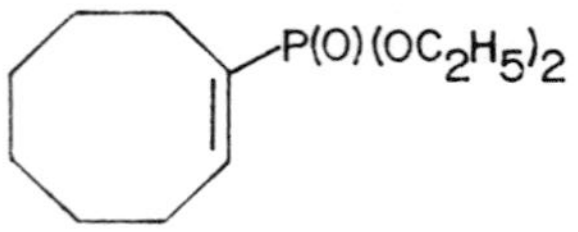

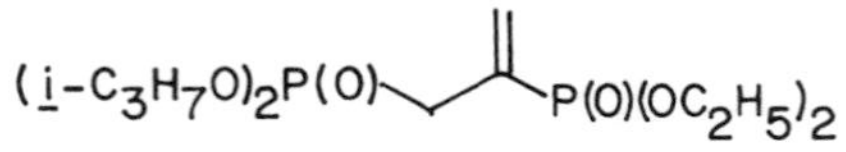

$(C_6H_5)_2C=CHP(O)(OC_2H_5)_2$

$(C_6H_5)_2C=CHP(O)(OC_3H_7-i)_2$

$NaOP(OC_3H_7-i)_2$	B	57	56
$(C_2H_5O)_3PCuBr$	A	95	57
$(i\text{-}C_3H_7O)_3PCuBr$	A	96	57

Note: A, metal halide-facilitated reaction B, thermal rearrangement; C, direct thermal reaction of vinylic halide.

REFERENCES

1. **Bratt, J. and Suschitzky, H.,** Reactions of polyhalogenopyridines and their *N*-oxides with benzenethiols, with nitrite, and with trialkyl phosphites, and of pentachloropyridine *N*-oxide with magnesium, *J. Chem. Soc. Perkin I,* p. 1689, 1973.
2. **Markovskii, L. N., Furing, G. G., Shermolovich, Y. G., and Yakobson, G. G.,** Phosphorylation of polyfluoroaromatic compounds. 3. Michaelis-Becker reaction in a series of polyfluoro-substituted benzene derivatives (transl.), *Izv. Akad. Nauk U.S.S.R.,* 646, 1981.
3. **Markovskii, L. N., Furin, G. G., Shermolovich, Y. G., and Yakobson, G. G.,** Phosphorylation of polyfluoroaromatic compounds. II. Reaction of triethyl phosphite with pentafluoro-substituted derivatives of benzene (transl.), *J. Gen. Chem. U.S.S.R.,* 49, 464, 1979.
4. **Kosolapoff, G. M.,** Isomerization of alkyl phosphites. VII. Reactions with chlorides of singular structure, *J. Am. Chem. Soc.,* 69, 1002, 1947.
5. **Plumb, J. B., Obrycki, R., and Griffin, C. E.,** Phosphonic acids and esters. XVI. Formation of dialkyl phenylphosphonates by the photoinitiated phenylation of trialkyl phosphites, *J. Org. Chem.,* 31, 2455, 1966.
6. **Obrycki, R. and Griffin, C. E.,** Phosphonic acids and esters. XIX. Syntheses of substituted phenyl- and arylphosphonates by the photoinitiated arylation of trialkyl phosphites, *J. Org. Chem.,* 33, 632, 1968.
7. **Issleib, K. and Vollmer, R.,** *o*-Substituted benzenephosphonic acid diethyl esters and *o*-amino, *o*-hydroxy, and *o*-mercaptophenylphosphine, *Z. Chem.,* 18, 451, 1978.
8. **Miles, J. A., Grabiak, R. C., and Beeny, M. T.,** Synthesis of novel phosphorus heterocycles: 2-aryl-1-methyl-2,3-dihydro-1*H*-2,1-benzazaphosphole 1-oxides, *J. Org. Chem.,* 46, 3486, 1981.
9. **Ohmori, H., Nakai, S., and Masui, M.,** Formation of dialkyl arylphosphonates *via* arylation of trialkyl phosphites, *J. Chem. Soc., Perkin I,* p. 2023, 1979.
10. **Maier, L.,** Herbicidally Active 2-Nitro-5-(2′-chloro-4′-trifluoromethylphenoxy)phenylphosphinic Acid Derivatives, U.S. Patent 4,434,108, 1984.
11. **Boduzek, B. and Wieczorek, J. S.,** A new method for the preparation of pyridine-4-phosphonic acids, *Synthesis,* p. 452, 1979.
12. **Tavs, P.,** Reaktion von Arylhalogeniden mit Trialkyl Phosphiten und Benzolphosphonigsaure-dialkylestern zu aromatischen Phosphonsaureestern unter Nickelsalz katalyse, *Chem. Ber.,* 103, 2428, 1970.
13. **Balthazor, T. M. and Grabiak, R. C.,** Nickel-catalyzed Arbuzov reaction; mechanistic observations, *J. Org. Chem.,* 45, 5425, 1980.
14. **Hechenbleikner, I. and Enlow, W. P.,** Arbuzov Rearrangement of Triphenyl Phosphite, U.S. Patent 4,113,807, 1978.
15. **Balthazor, T. M.,** Phosphindolin-3-one. A useful intermediate for phosphindole synthesis, *J. Org. Chem.,* 45, 2519, 1980.
16. **Horner, L. and Flemming, H. W.,** Fluorescing Naphthylphosphinates and Phosphonates, West German Patent 3,400,509, 1985.
17. **Balthazor, T. M., Miles, J. A., and Stults, B. R.,** Synthesis and molecular structure of 1,3-dihydro-1-hydroxy-3-methyl-1,2,3-benziodoxaphosphole 3-oxide, *J. Org. Chem.,* 43, 4538, 1978.
18. **Cristau, H.-J., Chene, A., and Christol, H.,** Arylation catalytique d'organophosphores. Produits de l'arylation, catalysee par les sels de nickel(II), de comnposes du phosphore tricoordine, *J. Organomet. Chem.,* 185, 283, 1980.
19. **Block, H.-D. and Dahmen, H.,** Process for the Production of Aryl Phosphonyl Compounds, U.S. Patent 4,391,761, 1983.
20. **Issleib, K., Balszuweit, K., Richter, S., and Koetz, J.,** Arylphosphonic or Arylphosphinic Acid Trimethylsilyl Esters, East German Patent 219,776, 1985.
21. **Connor, J. A. and Jones, A. C.,** Copper(II) enthanoate-assisted phosphonation of aryl halides, *J. Chem. Soc. Chem. Commun.,* p. 137, 1980.
22. **Hall, N. and Proce, R.,** The copper promoted reaction of *o*-halogenodiarylazo-compounds with nucleophiles. Part I. The copper-promoted reaction of *o*-bromodiarylazo-compounds with trialkyl phosphites. A novel method for the preparation of dialkyl arylphosphonates, *J. Chem. Soc., Perkin I,* p. 2634, 1979.
23. **Kukhar, V. P. and Sagina, E. I.,** Reactions of iodoarenes with triethyl phosphite in presence of cuprous chloride (transl.), *J. Gen. Chem. U.S.S.R.,* 47, 1523, 1977.
24. **Hirao, T., Masunaga, T., Oshiro, Y., and Agawa, T.,** A novel synthesis of dialkyl arenephosphonates, *Synthesis,* p. 56, 1981.
25. **Hirao, T., Masunaga, T., Yamada, N., Oshiro, Y., and Agawa, T.,** Palladium-catalyzed new carbon-phosphorus bond formation, *Bull. Chem. Soc. Jpn.,* 55, 909, 1982.
26. **Xu, Y., Li, Z., Xia, J., Guo, H., and Huang, Y.,** Palladium-catalyzed synthesis of unsymmetrical alkyl arylphenylphosphiniates, *Synthesis,* p. 377, 1983.
27. **Xu, Y. and Zhang, J.,** Palladium-catalyzed synthesis of functionalised alkyl alkylarylphosphinates, *Synthesis,* p. 778, 1984.

28. **Xu, Y., Li, Z., Xia, J., Guo, H., and Huang, Y.,** Palladium-catalysed synthesis of alkylarylphenyl-phosphine oxides, *Synthesis*, p. 781, 1984.

29. **Redmore, D.,** Phosphorus derivatives of nitrogen heterocycles. 2. Pyridinephosphonic acid derivatives, *J. Org. Chem.*, 35, 4114, 1970.

30. **Redmore, D.,** Phosphorus derivatives of nitrogen heterocycles. 3. Carbon-phosphorus bonding in pyridyl-2- and -4-phosphonates, *J. Org. Chem.*, 38, 1306, 1973.

31. **Redmore, D.,** Phosphonates of Full Aromatic Nitrogen Heterocycles, U.S. Patent 3,673,196, 1972.

32. **Redmore, D.,** Phosphorus derivatives of nitrogen heterocycles. 4. Pyridyl-4-phosphonates, *J. Org. Chem.*, 41, 2148, 1976.

33. **Redmore, D.,** Preparation of pyridyl-4-phosphonates, U.S. Patent 4,187,378, 1980.

34. **Katritzky, A. R., Keay, J. G., and Sammes, M. P.,** Regiospecific synthesis of dialkyl pyridin-4-yl, quinolin-4-yl, and isoquinolin-1-yl-phosphonates, *J. Chem. Soc., Perkin I,* p. 668, 1981.

35. **Akiba, K.-y., Matsuoka, H., and Wada, M.,** Regiospecific introduction of alkyl groups into 4-position of pyridine — novel synthesis of 4-substituted pyridines, *Tetrahedron Lett.*, p. 4093, 1981.

36. **Boduszek, B. and Wieczorek, J. S.,** Synthesis of 1-(4-pyridyl)-1,2-dihydropyridine-2-phosphonates and their derivatives, *Synthesis*, p. 454, 1979.

37. **Sweigart, D. A.,** Trialkyl phosphite addition to the bis(benzene)-iron(II) and -ruthenium(II) dications; catalyzed hydrolysis to dialkyl phosphites, *J. Chem. Soc., Chem. Commun.*, p. 1159, 1980.

38. **John, G. R. and Kane-Maguire, A. P.,** Kinetics of nucleophilic attack on co-ordinated organic moieties. Part 7. Mechanism of addition of tertiary phosphines and phosphites to tricarbonyl(dienyl)iron cations, *J. Chem. Soc., Dalton,* p. 873, 1979.

39. **John, G. R. and Kane-Maguire, A. P.,** Phosphite addition to organometallic cations to give phosphonium adducts, *J. Organomet. Chem.*, 120, C45, 1976.

40. **Bailey, N. A., Blunt, E. H., Fairhurst, G., and White, C.,** Reactions of the η^6-benzene(η^5-ethyltetra-methylene cyclopentadienyl) rhodium(III) cation and related species with nucleophiles, *J. Chem. Soc., Dalton,* p. 829, 1980.

41. **Chen, C. H. and Reynolds, G. A.,** Synthesis and reactions of (4H- and 2H-2,6-diphenylthiopyran-4-yl)phosphonates, *J. Org. Chem.*, 45, 2453, 1980.

42. **Fild, M. and Schmutzler, R.,** Halo and pseudohalophosphines, in *Organic Phosphorus Compounds*, Vol. 4, Kosolapoff, G. and Maier, L., Eds., Wiley-Interscience, 1972, 79.

43. **Photis, J. M.,** Preparation of arylphosphinic Acids, U.S. Patent 4,316,859, 1982.

44. **Simmons, K. A.,** Preparation of Arylphosphinic Acids, U.S. Patent 4,316,858, 1982.

45. **Neumaier, H.,** Process for making aryldichlorophosphines, U.S. Patent 4,536,351, 1985.

46. **Kormachev, V. V., Vasil'eva, T. V., and Korpov, R. D.,** Aryldichlorophosphines, Soviet Union Patent 1,151,540 1985.

47. **Buchner, B. and Lockhart, L. B.,** An improved method of synthesis of aromatic dichlorophosphines, *J. Am. Chem. Soc.*, 73, 755, 1951.

48. **Miles, J. A., Beeny, M. T., and Ratts, K. W.,** A general route to methoxy-substituted arylphosphonous dichlorides via mild Lewis acid catalysis, *J. Org. Chem.*, 40, 343, 1975.

49. **Kormachev, V. V., Vasil'eva, T. V., Abramov, I. A., Paramonov, V. I., Gradov, V. A., and Malovik, V. V.,** Diarylchlorophosphines, Soviet Union Patent 1,131,881, 1984.

50. **Grabiak, R. C., Miles, J. A., and Schwenzer, G. M.,** Synthesis of phosphonic dichlorides and correlation of their P-31 chemical shifts, *Phosphorus and Sulfur*, 9, 197, 1980.

51. **Piskunova, O. G., Yagodina, L. A., Kordev, B. A., and Bokanov, A. I.,** Phenophosphazines, IV. Electronic effects in 5,10-dihydro-5-methylphenophosphazines 10-oxides (transl.), *J. Gen. Chem. U.S.S.R.*, 48, 1205, 1978.

52. **Yagupol'skii, L. M., Pavlenko, N. V., Ignat'ev, N. V., Matyushecheva, G. I., and Semenii, V. Y.,** Aryl*bis*(heptafluoropropyl) phosphine oxides. Electronic nature of the $P(O)(C_3F_7)_2$ group (transl.), *J. Gen. Chem. U.S.S.R.*, 54, 297, 1984.

53. **Doak, G. O. and Freedman, L. D.,** The synthesis of arylphosphonic and diarylphosphinic acids by the Diazo reaction, *J. Am. Chem. Soc.*, 73, 5658, 1951.

54. **Melvin, L. S.,** An efficient synthesis of 2-hydroxyphenylphosphonates, *Tetrahedron Lett.*, p. 3375, 1981.

55. **Tavs, P. and Weitkamp, H.,** Herstellung und KMR-Spektren einiger α,β-ungesattigter phosphonsau-reestern, *Tetrahedron*, 26, 5529, 1970.

56. **Sturtz, G., Damin, B., and Clement, J.-C.,** Propene-1,3-and -2,3-diyldiphosphonates. Synthesis of diene-phosphonates by the Wittig-Horner reaction, *J. Chem. Res. (M)*, p. 1209, 1978.

57. **Axelrad, G., Laosooksathit, S., and Engel, R.,** A direct synthesis of vinylic phosphonates from vinylic halides, *Synthetic Commun.*, 10, 933, 1980.

58. **Axelrad, G., Laosooksathit, S., and Engel, R.,** Reactions of copper(I) halide complexes of trivalent phosphorus with vinylic halides, *J. Org. Chem.*, 46, 5200, 1981.

59. **Banerjee, S., Engel, R., and Axelrad, G.,** A direct synthesis of 1-cycloalkenylphosphonates, *Phosphorus and Sulfur*, 15, 15, 1983.

60. **Axelrad, G., Laosooksathit, S., and Engel, R.,** Halide exchange at vinylic position catalyzed by copper(I) halide complexes of trivalent phosphorus, *Synthetic Commun.,* 11, 405, 1981.
61. **Hirao, T., Masunaga, T., Oshiro, Y., and Agawa, T.,** Stereoselective synthesis of vinylphosphonates, *Tetrahedron Lett.,* p. 3595, 1980.
62. **Kreutzkamp, N. and Schindler, H.,** Ungesattigte Phosphonsaure-estern aus Hydroxymethylen-athern, *Chem. Ber.,* 92, 1695, 1959.
63. **Hammerschmidt, F. and Zbiral, E.,** Darstellung von 3-Oxo-1-alkenylphosphonsaure-dialkylestern, *Liebig's Ann. Chem.,* p. 492, 1979.
64. **Ohler, R., Haslinger, E., and Zbiral,** Synthesis and ¹NMR spectra of (3-acylbicyclo[2.2.1]hept-5-en-2-yl)phosphonates, *Chem. Ber.,* 115, 1028, 1982.
65. **Penz, G. and Zbiral, E.,** "Zur Synthese von (*rac*)-*cis*-1,2-Epoxy-3-oxo-alkyl-phosphonsaureestern Phosphonomycinanaloga," *Monatsh. Chem.,* 113, 1169, 1982.
66. **Ohler, E., El-Badawi, M., and Zbiral, E.,** Synthese von Hetaryl- und Hetarylvinylphosphonsaureestern aus 2-Brom-1-oxoalkylphosphonaten und 4-Brom-3-oxo-1-alkenylphosphonaten, *Chem. Ber.,* 117, 3034, 1984.
67. **Ohler, E. and Zbiral, E.,** Synthese, Reaktionen und NMR-Spektren von 2-Brom-3-oxo-1-alkenyl- und 3-Oxo-1-alkinyl-phosphonsaure-dialkylestern, *Monatsh. Chem.,* 115, 493, 1984.
68. **Ohler, E. and Zbiral, E.,** Cyclisierungsreaktionen von Diazoalkenyl-phosphonsaureestern Synthese von Pyrazolyl- und 2,3-Benzodiazepinylphosphonsaureestern, *Monatsh. Chem.,* 115, 629, 1984.
69. **Schneider, P. and Fischer, G. W.,** Reaktion vinyloger Chloromethylenimmoniumsalze mit Phosphorigsaure trialkylestern, *J. prakt. Chem.,* 322, 229, 1980.
70. **Aguilar, A. M. and Daigle, D.,** Vinyl halide displacement by metallo organophosphides. Preparation of *trans*-β-styryldiphenylphosphine oxide and sulfide, *J. Org. Chem.,* 30, 2826, 1965.
71. **Dittrich, R. and Hagele, G.,** Michaelis-Arbuzov perhalogenation reaction of olefins. III. The trifluorovinyl halide, CF_2CFX, *Phosphorus and Sulfur,* 10, 127, 1981.
72. **King, R. B. and Diefenbach, S. P.,** Transition-metal cyanocarbon derivatives. 5. Reactions of (1-Chloro-2,2-dicyanovinyl)manganese derivatives with trialkyl phosphites. A novel variant of the Michaelis-Arbuzov reaction leading to [2,2-Dicyanovinylphosphonato]metal complexes, *Inorg. Chem.,* 18, 63, 1979.
73. **Wong, Y. S., Paik, H. N., Chieh, P. C., and Carty, A. J.,** Two-carbon three-electron ligands. Phosphonium-betaine complexes *via* nucleophilic attack by phosphites on a σ-π-acetylide di-iron hexacarbonyl derivative, *J. Chem. Soc., Chem. Commun.,* p. 309, 1975.
74. **Malenko, D. M., Repina, L. A., and Sinitsa, A. D.,** Vinylphosphite-vinylphosphonate rearrangement (transl.), *J. Gen. Chem. U.S.S.R.,* 54, 2148, 1984.
75. **Korshin, E. E., and Mukhametov, F. S.,** Intramolecular rearrangement of vinyl phosphites to vinylphosphonates (transl.), *Bull. Acad. Sci. U.S.S.R.,* p. 1752, 1984.
76. **Pieper, W.,** Vinylphosphonic or Vinylpyrophosphonic Acid Derivatives, West German Patent 3,323,392, 1985.
77. **Kaesz, H. D., and Stone, F. G. A.,** Preparation and characterization of vinyldichlorophosphine, vinyldimethylphosphine, and ethyldimethylphosphine, *J. Org. Chem.,* 24, 635, 1959.
78. **Russell, G. A. and Hershberger, J.,** Substitution reactions of vinylmercurials by a free-radical chain mechanism, *J. Am. Chem. Soc.,* 102, 7603, 1980.

INDEX

BRISTOL-MYERS SQUIBB COMPANY